T0262357

Essentials of Molecular Photochemistry

Essentials of Molecular Photochemistry

Edited by **Todd Rollins**

New York

Published by NY Research Press,
23 West, 55th Street, Suite 816,
New York, NY 10019, USA
www.nyresearchpress.com

Essentials of Molecular Photochemistry
Edited by Todd Rollins

International Standard Book Number: 978-1-63238-192-7 (Hardback)

Printed in the United States of America.

Contents

Preface

Molecular photochemistry has garnered significant interest of researchers and scholars across the globe. This book has been compiled with the intention of addressing utilization of basic fundamentals and principles to more complex concepts in various fields of photochemistry. It is unique in its approach in comparison to various classical books on photochemistry which provide detailed accounts limited only to the basics of molecular photochemistry. There has been an overview on the core concepts used in diverse spheres of photochemistry which are not easily accessible. The aim of this text is to update academicians, students and experts actively involved in the field of molecular photochemistry. Latest developments have been highlighted and different functions of the technology in solution, metal oxides, biology, computational aspects and other applications have been dealt with. This book presents a unique overview on photochemistry.

After months of intensive research and writing, this book is the end result of all who devoted their time and efforts in the initiation and progress of this book. It will surely be a source of reference in enhancing the required knowledge of the new developments in the area. During the course of developing this book, certain measures such as accuracy, authenticity and research focused analytical studies were given preference in order to produce a comprehensive book in the area of study.

This book would not have been possible without the efforts of the authors and the publisher. I extend my sincere thanks to them. Secondly, I express my gratitude to my family and well-wishers. And most importantly, I thank my students for constantly expressing their willingness and curiosity in enhancing their knowledge in the field, which encourages me to take up further research projects for the advancement of the area.

Editor

Part 1

Photochemistry in Solution

Quinoline-Based Fluorescence Sensors

Xiang-Ming Meng, Shu-Xin Wang and Man-Zhou Zhu
Anhui University
China

1. Introduction

The human body is full of various ions, which play an important role in the normal physiological activities. For example, Zinc ion (Zn^{2+}) plays a vital role in protein organism and in many biochemical processes, such as inducing apotosis, enzyme regulation, and gene expression. Also, Ferrous ion (Fe^{2+}) is vital in the oxygen transporting. But there are some ions harmful to human body. When exposed to mercury, even at a very low concentration, they lead to kidney and neurological diseases. What's more, Cadmium (Cd^{2+}) could damage our tissues, resulting in renal dysfunction or even cancers. So far, we have known more about these ions' properties in metabolism, but little is known on mechanism.

We need a forceful instrument to study these mechanisms, need to know when and where ions are distributed, when ions are released, and so on.Therefore, traditional methods such as titration and electrochemistry are obviously unsuitable for in vivo detection. As a result, to accomplish the job, we need new tools and methods, among which fluorescence sensors are a good choice. So, what is a sensor? "Sensor" is a very broad concept, which accepts physical or chemical variables (input variables) information, and converts them into the same species of other kinds or converts their nature of the device output signal (Fig. 1) by following certain rules.

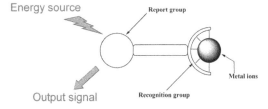

Fig. 1. Sensor structure (Once combination of ions, the output signal will change)

A chemosensor or a molecular sensor is a molecule that interacts with an analyte to produce a detectable change. Chemosenors consist of receptor and reporter, and after the receptor binds with a guest, the signal observed by the reporter will change. Fluorescent sensor is one of the most important chemosensors which uses fluorescence as the output signal, and also a powerful tool to monitor the metal ions in vivo system because of its simplicity, high sensitivity and real-time in situ imaging. In recent years, more and more chemosensors, especially the fluorescent sensors have been used to detect different ions, elbowing their way to center stage in the field of molecular recognition. Series of sensors based on fluorescein, coumarin, petide, quinoline, and proteins have been used to detect intracellular ions concentration, such as Zn^{2+} sensors of Zinpyr Family based on fluorescein designed by Woodroofe (2004) et al., Cadmium sensor based on boradiazaindacene synthesised by Xu (2007) et al., Cu^{2+} sensor based on rhodamine synthesised by Dujols (1997) et al., the benzimidazole sensor described by Henary (2004) et al., the protein sensor described by van Dongen (2006) et al., Hg^{2+} FRET sensor described by Joshi (2010) et al., and Fe^{3+} sensor based on 1,8-diacridylnaphthalene and synthesized by Wolf (2004) et al..

Different fluorophores bring different optic properties of sensors. For example, sensors based on rhodamine can be excited by visible light, but they get low Stock's shift. Benzofuran-based sensors get lower dissociation constant, but UV exiting with higher energy may damage cells. These disadvantages thus bring forward potential difficulties for quantitative determination and bioimaging, so how to solve these problems is still a challenge.

Quinoline sensors, especially Zn^{2+} and Cd^{2+}, have high selectivity and low detection limit (nM or pM). Modified quinoline chemosensors can also use low energy two-photon laser as the exciting source, which can reduce cell damage. Therefore, the current research of quinoline-based sensors attracts more and more attention.

Herein, the mechanism of quinoline-based fluorescence sensing, including PET (Photoinduced electron transfer), ICT (intermolecular charge transfer) and FRET (fluorescence resonance energy transfer), the synthetic strategies for functionalization of quinoline-based sensors will be reviewed, and the reasons for the choice of a particular synthetic pathway will be discussed. In order to contextualize the potential applications, a brief introduction of the photophysics property concerning quinoline-based sesnors is contained in the essay. At the same time, calculation method of sensor properties (eg, dissociation constant and quantum yield determination) is also included.

2. Mechanism of quinoline-based fluorescence sensing

Quinoline-based fluorescence sensors are usually used to measure intensity changes of fluorescence and/or shift of fluorescence wavelength. Photoinduced electron transfer (PET), intermolecular charge transfer (ICT) and fluorescence resonance energy transfer (FRET) are the three major mechanisms of fluorescence signal transduction in the design of quinoline-based fluorescence chemosensors (de Silva (1997) et al., Sarkar (2006) et al. and Banthia & Samanta (2006)). We will present the basic concepts of these mechanisms.

Chemosensors based on PET mechanism (Fig. 2) often use a atoms spacer less than three carbon atoms to connect a fluorescence group to a receptor containing a high-energy non-bonding electron pair, such as nitrogen or sulfur atom, which can transfer an electron to excited fluorescence group and result in fluorescence quench. But when the electron pair is coordinated by a metal ion (or other cation), the electron transfer will be prevented and the fluorescence is switched on. Most of quinoline-based fluorescence enhancement sensors can be explained by the PET type. Generally speaking, wavelengths of most PET chemosensors in Stokes shifts are less than 25 nm, which produces potential difficulties for quantitative determination and bioimaging. However, ratiometric chemosensors, which observe changes in the intensity ratio of the two wave bands in absorption and/or emission, would be more favorable in increasing the signal selectivity and can be widely used in vivo.

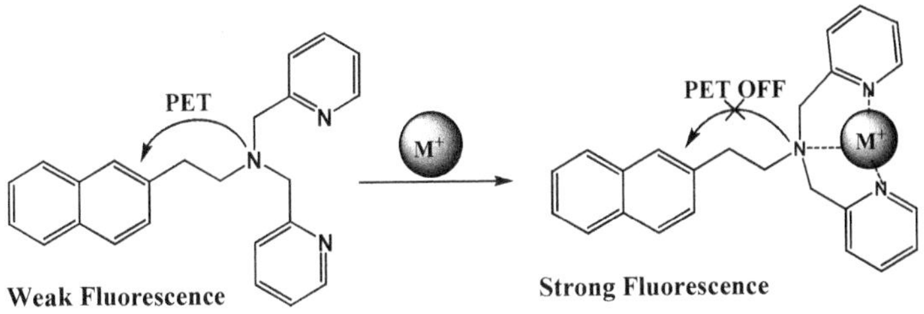

Fig. 2. PET mechanism (The intensity of Fluorescence will increase after combination of ions)

The ICT mechanism (Fig. 3) has been widely used in the design of ratiometric fluorescent chemosensors. Compared to PET mechanism, this type of chemosensor doesn't have any spacer. If a receptor (usually an amino group) is directly connected with a conjugation system and forms a new conjugation system with p-electron, resulting in electron rich and electron poor terminals, then ICT from the electron donor to receptor would be enhanced upon light excitation. When a receptor, as an electron donor within the fluorophore, is bound with a metal ion (or another cation), the cation will reduce the electrondonating capacity of the receptor and a blue shift of the emission spectrum is obtained. In the same way, if a receptor is an electron receptor, the coordination of the cation will further strengthen the push – pull effect. Then a red shift in emission will be observed. For example, the coordination of Zn^{2+} with quinoline derivatives can induce a red-shift ratiometric fluorescence signal.

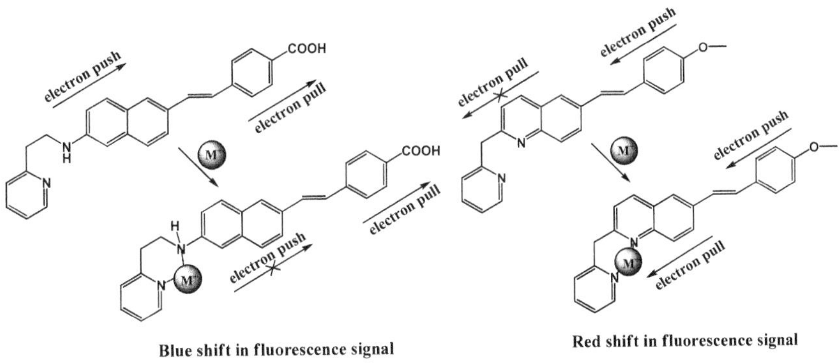

Fig. 3. ICT mechanism (Change in the intensity ratio of the two wave bands in emission)

Recently, the fluorescence resonance energy transfer (FRET Fig. 4), which involves the nonradiative transfer of excitation energy from an excited donor to a proximal ground-state acceptor, has been employed to design ratiometric sensors. The FRET-based sensors can be designed in the form of a small molecule, which usually contains two fluorophores connected by a spacer through covalent links. The following conditions must be satisfied for FRET: 1. The donor probe should have sufficient lifetime for energy transfer to occur. 2. The distance from the donor to the acceptor must be less than 10nm. 3. The absorption spectrum of the acceptor fluorophore must overlap with the fluorescence emission spectrum of the donor fluorophore (by approximately 30%). 4. For energy transfer, the donor and acceptor dipole orientations must be approximately parallel. Energy transfer is demonstrated by quenching of donor fluorescence with a reduction in the fluorescence lifetime, and an increase in acceptor fluorescence emission. FRET is very sensitive to the distance between fluorophores and can be used to estimate intermolecular distances. FLIM imaging can be used in association with FRET studies to identify and characterize energy transfer. Quinoline comprising another fluorophore (usually rhodamine) that will behave as FRET donor has been synthesized in order to produce FRET-based chemosensors.

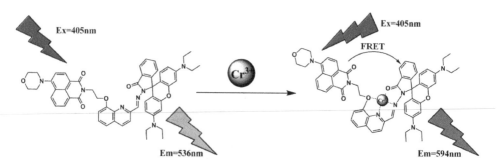

Fig. 4. A FRET chemosensor reported by Zhou (2008) et al.

It is worth mentioning that the combination of PET and ICT mechanisms in the design of chemosensors would be valuable, since a wavelength shift and fluorescence intensity enhancement can amplify the recognition event to a greater extent, for example, using decorated quinoline as mother nucleus, thus oxidizing methyl on 2 position, then connecting DPA group. Thereby excellent ICT effect and fluorescent shift can be obtained after nitrogen atom on quinoline is bound. Meanwhile, the binding N-atom on DPA can obstruct PET process, thus increasing fluorescent intensity. FRET process is also considerably flexible, which can be applied widely in double fluorescence group, and at the same time can be employed in the energy transfer between a single fluorescence group and nanoparticles. By using specific acceptor to separate fluorescent group from nanoparticles, FRET process will be blocked, and fluorescence is produced. By using acceptor to connect nanoparticles with fluorescent group, which was not formerly connected with nanoparticles, fluorescence vanishes. They are particularly significant to fluorescent sensors based on nanoparticles. These methods are extremely effective.

3. Structure and synthesis

The general structure of quinoline-based chemosensors is represented in Fig 5. Most quinoline-based sensors change the receptor group in the 2 (R_1) and 8 (R_5) positions, and the electron donating or withdrawing group in the 4 (R_2), 5 (R_3) and 6 (R_4) positions. Depending on the substituents R_1, R_2, R_3, R_4, R_5, the sensor will present different photophysical properties in solution, such as absorption and emission maxima (λ_{max} abs, λ_{max} em, and fluorescence quantum yield). Herein, the synthesis of different functionalized quinoline-based sensors will be discussed.

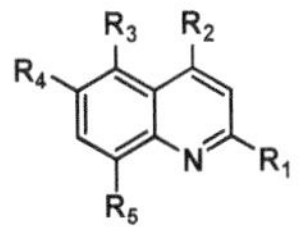

Fig. 5. Molecular structure of quinoline-based chemosensors.

Synthesis by Doebner-Von Miller (1996): According to this method, aniline and acetaldehyde are usually used as raw material in hydrochloric acid or zinc chloride. At the beginning, condensation acetaldehyde into crotonaldehyde, then crotonaldehyde reacts with aniline molecule, the intermediate product is produced, and then dehydrogenates into dihydroquinoline, which becomes 2-methylquinoline. The reaction formula is as follows:

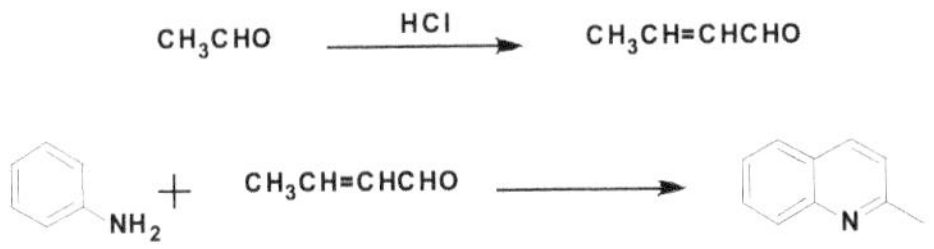

(1)

The improved method, which can also be applied to obtain larger conjugated system in other materials, uses crotonic aldehyde instead of methanal to get a higher yield. In this project, a sensor based on ICT and FRET mechanisms is designed and synthesized, which uses 4-bromo-phenylamine as raw materials. Besides, 4-methoxy styrene is introduced into the quinoline platform by applying the classic Heck reaction.

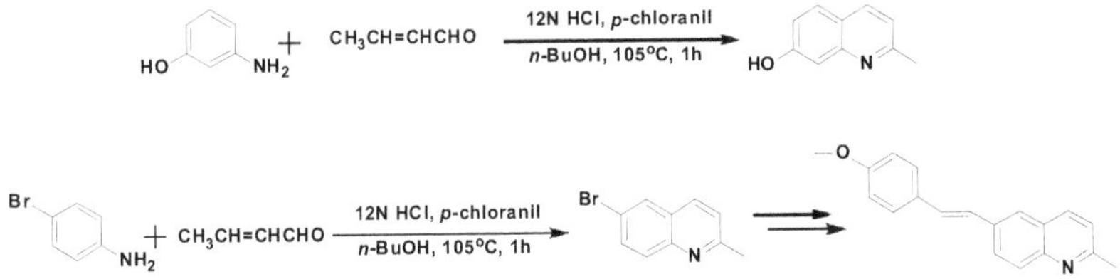

(2)

Matsubara (2011) et al. reported a new synthesis method of functionalized alkyl quinolines, which was based on sequential PdCl$_2$-catalyzed cyclization reactions of substituted anilines and alkenyl ethers. High efficiency and functional-group tolerance made this procedure widely applied in synthesis of a number of substituted 2-alkylquinolines and larger conjugated systems.

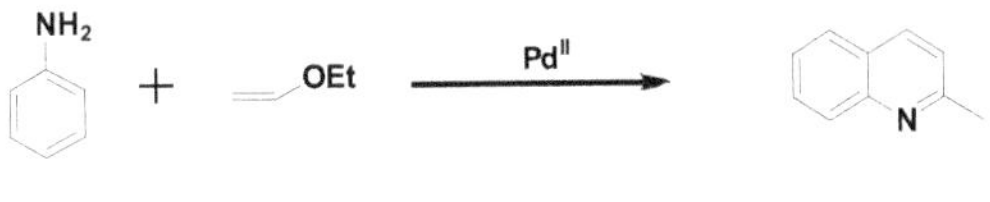

(3)

Guerrini (2011) group reported an innovative and convenient synthetic approach for synthesizing two important genres of heterocyclic scaffolds, which use the capability of the aromatic amides to rearrange photo-Fries. Quinolines from simple acetanilides derivatives have been obtained with satisfactory yield by using a single one-pot procedure.

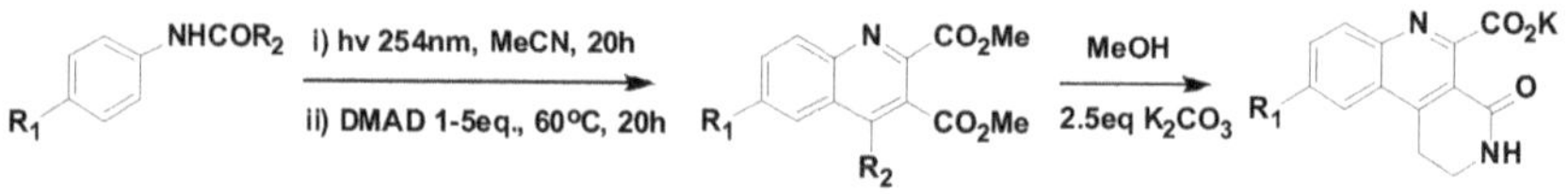

(4)

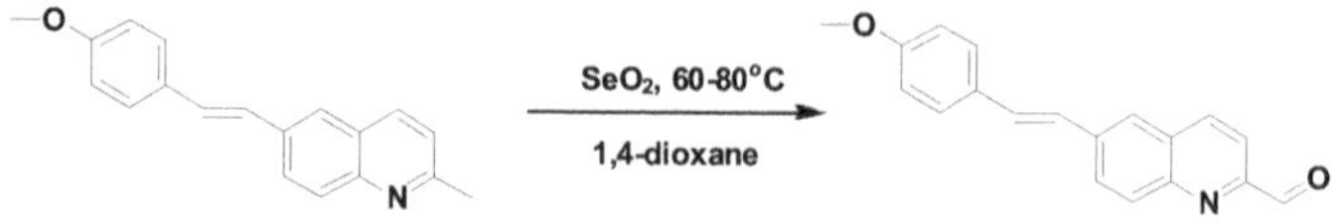

(5)

In order to introduce functional groups on the 2-position, we often oxidize the methyl to aldehyde. Using selenium dioxide as oxidant can gain very high yield. Generally dioxane is used as the reaction solvent at 60-80 degrees. Reaction usually ends within two hours.

The synthesis of 8-aminoquinoline: The cyclization reaction can be firstly adopted to synthesize quinoline, which is replaced by nitryl and its derivatives, then reduction is used to generate 8-aminoquinoline. Classic reactions to synthesize quinoline ring include Friedlander, Skraup, Dobner-Miller and so on. The reaction equation is as follows:

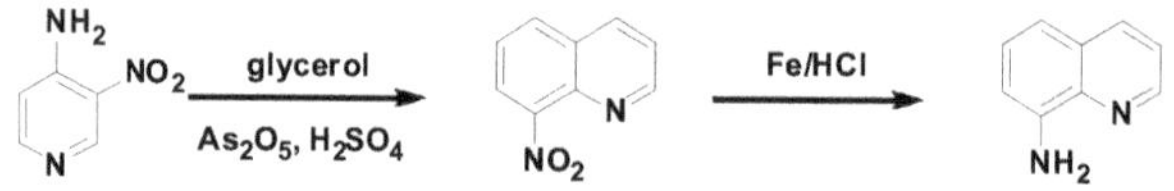

(6)

4. Dissociation constant and quantum yield determination

The dissociation constant is commonly used to describe the degree of affinity between sensor and metal ion. It is a key parameter used to describe the sensor's selectivity. It can be calculated by eq 1,

$$K_d = \frac{[M^{n+}]_{free}(F_{max} - F)}{F - F_0} \quad (1)$$

where F=normalized fluorescence intensity, K_d=dissociation constant, F_{min}=fluorescence intensity without metal ions (M^{n+}), F_{max}=fluorescence intensity of bound sensor and $[M^{n+}]_{free}$ is the concentration of the free M^{n+}. Free metal ion concentrations are controlled by metal ion buffers (e.g., NTA (nitrilotriacetic acid), EDTA or other chelating agent.). As for log K of different metals with NTA and EDTA, see Tab. 1 (log K (ML), I=0.1mol/L, 25°C, 0.1mol/L).

Quinine sulfate is widely used as the standard in the calculation of fluorescence quantum yields of quinoline-based chemosensors (in 0.1N H_2SO_4, Φ =0.55, λ_{ex} = 320 nm). The quantum yields are calculated by eq 2.

$$\Phi u = \frac{\Phi_S (F_u A_S)}{F_S A_u} \qquad (2)$$

A_u is the UV absorption of unbound sensor or bound sensor, with A_s being the standard. F_u is integrated fluorescence emission corresponding to sensor or metal complex, and F_s is the standard.

	Ca^{2+}	Cd^{2+}	Zn^{2+}	Co^{2+}	Cu^{2+}	Fe^{3+}	Hg^{2+}	Pb^{2+}	Mg^{2+}
EDTA	10.61	16.36	16.44	16.26	18.70	25.0	21.5	17.88	8.83
NTA	6.39	9.78	10.66	10.38	12.94	15.9	14.6	11.34	5.47

Table 1. Log K of different metals with NTA and EDTA.

5. Quinoline used for detecting different metal ions

5.1 Quinoline used for detecting Zn^{2+} ion

Quinoline and its derivatives, especially 8-hydroxyquinoline and 8-aminoquinoline, are very important fluorogenic chelators for metal ions transition. Derivative of 8-aminoquinoline with an aryl sulfonamide is the first and most widely applied fluorescent chemosensor for imaging Zn^{2+} in biological samples. It was first reported by Toroptsev and Eshchenko. In 1987, Frederickson (1987) et al. reported a new quinoline-based sensor 1, which showed 100 folds in fluorescence enhancement after being bound with Zn^{2+}. And it is the first high-sensitive sensor to detect Zn^{2+} in high concentrations of Ca^{2+} and Mg^{2+}, which is very important for application in vivo. But low water solubility limits its application, so Zalewski (1994) led in a water-soluble group at the 6-position of quinoline, chemosensors 2 and 3 were synthesized. The research showed that this improvement made these two chemosensors much more water-soluble, and also showed a large increase in fluorescence upon Zn^{2+}addition. Ca^{2+} and Mg^{2+} had little effect on the fluorescence whereas Fe^{2+} and Cu^{2+} quenched the fluorescence. Recently, Zhang (2008) et al. reported a new high-selective water-soluble and ratiometric chemosensor 4, based on 8-aminoquinoline for Zn^{2+} ion, which showed 8-fold increase in fluorescence quantum yield and a 75 nm red-shift fluorescence emission from 440 to 515 nm. But its excited source's energy is too high to be applied in vivo. Except the ability to be the fluorescence report group, quinoline's capability of binding Zn^{2+} enables it to be used as merely a binding group, so that high selectivity recognition of Zinc ion can be achieved. Nolan and Lippard (2005) et at., use ethyl 8-aminoquinoline to synthesize chemosensor 5. In addition, 5 exhibits 150-fold increase in fluorescence upon Zn^{2+} binding because of the low background fluorescence and high emission when binding with Zn^{2+}. This binding is selective for Zn^{2+} from other biologically relevant metal cations, toxic heavy metals, and most first-row transition metals and is of appropriate affinity (K_d=41uM) to reversibly bound Zn^{2+} at physiological levels, and the quantum yield for the Zn^{2+}-bound complex is 0.7 (λ_{ex}=518nm). In this job, quinoline's recognition of Zinc ions is utilized. Meanwhile, rhodamine's high yield of fluorescent quantum and high sensitivity are taken full advantage of. So we can see that excellent properties such as light excitation provide superior possible ways for designing quinoline sensors.

8-Hydroxyquinoline, also a traditional fluorogenic agent for analyzing Zn^{2+} and other metal ions, was used as a reporter group in the chemosensor. Di-2-picolylamine (DPA) is a classic chelator with high selectivity for Zn^{2+} over other metal ions that can not be influenced by higher concentration of Ca^{2+}, Na^+ and K^+ ions in biological samples. Xue (2008) et al. incorporated DPA into 8-hydroxy-2-methylquinoline at the 2-position to prepare a series of chemosensors 6, 7 and 8. The NMR studies and crystal structures of Sensor–Zn^{2+} complexes indicated that oxygen at the 8-position participated in the coordination of Zn^{2+} along with the quinoline nitrogen atom, and that DPA group endow the sensor with a high affinity (7, $K_d = 0.85$ pM). The fluorescence intensities of sensors showed a 4 to 6 fold enhancement and the quantum yields were also remarkably enhanced (Fig. 6a). According to the study of sensor's selectivity, the emissions of sensors showed slight enhancement upon addition of K^+, Mg^{2+} and Ca^{2+} in the millimolar range, whereas the fluorescence intensities were slightly quenched by 1 equiv. of transition metals such as Mn^{2+}, Co^{2+}, Fe^{2+}, Ni^{2+} and Cu^{2+}, with the exception of Cd^{2+} showing enhanced fluorescence. Xue (2009) improved the sensor via choosing ICT process instead of PET process. By adding a cation which interacted with a receptor, the electron-withdrawing ability of the expanded conjugated system was enhanced, and 8 was designed. This results in a larger red-shift emission and Stokes shift (Fig. 6b). 8 shows a maximum emission at 545 nm with a large Stokes shift of 199 nm in the absence of Zn^{2+}. The ratio of emission intensity (I_{620} nm/I_{540} nm) increases linearly with increased Zn^{2+} concentration. Ratiometric has brought about higher sensitivity, and other background disturbance. Only Zn^{2+} and Cd^{2+} show distinct ratiometric responses. Cell staining experiments demonstrate that 8 can readily reveal changes in intracellular Zn^{2+}. Dual emissions and cell-permeable nature of 8 make it possible to study cellular Zn^{2+} in hippocampus in a ratiometric approach. The same problem appears here too: high-energy excitation puts cells vulnerable to harm.

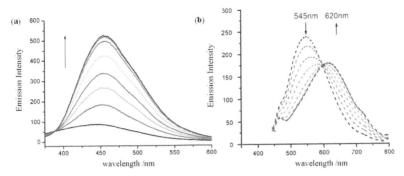

Fig. 6. (a) Fluorescence spectra (λ_{ex}=320 nm) of 5μM 6 upon the titration of Zn^{2+} (0–1.6 equiv.) in a HEPES buffer. (b) Fluorescence response upon titration of 8 (5 mM) with Zn^{2+} (0–1.6 equiv.), λ_{ex}=405 nm.

One year later, a both visual and fluorescent sensor 9 for Zn^{2+} was synthesized by Zhou (2010) et al., it displays high selectivity for Zn^{2+} and can be used as a ratiometric Zn^{2+} fluorescent sensor under visible light excitation. The strong coordination ability of Zn^{2+} with 9 leads to approximately 14-fold Zn^{2+} enhancement in fluorescence response and more than 7-fold increase in quantum yield (form 0.006 to 0.045) in THF-H_2O solution. It is important that 9 have little or no effect on Cd^{2+}, whereas Cu^{2+} and Co^{2+} quench the fluorescence. The quenching is not due to the heavy-atom effect, for, other heavy-atom did not quench the fluorescence.

In recent years, two-photon microscopy (TPM) imaging has gained much interest in biology because this method leads to less phototoxity, better three dimensional spatial localization, and greater penetration into scattering or absorbing tissues. Sensor 10 for monitoring Zn^{2+} was synthesized by Chen (2009) et al. based on the structure of 7-hydroxyquinoline. Its fluorescence enhancement (14-fold) and nanomolar range sensitivity (K_d=0.117 nM) were favorable in biological applications. JOB'S plot, NMR study and X-ray crystal structure indicated the binding model between sensor and Zn^{2+} is 1:1. Moreover, 10 also showed high selectivity for Zn^{2+} toward other first row transition metal ions including Fe^{2+}, Co^{2+}, and

Cu^{2+}, but it was slightly enhanced by Cd^{2+}. Furthermore, 10 can be used for imaging Zn^{2+} in living cells with two-photon microscopy (Fig. 7). This is also one of the directions of designing fluorescence sensors, that is, using widely used low-energy 800nm laser as excitation source so as to avoid the harm to cells caused by ultraviolet rays.

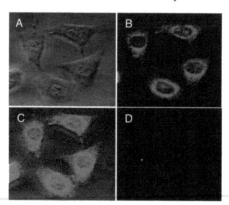

Fig. 7. (A) Bright-field image of A431 cells labeled with 30 μM 10 after 30 min of incubation, λ_{ex}=800 nm. (B) TP image after a 30 min treatment with zinc(II)/pyrithione (50 μM, 1:1 ratio). (C) The overlay of (A) and (B). (D) TP image of cells that are further incubated with 50 μM TPEN for 10 min.

Aoki, S. (2006) et al. have designed and synthesized new cyclen-based Zn^{2+} chemosensor 11 having an 8-hydroxy-5-N and N-dimethylaminosulfonylquinoline unit on the side chain. In the study, they found that using deprotonation of the hydroxyl group of 8-HQ and chelation to Zn^{2+} at neutral pH allows more sensitive detection of Zn^{2+} than dansylamide-pendant cyclen and (anthrylmethylamino) ethyl cyclen. They also introduced deprotonation behavior and fluorescence behavior, which was different by modifying the 5-position. This was very important in designing Quinoline-based chemosensors.

12   **13 n=0,1,2**

A space comprised of nitrogen atoms with quinoline fragments at both ends is often used in detection of Zn^{2+}. Sensor 12 was synthesized by Liu (2010) et al, by using ethidene diamine to connect two 2-oxo-quinoline-3-carbaldehydes, thus schiff-base was composed to achieve Zn^{2+} detection. Compared with other metal ions, chemosensor 12 exhibits high selectivity and sensitivity for Zn^{2+} in acetonitrile solution compared with Cd^{2+} and other metal ions (Fig. 8a). The single crystal was taken for demonstrating the binding model of sensor and Zn^{2+}. A simple-structured sensor 13 was reported by Shiraishi (2007) et al. 13 was easily synthesized by one-pot reaction in ethanol via condensation of diethylenetriamine and 2-quinolinecarbaldehyde followed by reduction with $NaBH_4$. 13 is a new member of the water-soluble fluorescent Zn^{2+} sensor capable of showing linear and stoichiometrical response to Zn^{2+} amount without background fluorescence. 13 also shows high Zn^{2+}

selectivity and sensitivity in water solution (Fig. 8b). Cd^{2+} induces slight enhancement of fluorescence emission intensity. It is a easily synthesized recognition to connect two 2-position quinoline sensors through a bridge that comprises heteroatoms (usually N, O, S atoms). In addition, the size of the cavity after bridge connection can be controlled in order to recognize specific ions. This is a very good choice to recognize sensors of different ions.

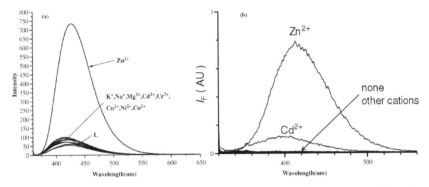

Fig. 8. (a) Fluorescence intensity of 12 (10 µM) in the presence of various metal ions (20 µM) in acetonitrile solution (λ_{ex}=305 nm, λ_{em}=423 nm). (b) Fluorescence spectra and intensity (monitored at 410 nm) of 13 (50 µ M) measured with respective metal cations (1 equiv) at pH 7.0 (KH_2PO_4–NaOH buffered solution).

Although quinoline based chemosensors can serve as both the metal ligand and the fluorophore, their optical properties limit the application in vivo. The main disadvantage of these chemosensors is high-energy UV excitation which is detrimental to cells. Fluorescence at short wavelengths (most of the emission wavelengh is under 500nm) and most of the fluorescent sensors, based on quinoline with DPA as receptor, are more or less affected by Cd^{2+}. So how to improve quinoline-based sensor is still a challenge.

5.2 Detection of Cd^{2+}

The interference of Cd^{2+} is a well-known problem for zinc fluorescence sensors and cadmium fluorescence sensors. Xue (2009) et al. reported an chemosensor that modulated the 8-position oxygen of the quinoline platform on sensor 14, while bound Zn^{2+} in 14 can be displaced by Cd^{2+}, resulting in another ratiometric sensing signal output (Fig. 9a). 14 shows a blue-shift of 33nm in emission spectrum. ^{1}H-NMR and optical spectra studies indicate that that 14 has higher affinity for Cd^{2+} than for Zn^{2+}, which consequently incurs the ion displacement process. Recently Xue (2011) et al., synthesized a new cadmium sensor 15 based on 4-isobutoxy-6-(dimethylamino)-8-methoxyquinaldine in line with the ICT mechanism. Sensor 15 exhibits very high sensitivity for Cd^{2+} (K_d=51pM) and excellent selectivity response for detection Cd^{2+} from other heavy and transition metal ions, such as Na^+, K^+, Mg^{2+}, and Ca^{2+} at millimolar level. They also established a single-excitation, dual-emission ratiometric measurement with a large blue shift in emission ($\Delta\lambda$ = 63 nm) and remarkable changes in the ratio (F_{495} nm/F_{558} nm) of the emission intensity (R/R_0 up to 15-fold, Fig. 9b). The crystal structures data of 15 binding with Cd^{2+} and Zn^{2+} demonstrate that the DPA moiety plays the main function of grasping the metal ions, while the 8-position methoxy oxygen can be used to tune the selectivity of the sensor. Furthermore, confocal experiments in HEK 293 cells were carried out with 15, demonstrating 15 to be a ratiometric chemosensor to image intracellular, which is obviously superior to intensity-based images of the sole emission channel. This job is a guide to design Quinoline-based Cd^{2+} sensor.

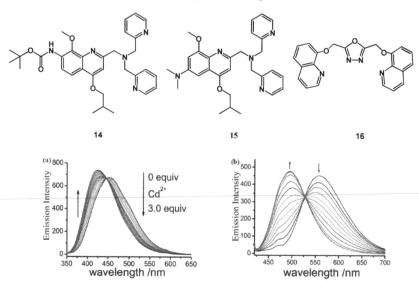

Fig. 9. (a) Fluorescence spectra (λ_{ex}=295 nm) of 10 µM 14 + 1 equiv of Zn^{2+} upon the titration of Cd^{2+} (0-3.0 equiv) in buffer solution. (b) Fluorescence spectra (λ_{ex} = 405 nm) of 10 µM 15 upon titration of Cd^{2+} (0-20 µM) in aqueous buffer.

Tang (2008) et al. merged 8-hydroxyquinoline with oxadiazole to develop a ratiometric chemosensor 16 for Cd^{2+}. If 1,3,4-oxadiazole subunit contained lone electron pairs on N, the semirigid ligand could effectively chelate Cd^{2+} according to the ionic radius and limit the geometric structure of the complex; thus 16 showed very high selectivity over other heavy and transition metal ions. This is also a designing method of bridge connection, that is, to make the detection group to form half heterocycle structure through the bridge, control the size of the heterocycle, and use the affinity of different heteroatoms to different ions, so that the selective recognition of different ions can be reached.

5.3 Detection of Cu^{2+} and Ag^+

Calixarenes are an important class of macrocyclic compounds, and they have been widely used as an ideal platform for the development of fluorescence chemosensors for alkali and alkaline-earth metal ions. Li (2008) et al. reported a turn on fluorescent sensor 17 for detecting Cu^{2+} based on calyx[4]arene bearing four iminoquinoline, which showed a largely enhanced fluorescent signal (1200-fold) upon addition of Cu^{2+} and a high selectivity toward Cu^{2+} over others. The 1:1 binding mode between sensor and Cu^{2+} was indicated by JOB's plot and mass spectrum. In Moriuchi-Kawakami's (2009) study, Cyclotriveratrylene can also act as a host analogous to calixarenes, a new C3-functionalized cyclotriveratrylene (CTV) bearing three fluorogenic quinolinyl groups. Sesnor 18 was synthesized, meanwhile, the fluorescence emission was remarkably increased by the addition of Cu2+ with 1332% efficiency. In the two operations, the main function of quinoline is reflected in fluorescence changes before and after its N atom coordination. With regard to the selectivity of ions, it is decided by the cavity size formed by the middle cyclocompounds. The design is instructional, because compared with the bridge mentioned before, it produces a three-dimensional bridge, and brings about better selectivity of particular ions.

17   18

Although quinoline moiety can be used both as the metal binding site and the fluorophore, the application of quinoline-based chemosensors in biological systems is limited by their optical properties. The main disadvantage of these chemosensors is high-energy UV excitation, which is possibly detrimental to biological tissues. It can induce autofluorescence from endogenous components and fluorescence at short wavelengths. Chemosensors 19 developed by Ballesteros (2009) et al. enlarged the conjugated system with 5-bromoindanone at the 8-position by Suzuki reaction. With this improvement, after being excited by visible light (λ_{ex}=495nm), 19 showed a 5-fold increase in the intensity of emission centered at 650 nm after Cu2+ was added.

19   20

Receptor is a key role in the design of chemosensors, by choosing azacrown[N,S,O] instead of DPA. Wang (2010) et al. synthesized a new chemosensor 20 based on ICT mechanism. Chemosensor 20 is an effective ratiometric fluorescent sensor for silver ion and bears the features of a large Stokes shift at about 173 nm, with red-shift up to 50 nm in the emission spectra, and brings high affinity for silver ions (log K = 7.21) in ethanol in comparison with other competitive d^{10} metal ions. Crown compounds are all along used as highly recognized receptor groups for particular ions. As for groups which are not easily bounded, for instance, K+, 18-crown-6 can be used to recognize it. However, crown compounds have some application limits. At the beginning, crown compounds comprised of different heteoatoms can not achieve highly efficient synthesis. Then, they are considerably poisonous to in vivo cells. This imposes restrictions on their application in organisms to some extent.

5.4 Detection of Hg^{2+}

Modified quinoline can also become very good binding group for Hg^{2+} ion. Han (2009) et al. reported highly selective and highly sensitive Hg^{2+} chemosensor 21 based on quinoline and porphyrin ring. The 21 complexation quenches the fluorescence of porphyrin at 646 nm and induces a new fluorescent enhancement at 603 nm. The fluorescent response of 21 towards Hg^{2+} and ^{1}H-NMR indicates Hg^{2+} ion is binding with the quinoline moiety. Yang (2007) et al. reported chemosensor 22, which connect 8-hydroxyquinoline with rhodamine and ferrocene, and recognize Hg^{2+} through opening and closing rhodamine ring before and after binding. At the same time, because the density of the interior electron cloud changes before and after binding, the electrochemistry signal of ferrocene varies, so that the detection accuracy is improved. Concerning Hg^{2+} recognition, there is another method that receives considerable attention, that is, modified thioamides on quinoline, using the sulfur addicting feature of Hg^{2+}, will be transformed into amides by Hg^{2+}, which will change the PET process, thus inducing the production of fluorescence.

21   22   23

Song (2006) et al. reported 8-hydroxyquinoline derivative chemosensor 23. In 23, the fluorescence background is very weak. 23 is demonstrated to be highly sensitive to the detection of Hg^{2+}, because the hydrolytic conversion of thioamides into amides catalyzed by Hg^{2+} is very efficient. NMR, IR, and mass studies indicate the Hg^{2+} ion induced the transformation of thioamide into amide.

5.5 Detection of Cr^{3+} and Fe^{3+}

Because paramagnetic Fe^{3+} and Cr^{3+} are reported as two of the most efficient fluorescence quenchers among the transition metal ions, the development of Fluorescence chemosensors working with these inherent quenching metal ions is a challenging job. Zhou (2008) et al. reported a FRET-based Cr^{3+} chemosensor 24. With increased FRET from 1,8-naphthalimide (donor) to the open, colored form of rhodamine (acceptor), the intensity of the fluorescent peak at 544 nm gradually decreased and that of new fluorescent band centered at 592 nm increased, 24 showed an 7.6-fold increase in the ratio of emission intensities (F_{592} nm/F_{544} nm). We (2011) reported a turn-on fluorescent probe 25 for Fe^{3+} based on the rhodamine platform. An improved quinoline fluorescent group, which could be excited by about 400 nm wavelength of light, was linked to the rhodamine platform. The emission of conjugated quinoline was partly in the range of rhodamine absorption, so 1,8-naphthalimide could be removed. And we also removed the hydroxy group at 8-position to reach a different coordination mode choice on the Fe^{3+} ions. 25 shows high selectively for Fe^{3+} over Cr^{3+} both in fluorescence and visible light (Fig. 10). JOB's plot indicate that, the 1:2 binding model between Fe^{3+} and 25, pH and cytotoxic effect also suggest that the new sensor is suitable for bioimaging. More importantly, the improved quinoline group can be excited by 800nm two-photon laser source, which is more suitable for bioimaging. We can utilize the FRET processes of quinoline and other fluorescent groups to design sensors, so as to take advantage of their respective merits. For example, we can make use of other groups' visible light changes, water solubility, and high sensitivity brought about by switch construction and so on to compensate demerits of quinoline groups. This will be one of the directions for designing sensors.

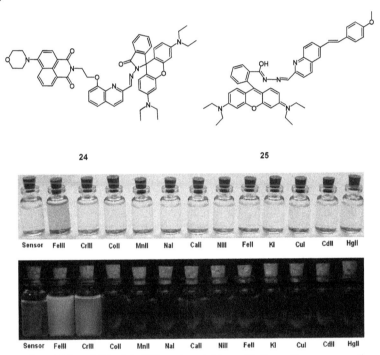

24   25

Fig. 10. Top: color of 25 and 25 with different metal ions. Bottom: fluorescence (λ_{ex} = 365 nm) change upon addition of different metal ions.

6. Conclusion

In this review, we cover quinoline-based chemosensors for detection of different metal ions. There has been tremendous interest in improving quinoline-based chemosensors due to its easy synthesis method, high sensitivity and stability. However, there is still much room for progress in its application in vivo such as water solubility, high selectivity, and fluorescence bio-imaging capacity. Accordingly, the design of receptor for different ions is very important. For example, 15 adopted the different bonding model to distinguish between

Zn^{2+} and Cd^{2+}. 24, through another fluorophore, thus achieved FRET process. Extended conjugated system of quinoline can be excited by two-photon laser source, and so on. We anticipate that more and more quinoline-based fluorescence chemosensors can be synthesized which are useful for detection of metal ions.

7. Acknowledgements

We acknowledge financial support by NSFC(21072001, 21102002), 211 Project of Anhui University for supporting the research.

8. References

Aoki, S.; Sakurama, K.; Matsuo, N.; Yamada, Y.; Takasawa, R.; Tanuma, S.; Shiro, M.; Takeda, K. & Kimura, E. (2006). A New Fluorescent Probe for Zinc(II): An 8-Hydroxy -5-N,Ndimethylaminosulfonylquinoline-Pendant   1,4,7,10-Tetraazacyclododecane. *Chemistry –A European journal* 12, 9066-9080

Ballesteros, E.; Moreno, D.; Gomez, T.; Rodriguez, T.; Rojo, J.; Garcia-Valverde, M. & Torroba, T. (2009). A New Selective Chromogenic and Turn-On Fluorogenic Probe for Copper(II) in Water-Acetonitrile 1:1 Solution. *Organic Letters* 11, 1269-1272

Banthia, S. & Samanta, A. (2006). A New Strategy for Ratiometric Fluorescence Detection of Transition Metal Ions. *The Journal of Physical Chemistry B* 110, 6437-6440

Bhalla, V.; Tejpal, R.; Kumar, M. & Sethi, A. (2009). Terphenyl Derivatives as "Turn On" Fluorescent Sensors for Mercury. *Inorganic Chemistry* 48, 11677-11684

Burdette, A. C. & Lippard, S. J. (2003). A Selective Turn-On Fluorescent Sensor for Imaging Copper in Living Cells. *Proceeding of the national academy of Science of the United States of America* 100, 3605-3610

Chen, X. –Y.; Shi, J.; Li, Y. –M.; Wang, F. –L.; Wu, X.; Guo, Q. –X. & Liu, L. (2009). Two-Photon Fluorescent Probes of Biological Zn(II) Derived from 7-Hydroxyquinoline. *Organic Letters* 11, 4426-4429

de Silva, A. P.; Nimal Gunaratne, H. Q.; Gunnlaugsson, T.; Huxley, A. J. M.; McCoy, C. P.; Rademacher, J. T. & Rice, T. E. (1997). Fluorescence-Based Sensing of Divalent Zinc in Biological Systems. *Chemical Reviews* 97, 1515-1566

Dobson, S. (1992) Cadmium: EnVironmental Aspects; World Health Organization:Geneva. ISBN 92-4-157135-7

Doebner, O.; Von Miller, W. & Ber, D. T. (1996). Synthesis of Pyrroloquinolinequinone Analogs Molecular Structure and Moessbauer and Magnetic Properties of Their Iron Complexessch. *Chem. Ges* 34, 6552-6555

Dujols, V.; Ford, F. & Czarnik, A. W. (1997). A Long-Wavelength Fluorescent Chemodosimeter Selective for Cu(II) Ion in Water. *Journal of American Chemistry Society* 119, 7386-7387

Falchuk, K. H. (1998). The molecular basis for the role of zinc in developmental biology. *Molecular and Cellular Biochemistry* 188, 41-48

Frederickson, C. J.; Kasarskis, E. J.; Ringo, D. & Frederickson, R. E. (1987). A quinoline fluorescence method for visualizing and assaying the histochemically reactive zinc (bouton zinc) in the brain. *Journal of Neuroscience Methods* 20, 91-103

Guerrini, G.; Taddei, M. & Ponticelli, F. (2011). Synthesis of Functionalized Quinolines and Benzo[c][2,7]naphthyridines Based on a Photo-Fries Rearrangement. *The Journal of Organic Chemistry* 76, 7597-7601

Han, Z. –X.; Luo, H. –Y.; Zhang, X. –B.; Kong, R. –M.; Shen, G. –L. & Yu, R. –Q. (2009). A ratiometric chemosensor for fluorescent determination of Hg2+ based on a new porphyrin-quinoline dyad. Spectrochimica Acta Part A: *Molecular and Biomolecular Spectroscopy* 72, 1084-1088

Hanaoka, K.; Kikuchi, K.; Kojima, H.; Urano, Y. & Nagano, T. (2004). Development of a Zinc Ion-Selective Luminescent Lanthanide Chemosensor for Biological Applications. *Journal of the American Chemical Society* 126, 12470-12476

Henary, M. M.; Wu, Y. & Fahrni, C. J. (2004). Zinc(II)-Selective Ratiometric Fluorescent Sensors Based on Inhibition of Excited-State Intramolecular Proton Transfer. *Chemistry –A European Journal* 10, 3015-3025

Hirano, T.; Kikuchi, K.; Urano, Y. & Nagano, T. (2002). Improvement and Biological Applications of Fluorescent Probes for Zinc, ZnAFs. *Journal of the American Chemical Society* 124, 6555-6562

Joshi, B. P.; Lohani, C. R. & Lee, K. H. (2010). A highly sensitive and selective detection of Hg(II) in 100% aqueous solution with fluorescent labeled dimerized Cys residues. *Organic & Biomolecular Chemistry* 8, 3220-3226

Jung, H. S.; Ko, K. C.; Lee, J. H.; Kim, S. H.; Bhuniya, S.; Lee J. Y.; Kim, Y.; Kim, S. J. & Kim, J. S. (2010). Rationally designed fluorescence turn-on sensors: a new design strategy based on orbital control. *Inorganic Chemistry* 49, 8552-8557

Komatsu, K.; Urano, Y.; Kojima, H. & Nagano, T. (2007). Development of an Iminocoumarin-Based Zinc Sensor Suitable for Ratiometric Fluorescence Imaging of Neuronal Zinc. *Journal of the American Chemical Society* 129, 13447-13454

Kimber, M. C.; Mahadevan, I. B.; Lincoln, S. F.; Ward, A. D. & Tiekink, E. R. T. (2000). The Synthesis and Fluorescent Properties of Analogues of the Zinc (II) Specific Fluorophore Zinquin Ester. *Journal of Organic Chemistry* 65, 8204-8209

Lee, D. Y.; Singh, N.; Kim, M. J. & Jang, D. O. (2010). Ratiometric fluorescent determination of Zn(II): a new class of tripodal receptor using mixed imine and amide linkages. *Tetrahedron* 66, 7965-7969

Li, G. –K.; Xu, Z. –X.; Chen, C. –F. & Huang, Z. –T. (2008). A highly efficient and selective turn-on fluorescent sensor for Cu2+ ion based on calix[4]arene bearing four iminoquinoline subunits on the upper rimw. *Chemical Communications* 1774-1776

Lim, N. C.; Schuster, J. V.; Porto, M. C.; Tanudra, M. A.; Yao, L.; Freake, H. C. & Brückner, C. (2005). Coumarin-Based Chemosensors for Zinc(II): Toward the Determination of the Design Algorithm for CHEF-Type and Ratiometric Probes. *Inorganic Chemistry* 44, 2018-2030

Liu, Z. –C.; Wang, B. –D.; Yang, Z. –Y.; Li, T. –R. & Li, Y. (2010). A novel fluorescent chemosensor for Zn(II) based on 1,2-(2'-oxoquinoline-3'-yl-methylideneimino) ethane. *Inorganic Chemistry Communications* 13, 606-608

Marct, W.; Jacob, C.; Vallee, B. L. & Fishcher, E. H. (1999). Inhibitory sites in enzymes: zinc removal and reactivation by thionein. *Proceeding of the national academy of Science of the United States of America* 96, 1936-1940

Mason, W. T. (1999). Fluorescent and Luminescent Probes for Biological Activit., 2nd cd.; Academic Press: New York. ISBN 0-12-447836-0

Anthea, M.; Hopkins, J.; McLaughlin, C. W.; Johnson, S.; Warner, M. Q.; LaHart, D. & Wright J. D. (1993). Human Biology and Health. Englewood Cliffs, New Jersey, USA: Prentice Hall. ISBN 0-13-981176-1

Matsubara, Y.; Hirakawa, S.; Yamaguchi, Y. & Yoshida, Z. (2011). Assembly of Substituted 2-Alkylquinolines by a Sequential Palladium-Catalyzed C-N and C-C Bond Formation). *Angew Angewandte Chemie International Edition* 50, 7670-7673

Meng, X. M.; Zhu, M. Z.; Liu, L. & Guo, Q. X. (2006). Novel highly selective fluorescent chemosensors for Zn(II). *Tetrahedron Letters* 47, 1559-1562

Mizukami, S.; Okada, S. & Kimura, S. (2009). Design and Synthesis of Coumarin-based Zn2+ Probes for Ratiometric Fluorescence Imaging. *Inorganic Chemistry* 48, 7630-7638

Moriuchi-Kawakami, T.; Sato, J. & Shibutani, Y. (2009). C3-Functionalized Cyclotriveratrylene Derivative Bearing Quinolinyl Group as a Fluorescent Probe for Cu2+. *Analytical Sciences* 25, 449-452

Nolan, E. M.; Jaworski, J.; Okamoto, K. -I.; Hayashi, Y.; Sheng, M. & Lippard, S. J. (2005). QZ1 and QZ2: Rapid, Reversible Quinoline-Derivatized Fluoresceins for Sensing Biological Zn(II). *Journal of the American Chemical Society* 127, 16812-16823

Pearce, D. A.; Jotterand, N.; Carrico, I. S. & Imperiali, B. (2001). Derivatives of 8-Hydroxy-2-methylquinoline are Powerful Prototypes for Zinc Sensors in Biological Systems. *Journal of the American Chemical Society* 123, 5160-5161

Peng, X. J.; Du, J. J.; Fan, J. L.; Wang, J. Y.; Wu, Y. K.; Zhao, J. Z.; Sun, S. G. & Xu, T. (2007). A Selective Fluorescent Sensor for Imaging Cd2+ in Living Cells. *Journal of American Chemistry Society* 129, 1500-1501

Sarkar, M.; Banthia, S.; Patil, A.; Ansari, M. B. & Samanta, A. (2006). pH-Regulated "Off–On" fluorescence signalling of d-block metal ions in aqueous media and realization of molecular IMP logic function. *New Jornal of Chemistry* 30, 1557-1560

Sarkar, M.; Banthia, S. & Samanta, A. (2006). A highly selective 'off–on' fluorescence chemosensor for Cr(III). *Tetrahedron Letters* 47, 7575-7578

Shiraishi, Y.; Ichimura, C. & Hirai, T. (2007). A quinoline–polyamine conjugate as a fluorescent chemosensor for quantitative detection of Zn(II) in water. *Tetrahedron Letters* 48, 7769-7773

Shults, M. D.; Pearce, D. A. & Imperiali, B. (2003). Versatile Fluorescence Probes of Protein Kinase Activity. *Journal of the American Chemical Society* 125, 14248-14249

Song, K. C.; Kim, J. S.; Park S. M.; Chung, K. –C.; Ahn, S. & Chang, S. –K. (2006). Fluorogenic Hg2+-Selective Chemodosimeter Derived from 8-Hydroxyquinoline. *Organic Letters* 8, 3413-3416

Tang, X. –L.; Peng, X. –H.; Dou, W.; Mao, J.; Zheng, J. –R.; Qin, W. –W.; Liu, W. –S.; Chang, J. & Yao, X. –J. (2008). Design of a Semirigid Molecule as a Selective Fluorescent Chemosensor for Recognition of Cd(II). *Organic Letters* 10, 3653-3656

Teolato, P.; Rampazzo, E.; Arduini, M.; Mancin, F.; Tecilla, P. & Tonellato, U. (2007). Silica Nanoparticles for Fluorescence Sensing of ZnII:Exploring the Covalent Strategy. *Chemistry –A European journal* 13, 2238-2245

Vallee, B. L. & Falchuk, K. H. (1993). The biochemical basis of zinc physiology. *Physiological Reviews* 73, 79-118

van Dongen, E. M. W. M.; Dekkers, L. M.; Spijker, K.; Meijer, E. W.; Klomp, L.W. W. J. & Merkx, M. (2006). Ratiometric Fluorescent Sensor Proteins with Subnanomolar

Affinity for Zn(II) Based on Copper Chaperone Domains. *Journal of the American Chemical Society* 128, 10754-10762

Walkup, G. K.; Burdette, S. C.; Lippard, S. J. & Tsien, R. Y. (2000). A New Cell-Permeable Fluorescent Probe for Zn2+. *Journal of the American Chemical Society* 122, 5644-5645

Wang, H. –H.; Xue, L.; Qian, Y. –Y. & Jiang, H. (2010). Novel Ratiometric Fluorescent Sensorfor Silver Ions. *Organic Letters* 12, 292-295

Wang, S. X.; Meng, X. M. & Zhu, M. Z. (2011). A naked-eye rhodamine-based fluorescent probe for Fe(III) and its application in living cells. *Tetrahedron Letters* 52, 2840-2843

Wang, X. H.; Cao, J. & Chen, C. F. (2010). A Highly Efficient and Selective Turn-on Fluorescent Sensor for Cu2+ Ion. *Chinese Journal of Chemistry* 28, 1777-1779

Wolf, C.; Mei, X. F. & Rokadia, H. K. (2004), Selective detection of Fe(III) ions in aqueous solution with a 1,8-diacridylnaphthalene-derived fluorosensor. *Tetrahedron Letters* 45, 7867-7871

Woodroofe, C. C.; Masalha, R.; Barnes, K. R.; Frederickson, C. J. & Lippard, S. J. (2004). Membrane-Permeable and -Impermeable Sensors of the Zinpyr Family and Their Application to Imaging of Hippocampal Zinc In Vivo. *Chemistry & Biology* 11, 1659-1666

Woodroofe, C. C. & Lippard, S. J. (2003). A Novel Two-Fluorophore Approach to Ratiometric Sensing of Zn2+. *Journal of the American Chemical Society* 125, 11458-11459

Xue, L.; Wang, H. –H.; Wang, X. –J. & Jiang, H. (2008). Modulating Affinities of Di-2-picolylamine (DPA)-Substituted Quinoline Sensors for Zinc Ions by Varying Pendant Ligands. *Inorganic Chemistry* 47, 4310-4318

Xue, L.; Liu, C. & Jiang, H. (2009). A ratiometric fluorescent sensor with a large Stokes shift for imaging zinc ions in living cellsw. *Chemical Communications* 1061-1063

Xue, L.; Liu ,Q. & Jiang, H. (2009). Ratiometric Zn2+ Fluorescent Sensor and New Approach for Sensing Cd2+ by Ratiometric Displacement. *Organic Letters* 11, 3454-3457

Xue, L.; Li, G. P.; Liu, Q.; Wang, H. H.; Liu, C.; Ding, X.; He, S. & Jiang, H. (2011). Ratiometric Fluorescent Sensor Based on Inhibition of Resonance for Detection of Cadmium in Aqueous Solution and Living Cells. *Inorganic Chemistry* 50, 3680-3690

Yang, H.; Zhou, Z.; Huang, K.; Yu, M.; Li, F.; Yi, T. & Huang, C. (2007). Multisignaling Optical-Electrochemical Sensor for Hg2+ Based on a Rhodamine Derivative with a Ferrocene Unit. *Organic Letters* 9, 4729-4732

Zalewski, P. D.; Forbes, I. J. & Betts, W. H. (1993). Correlation of apoptosis with change in intracellular labile Zn(II) using zinquin [(2-methyl-8-p-toluenesulphonamido-6-quinolyloxy)acetic acid], a new specific fluorescent probe for Zn(II). *Biochemical Journal* 296, 403-408

Zalewski, P. D.; Forbes, I. J.; Borlinghaus, R.; Betts, W. H.; Lincoln, S. F. & Ward, A. D. (1994). Flux of intracellular labile zinc during apoptosis (gene-directed cell death) revealed by a specific chemical probe, Zinquin. *Chemistry & Biology* 1, 153-161

Zalewski, P. D.; Millard, S. H.; Forbes, I. J.; Kapaniris, O.; Slavotinek, A.; Betts, W. H.; Ward, A. D.; Lincoln S. F. & Mahadevan, I. (1994). Video image analysis of labilc zinc in viable pancreatic islet cells using a specific fluorescent probe for zinc. *Journal of Histochemistry and Cytochemistry* 42, 877-884

Zhang, Y.; Guo, X. F.; Si, W. X.; Jia, L. & Qian, X. H. (2008). Ratiometric and Water-Soluble Fluorescent Zinc Sensor of Carboxamidoquinoline with an Alkoxyethylamino Chain as Receptor. *Organic Letters* 10, 473-476

Zhou, X. Y.; Yu, B. R.; Guo, Y. L.; Tang, X. L.; Zhang, H. H. & Liu, W. S. (2010). Both Visual and Fluorescent Sensor for Zn2+ Based on Quinoline Platform. *Inorganic Chemistry* 49, 4002-4007

Zhou, Z.; Yu, M.; Yang, H.; Huang, K.; Li, F.; Yi, T. & Huang C. (2008). FRET-based sensor for imaging chromium(III) in living cellsw. *Chemical Communications* 3387-389

Part 2

Photochemistry on Metal Oxides

2

Photocatalytic Deposition of Metal Oxides on Semiconductor Particles: A Review

Miguel A. Valenzuela, Sergio O. Flores, Omar Ríos-Bernÿ,
Elim Albiter and Salvador Alfaro
Lab. Catálisis y Materiales, ESIQIE-Instituto Politécnico Nacional Zacatenco,
México D.F.,
México

1. Introduction

As it has been well recognized in the last decade, heterogeneous photocatalysis employing UV-irradiated titanium dioxide suspensions or films in aqueous or gas media, is now a mature field [Chong et al. 2010, Ohtani B. 2010, Paz Y., 2010]. Semiconductor photocatalysis is considered as a green process that focuses basically on exploiting solar energy in many ways. Its investigations have been mainly targeted to the degradation/mineralization of organic pollutants and water splitting solar energy conversion, among others. However, there are other exciting applications such as metal photodeposition, organic synthesis, photoimaging, antibacterial materials, which have now an intense investigation [Wu et al. 2003, Chan S. & Barteau M. 2005, Litter M. 1999, Fagnoni et al. 2007, Choi W. 2006, Valenzuela et al. 2010, Zhang et al. 2010].

In particular, the photodeposition has been used since the decade of 70's by the pioneer work of Bard [Kraeutler and Bard, 1978] to prepare supported-metal catalysts and photocatalysts as well as to recover noble metals and to remove metal cations from aqueous effluents [Ohyama et al. 2011]. In this case, the reduction of each adsorbed individual metal ions occurs at the interface by acceptance of electrons from the conduction band forming a metallic cluster. A variant of metals deposition is the reductive deposition of metal oxides and a clear example of this route is the photocatalytic reduction of Cr (VI) which is transformed to Cr(III), so that in acidic environment, chromates are easily converted to Cr_2O_3 [Lin et al. 1993 and Flores et al. 2008].

The oxidative deposition of metal oxides is less frequently reported and it has been demonstrated that proceeds via the oxidative route [Tanaka et al. 1986]. For instance, checking the electrochemical potentials of Mn and Pb (Table 1), they are more easily oxidized by the valence band holes than reduced by the conduction band electrons in presence of TiO_2 as follows [Wu et al. 2003]:

$$Mn^{2+} + 2H_2O \xrightarrow{\quad TiO_2 \quad} MnO_2 + 4H^+ \tag{1}$$

$$Pb^{2+} + 2H_2O \xrightarrow{TiO_2} PbO_2 + 4H^+ \tag{2}$$

Half cell	E° (V)
MnO_2/Mn^{2+}	1.23
Mn^{2+}/Mn	-1.18
PbO_2/Pb^{2+}	1.46
Pb^{2+}/Pb	-0.12
Tl_2O_3/Tl^{1+}	0.02
Tl^{1+}/Tl	-0.336
Co^{2+}/Co	-0.28
Cr^{6+}/Cr^{3+}	1.232
Cr^{3+}/Cr	-0.744

Table 1. Half wave potentials of different couples.

These two reactions represent a good example of the photocatalytic deposition of metal oxides in aqueous solution onto titanium dioxide. This means that the complete photocatalytic cycle should consider the photoredox couple in which one metallic ion (single component) in solution is oxidized and the oxygen of the media is reduced. The deposition is driven by particle agglomeration after reaching their zero point charge and a critical concentration to be deposited on the surface of the semiconductor. It has been reported that single component metal oxides for example, PbO_2, RuO_2, U_3O_8, SiO_2, SnO_2, Fe_2O_3, MnO_2, IrO_2 and Cr_2O_3 can be deposited on semiconductor particles following a photo-oxidative or photo-reductive route [Maeda et al. 2008].

On the other hand, when a semiconductor is irradiated with UV light in presence of aqueous solutions containing dissolved Ag^+ or Pb^{2+} cations a redox process is undertaken giving rise to the reduction of silver ions or the oxidation of lead ions, according to the following reactions [Giocondi et al. 2003]:

$$Ag^+ + e^- \rightarrow Ag^{\circ} \tag{3}$$

$$Pb^{2+} + 2H_2O + 2h^+ \rightarrow PbO_2 + 4H^+ \tag{4}$$

Lately, it has been reported the photocatalytic deposition of mixed-oxides such as $Rh_{2-y}Cr_yO_3$ dispersed on a semiconductor powder with applications in the water splitting reaction [Maeda et al. 2008]. Hence, we intend to offer the reader a condensed overview of the work done so far considering photocatalytic deposition of a single or mixed oxide on semiconductor materials by either oxidative or reductive processes.

2. Photocatalytic oxidation of a single component

2.1 Pb^{2+}

In regard with the very negative impact of lead on environment and population, many efforts are conducted to remove it from water of distinct origins. It is commonly removed by precipitation as carbonate or hydroxide; besides other physicochemical methods are available to lead elimination. The maximun contaminant level in drinking water established by EPA is 15µg/l. However, it is desirable the total elimination of lead due to its extreme potential toxicity [Murruni et al., 2007]. In a first report concerning to the photodeposition of Pb^{2+} ions on TiO_2 and metallized TiO_2, it was found that the former only produces PbO, whereas the last converts efficiently Pb ions to PbO_2 [Tanaka et al., 1986]. In the same work, it was proposed a reaction mechanism in two steps involving the reduction of oxygen to form the superoxide ion and the subsequent oxidation of Pb ions:

$$2O_2 + 2e^- \rightarrow 2O_2^- \tag{5}$$

$$Pb^{2+} + 2O_2^- \rightarrow PbO_2 + O_2 \tag{6}$$

Their mechanism was supported by experiments carried out in different atmospheres: nitrogen, argon and oxygen at several partial pressures. In N_2 and Ar, irradiation of TiO_2 suspensions did not result in lead oxide formation. It is worth noting that a high pressure Hg lamp (500 W) was used for all their photocatalytic deposition reactions.

Litter et al. 1999, have proposed a different mechanism which involves two consecutive electron transfer reactions. Lead ions are oxidized by holes or by hydroxyl radicals passing through the divalent to the tetravalent state, equations 7 and 8:

$$h^+/HO^{\bullet} + Pb^{2+} \rightarrow Pb^{3+} + OH^- \tag{7}$$

$$h^+/HO^{\bullet} + Pb^{3+} \rightarrow Pb^{4+} + OH^- \tag{8}$$

A further enhancement was achieved with platinized TiO_2 by decreasing the overpotential of oxygen. In fact, the role of oxygen is crucial to carry out the photocatalytic cycle and it has found a linear dependence of oxygen partial pressure based on a Langmuir-Hinshelwood mechanism [Torres & Cervera-March, 1992].

2.2 Co^{2+}

Recently, it has been highlighted many applications of cobalt compounds deposited on semiconductors such as: catalysts for solar oxygen production, gas sensors, batteries, electrochromic devices, among others [Steinmiller & Choi, 2009; Tak & Yong, 2008]. In particular, the photodeposition of Co_3O_4 spinel phase on ZnO has been prepared by two routes, one consisting in the direct photo-oxidation of Co^{2+} ions to Co^{3+} ions and the other by an indirect procedure involving the reduction of Co^{2+} to Co° and the oxidation of metallic cobalt to Co_3O_4 by means of the oxygen coming from the photo-oxidation of water.

By using the direct deposition route, a ZnO electrode was immersed in an aqueous solution of $CoCl_2$ mantaining the pH constant at 7 and illuminating with UV light ($\lambda= 302$ nm). Due to the oxidation potential of Co^{2+} to form Co_3O_4 is 0.7 V at pH= 7 at low concentrations of Co^{2+} (10^{-3} M) and the valence band edge of ZnO is located at around 2.6 V vs NHE, the photogenerated holes can easily oxidize Co^{2+} ions to Co^{3+} ions. The complete photocatalytic cycle must also include the reduction of water or dissolved oxygen in the solution to have an efficient Co^{2+} photo-oxidation [Steinmiller & Choi, 2009]:

$$2 Co^{2+} + 3H_2O \rightarrow Co_2O_3 + 6H^+ + 2e^- \tag{9}$$

From a thermodynamic point of view, Co^{3+} ions can be deposited on any semiconductor that has a valence band edge located at a more positive potential than that of the Co^{2+} ions, as shown in Figure 1.

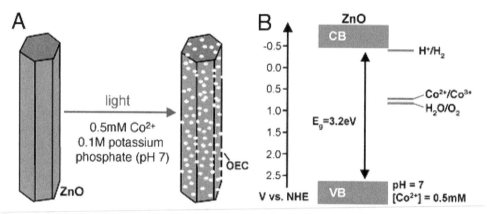

Fig. 1. Schematic representation. (A) Photochemical deposition of the Co-based catalyst on ZnO and (B) relevant energy levels. [Reproduced from Steinmiller and Choi, with permission from PNAS, copyright 2009 by National Academy of Sciences].

By the second route, the ZnO nanowires were grown by ammonia solution hydrothermal method and then coated with Co_3O_4 using a photocatalytic reaction. This last method was selected considering that the redox reactions of aqueous chemical species on irradiated semiconductor surfaces has characteristics of site-specific growth. Briefly, the ZnO nanowire array was immersed in an aqueous solution of $Co(NO_3)_2$ and was irradiated with UV-light of 325 nm from minutes to 24 h. According to the results of this work, the morphology of the heterostructures depended on the photocatalytic reaction parameters such as the concentration of Co^{2+} in solution, UV irradiation time and the geometrical alignment of the ZnO nanowires. The photocatalytic process was explained in terms of redox cycle which includes the reduction of Co^{2+} species into $Co°$ and the oxidation of water to produce O_2. In fact, after irradiation of ZnO with photon energy larger than the band gap of ZnO (3.4 eV) generates the charge carriers (electron-hole pairs). The photogenerated electrons in the conduction band reduce Co^{2+} to $Co°$ favoring the accumulation of holes in the valence band. In addition, the holes oxidizes water to molecular oxygen which carries out the partial oxidation of $Co°$ to $Co^{2+}Co_2^{3+}O_4$ spinel, as outlined in Figure 2. It seemed that this simple, room temperature and selective photodeposition process can be applicable to other semiconductors (e.g. TiO_2, CdS, SnO_2...) or to other shapes of nanomaterials.

2.3 Hg°

Mercury is a neurotoxic heavy metal frecuently found in industrial wastewaters at concentrations higher than 0.005 ppm and unfortunately it cannot be bio- or chemically degraded [Clarkson & Magos, 2006]. It is released to the environment by coal combustion and trash incineration, mainly as gaseous mercury producing methyl mercury in the aquatic ecosystem by the action of sulfate-reducing bacteria. Certainly, due to its multiple industrial applications (e.g. pesticides, paints, catalysts, electrical device etc.) it can also be found in

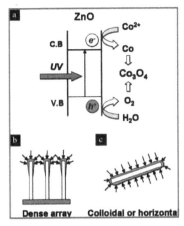

Fig. 2. (a) photocatalytic reaction scheme showing the reduction of Co ions to metallic Co and its oxidation to the spinel Co_3O_4, (b) Co_3O_4 deposited on the tip of ZnO, (c) Co_3O_4 deposited on the whole surface of ZnO. [Reproduced from Tak and Yong with permission from The Journal of Physical Chemistry; copyright 2008].

solution as Hg (II). Several methods have been investigated for its removal or control, such as, precipitation, ion exchange, adsorption, coagulation and reduction. However, the photocatalytic oxidation (PCO) of gaseous mercury by UVA-irradiated TiO_2 surfaces has been reported as a good option for its capture [Snider and Ariya, 2010].

For instance, an enhanced process including adsorption of gaseous mercury on silica-titania nanocomposites and then its photocatalytic oxidation has been published [Li and Wu, 2007]. However, some problems of reactivation of the nanocomposite as well as pore structure modification during Hg and HgO capture and deposition have to be solved. In the same work, it has been proposed the use of pellets of silica-titania composites and it was found that a decrease of contact angle was likely responsible for mercury capture for long periods. Usually, the experimental systems to evaluate the PCO of gaseous mercury include water vapor to supply the OH radicals required for the oxidation and a source of UVA irradiation (320-400 nm, 100 W Hg lamp). Figure 3 shows a typical schematic diagram for the PCO using titania-silica pellets.

According to the results obtained for the PCO of Hg in gas phase using a titania-silica nanocomposite [Li and Wu, 2007] , it has been proposed the following reaction mechanism:

$$TiO_2 + h\upsilon \rightarrow e^- + h^+ \tag{10}$$

$$H_2O \leftrightarrow H^+ + OH^- \tag{11}$$

$$h^+ + OH^- \rightarrow {}^\bullet OH \tag{12}$$

$$h^+ + H_2O \rightarrow {}^\bullet OH + H^+ \tag{13}$$

$${}^\bullet OH + Hg^\circ \rightarrow HgO \tag{14}$$

which was successfully expressed by the Langmuir-Hinshelwood model. The rate of photo-oxidation of Hg was significantly inhibited by the presence of water vapor explained in terms of a competitive adsorption of water and mercury on the surface of TiO_2.

Efforts to gas mercury oxidation in air are now focused by using immobilized semiconductors irradiated with visible light looking for a potentially safe, low-cost process [Snider and Ariya, 2010].

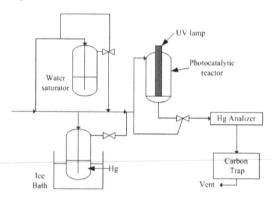

Fig. 3. Schematic diagram for the photocatalytic oxidation of mercury gas. After [Li and Wu, 2007].

2.4 Mn^{2+}

Manganese (II) in aquatic systems is a problem of environment concern due to its slow oxidation to MnO_2 which is responsible for the formation of dark precipitates. The photocatalytic oxidation of Mn^{2+} to Mn^{4+} in the presence of irradiated titanium dioxide has been scarcely studied since the 80's [Tanaka et al. 1986, Lozano et al. 1992 and Tateoka et al. 2005, Matsumoto et al., 2008]. This process represents an alternative route for its removal and the resulting material could be used as supported metal oxides catalysts [Tateoka et al. 2005, Matsumoto et al., 2008]. In the first publication, it was used concentrations ranging from 10^{-4}-10^{-3} mol/L aqueous solutions of Mn^{2+} with irradiated TiO_2 and Pt/TiO_2 photocatalysts using a high pressure Hg lamp of 500 W. Mn^{2+} conversion to Mn^{4+} was 98 and 78% from low to high concentrations onto Pt-loaded TiO_2 in 1 h of irradiation time. In the second work published in 1992, the oxidation of Mn^{2+} was carried out in acidic conditions using TiO_2 Degussa P-25 and irradiating with a Hg vapor lamp of 125 W at initial concentration of Mn^{2+} of 10^{-4} mol/L. One of the visual evidence of the photocatalytic oxidation of Mn^{2+} to Mn^{4+} is the appearance of a slight dark coloration over the TiO_2. The overall reaction scheme for the photo-oxidation was presented as follows:

$$Mn^{2+} + \frac{1}{2} O_2 + H_2O \rightarrow MnO_2 + 2H^+ \tag{14}$$

In a recent work, it was studied the photodeposition of metal and metal oxide at the TiO_x nanosheet to observe the photocatalytic active site (Matsumoto et al., 2008). It was investigated the photodeposition of Ag, Cu, Cu_2O and MnO_2 at a TiO_x nanosheet with a lepidocrocite-type structure prepared from K-Ti-Li mixed oxide. As expected, the photoreduction of Ag, Cu and Cu_2O, occurred mainly at edges where the photoproduced electrons move in the network of Ti^{4+} ions in the nanosheet. On the other hand, the photo-oxidation of Mn^{2+} ions occurred on all over the surface of the nanosheet, which is indicative of the presence of holes at the O^{2-} ion. The photo-oxidation reaction was carried out in a diluted solution of $MnSO_4$ (10^{-6} M) in air at room temperature irradiating with UV light of 265 nm. It was found that pH played an important role in the photodeposition: the amount of formed MnO_2 increased at higher values of pH, and no metallic Mn was observed at lower values of pH (pH=2.1). The coupled reactions that takes place during the photodeposition of MnO_2 are described by eqs. (15,16):

$$Mn^{2+} + 2H_2O + 4h^+ \text{ (produced in VB)} \rightarrow MnO_2 + 4H^+ \text{ (on the surface)} \qquad (15)$$

$$O_2 \text{ (in air)} + 4H^+ + 4e^- \text{ (produced in CB)} \rightarrow 2H_2O \text{ (at the edge)} \qquad (16)$$

A complete model explaining the photodeposition process and charge mobility are illustrated in Figure 4. In other words, according to the results reported by (Matsumoto et al., 2008) the photoproduced electrons move at the 3d orbital conduction band of the Ti^{4+} network in the nanosheet, whereas the photoproduced holes are located at the 2p orbital as O^{2-} species at the surface. Finally, the charge carriers recombination is favored under low pH which was found as a key parameter to control the photoprocess on the oxide nanosheet.

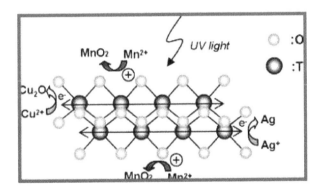

Fig. 4. Model of the movements of the photoproduced electron and hole at the TiO_x nanosheet with a lepidocrocite-type structure. The electron moves in the 3d CB consisting of the Ti^{4+} network in the nanosheet and then reduces Ag^+ and Cu^{2+} at the edge, while the hole exists at the 2p VB consisting of the O^{2-} surface and oxidizes Mn^{2+} on the surface. Reproduced from Matsumoto et al., with permission from The Journal of Physical Chemistry C, copyright 2008.

2.5 Fe^{2+}

One of the main drawbacks to commercialize the TiO_2 photocatalytic process at large scale is the use of UV light as irradiation source. Then, many efforts have been done during the past two decades to develop new photocatalysts active under visible light [Choi, 2006]. For instance, the presence of Fe^{3+} ions on TiO_2 favors the absorption of photons in the visible region as well as accelerates the photocatalytic oxidation of organic compounds. In this case, Fe^{3+} ions reduce to Fe^{2+} by the photoelectrons of the conduction band avoiding the charge recombination and increasing the photonic efficiency. However, the reverse process, this means the photo-oxidation of Fe^{2+} has been scarcely studied. A photoelectrochemical oxidation of Fe^{2+} ions on porous nanocrystalline TiO_2 electrodes was studied by using in situ EQCM (electrochemical quartz crystal microbalance) technique [Si et al., 2002]. In this work, it was found that the pH of iron precursor solution plays an important role in terms of the amount of adsorbed Fe^{2+} ions. The maximum value was 1.1 mmol Fe^{2+} at pH 4. The stability and the adsorption process was studied by the EQCM technique and it was found that the adsorption amount of Fe^{2+} ions on TiO_2 support was not affected by bias potential drop. The above result was attributed to Fe^{2+} ions are coordinated with hydroxyl groups of TiO_2 surface by the following reaction:

$$\text{Ti-OH} + Fe^{2+} \rightarrow \text{Ti-OH} \ldots Fe^{2+} \tag{17}$$

As is well known at low pH values, TiO_2 has negative surface charge favoring the electrostatic attraction of Fe^{2+} ions. Therefore, the adsorption-desorption behavior of Fe^{2+} ions on TiO_2 surface is strongly affected by pH changes. After irradiation of the adsorbed Fe^{2+} ions on TiO_2 the following photochemical reactions can be expected:

$$TiO_2 + h\nu \rightarrow h^+_{vb} + e^-_{cb} \tag{18}$$

$$\text{Ti-OH} \ldots Fe^{2+} + h^+_{vb} \rightarrow \text{Ti-OH} \ldots Fe^{3+} \tag{19}$$

$$(H_2O)_{surf} + h^+_{vb} \rightarrow (\bullet OH)_{surf} + H^+ \tag{20}$$

$$2Fe^{2+}_{aq} + 2(\bullet OH)_{surf\ or\ aq} + H_2O \rightarrow Fe_2O_3 + 4H^+ \tag{21}$$

2.6 Ce^{3+} and Sn^{2+}

Nowadays, the preparation of semiconductor nanoparticles with precise control of size and morphology has found new applications as ion-conducting,sun-screening, anti-corrosion and electro-catalytic properties [Kamada & Moriyasu 2011]. For instance, CeO_2 and SnO_2 have been synthesized as semiconducting oxide films by a photodeposition method [Kamada & Moriyasu 2011]. This method has the advantage of depositing homogeneously a thin film of the respective semiconductor by manipulating certain parameters such as concentration of the precursor, time and intensity of the irradiation, etc.

In the work reported by Kamada and Moriyasu, a photo-excited electroless deposition was carried out by the irradiation with UV light of an aqueous solution of cerium triacetate in a platinum substrate. It was observed an enhancement of the deposition of CeO_2, which was explained in terms of an electron transfer local cell mechanism. In this case, Ce^{3+} was oxidized by dissolved oxygen through an electron transfer in the Pt substrate and then transformed in a CeO_2 thin film, as shown in Figure 5. Surprisingly, the deposition rate was detrimentally affected by increasing the concentration of Ce^{3+} ions.

In a similar way, Sn^{2+} ions were anodically oxidized to Sn^{4+} and deposited on a Pt electrode with UV light irradiation. This process was followed through a different reaction mechanism than that of cerium. Tin oxide deposition proceeded by a photochemical reaction started with the disproportionation of Sn^{2+} and the further production of Sn^o and

Sn⁴⁺. Then Sn^{4+} was hydrolized to the insoluble H_2SnO_3, which finally is decomposed to SnO_2.

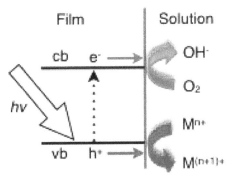

Fig. 5. A schematic representation of the photocatalytic oxidation of metal oxides in presence of a conductive substrate (Pt film) and dissolved oxygen. After [Kamada and Moriyasu 2011].

3. Photocatalytic reduction of single component

3.1 MnO₄⁻/MnO₂

This particular route for depositing metal oxides on semiconductors, also called reductive deposition, has been studied intensively for its potential in environmental remediation: for instance in the partial reduction of chromates (Cr^{6+} are extremely toxic) to the much less toxic Cr^{3+} or for UO_2^{2+} to UO_2 or but also in the preparation of special catalysts containing Cu_2O obtained by the partial reduction of Cu^{2+} to Cu^{1+} [Wu et al., 2003].

Lately, it has been reported works devoted to the reductive deposition of Mn_3O_4 or RuO_2 on titanium dioxide by using $KMnO_4$ contained in waste water or pure aqueous solutions of $KRuO_4$. The reaction mechanism involves a cathodic process where anions (e.g. CrO_4^{2-}, MnO_4^-, etc.) having strong oxidation power effectively accept the photogenerated electrons of the conduction band of TiO_2 after irradiation with UV light and the deposition of the corresponding oxide. On the other hand, in the anodic process the holes found in the valence band oxidize the sacrificial oxidant agent to produce the proton required for the photoreduction of the anion.

In this sense, Nishimura et al., 2008, have prepared coupled catalysts nanoparticles of MnO_2/TiO_2 by the photoreduction of harmful MnO_4 anions in water, see Fig. 6, and applied to the decomposition of hydrogen peroxide in the dark or irradiated with UV light. This coupled semiconductors can improve the charge separation efficiency through interfacial electron transfer. In addition, it is well known the catalytic properties of MnO_2 for the oxidation of organic pollutants which coupled with TiO_2 could have a special synergism in conventional catalytic or photocatalytic reactions. It was used a 10^{-3} M aqueous solution of $KMnO_4$ at pH 7, UV light ($\lambda > 300$ nm) and inert atmosphere to carry out the photoreduction reaction of manganate ions. In a blank experiment during the irradiation of the solution of $KMnO_4$ (without TiO_2) it was only found a partial decomposition of MnO_4^- ions to MnO_4^{2-} and O_2. The photodeposition of Mn_3O_4 on TiO_2 was confirmed by XPS and these stick-shaped nanoparticles were converted to cubic β-MnO_2 by heating a 600°C.The overall photodeposition reaction of Mn_3O_4 was as follows:

$$3MnO_4^- + 12H^+ \xrightarrow[\text{hv>300 nm}]{TiO_2} Mn_3O_4 + 6H_2O + O_2 \qquad (22)$$

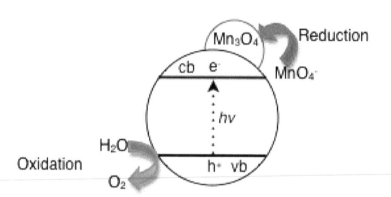

Fig. 6. Scheme showing the simultaneous photocatalytic reduction of permanganate anions in aqueous solution forming Mn_3O_4 and the photooxidation of water forming oxygen during UV illumination of TiO_2. After Nishimura et al., 2008.

3.2 RuO_4^-/RuO_2

The photocatalytic decomposition of water strongly requires the presence of effective catalysts for hydrogen and oxygen evolution. Usually, most published works are focused to the overall water splitting and a few have independently tested the water photo-oxidation reaction. In particular, the water photo-oxidation has been successfully studied with partially dehydrated RuO_2. However, its loading onto substrate surfaces by the conventional thermal methods lead to deep dehydration and sintering, reducing dramatically its activity and stability. An early work of Mills et al., 2010, has achieved the photodeposition of RuO_2 on titanium dioxide by a simple reaction of an aqueous solution of $KRuO_4$ mixed with TiO_2 and irradiation with a Xe or Hg lamp and Ce^{4+} ions as sacrificial electron donor. The following reaction scheme was proposed:

$$4RuO_4^- + 4H^+ + 4TiO_2 \rightarrow 4TiO_2/RuO_2 + 3O_2 + 2H_2O \qquad (23)$$

The photoreduction of ruthenate ion (RuO_4^-) in the absence of the titania photocatalyst remain unchanged.

4. Photocatalytic oxidation to obtain mixed oxides

4.1 $Rh_{2-y}Cr_yO_3$

The direct photodeposition of nanoparticulate mixed oxides on semiconductors was firstly reported by Maeda et al. [Maeda et al. 2008] supported in the pioneer work of Kobayashi et al. 1983, who studied the simultaneous photodeposition of Pd/PbO_2 and

Pt/RuO_2 on single crystals of TiO_2. Searching a good photocatalyst for overall water splitting, Maeda et al., 2006, developed a complex semiconductor $(Ga_{1-x}Zn_x)(N_{1-x}O_x)$ as a promising stable material active under visible light irradiation. However, this semiconductor only presented activity for water oxidation and its activity for water reduction was very low. Therefore, an effective modification of the GaN:ZnO semiconductor to promote the water reduction photoactivity was required. As is well known, noble metals or transition-metal oxides are often employed as cocatalysts to facilitate the water reduction reaction. Then, it was proposed the preparation of a noble-metal/mixed oxide (core/shell) supported on the GaN:ZnO solid solution by in situ photodeposition method [Maeda et al., 2006]. A two steps procedure was employed, Rh nanoparticles were firstly deposited on the mixed support with an aqueous precursor of $Na_3RhCl_6.H_2O$ and then Cr_2O_3 was deposited from a K_2CrO_4 solution, in both steps visible light irradiation was employed ($\lambda>400$ nm), as shown in Fig. 7. The authors confirmed the formation of a Rh/Cr_2O_3 core/shell nanoparticle with an average size of the ensemble of 12 nm and found a dramatical change in photocatalytic activity for overall water splitting in comparison with Rh or Cr_2O_3/GaN:ZnO supported systems.

In a second similar work of Maeda et al., 2008, it was reported a method to prepare mixed oxides of rhodium and chromium on five different semiconductors. They used aqueous solutions of $(NH_4)RhCl_6$ and K_2CrO_4 containing dispersed semiconductor powders and irradiated them during 4 hours with wavelengths whose energy exceeded those of each semiconductor band gap, as shown in Table 2.

Semiconductor	Rhodium (%wt)	Chromium (%wt)	Irradiation wavelength (nm)
$(Ga_{1-x}Zn_x)(N_{1-x}O_x)$ (x=0.12)	1	1.5	>400
$(Zn_{1+x}Ge)(N_2O_x)$ (x=0.44)	1	1.5	>400
$SrTiO_3$	0.5	0.75	>200
$Ca_2Nb_2O_7$	0.5	0.75	>200
ß-Ga_2O_3	0.5	0.75	>200

Table 2. Semiconductor powders and Rh-Cr content for mixed oxide photodeposition.

Based on XPS characterization, authors concluded that photodeposited mixed oxides have the composition $Rh_{2-y}Cr_yO_3$ and explained that the photoreduction of both, Rh^{3+} and Cr^{6+} proceeds via a band-gap transition of the semiconductor powder.

Furthermore, it was found that this mixed oxide is only formed when Rh and Cr are simultaneously present in the precursor solution. The photocatalytic performance of the materials was investigated for the evolution of H_2/O_2 in water splitting displaying different photocatalytic activity values depending of the support employed. In particular, photocatalyst containing the mixed oxide $Rh_{2-y}Cr_yO_3$ exhibited a two fold activity compared to that of semiconductor alone.

4.2 NiCoO_x

In 2006, Buono-Core et al. 2006 reported the photodeposition of $NiCoO_x$ on Si (100). Interest in this mixed oxide system regards on its antiferromagnetic characteristics. Authors synthesized $NiCo(DBA)_2$ as a single source precursor for the preparation of NiCo mixed oxide thin films, Figure 8. A solution of precursor in chloroform was prepared and then spin coated onto Si (100) chips. The films were irradiated under a 200 W Hg-Xe lamp ($\lambda \sim 254$ nm) until no ligand absorptions were observed in FT-IR. Characterization by AFM and XRD lead the authors to conclude about the amorphous nature of the mixed oxide films. After annealing at 600 °C of those films, XRD evidenced individual NiO and CoO oxides confirming the metastable nature of $NiCo_x$ films. Furthermore, EDAX analyses demonstrated homogeneity of Ni and Co dispersion throughout Si (100) surface. Finally, authors suggest the extendibility of this technique for rendering a wide range of binary metal oxide phases.

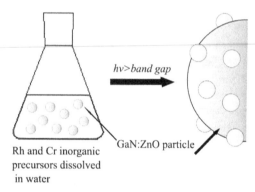

$hv > band\ gap$

Rh and Cr inorganic
precursors dissolved
in water

GaN:ZnO particle

Fig. 7. Scheme showing the photoconversion of inorganic precursors of Rh and Cr and their deposition of particles of GaN:ZnO. After from Maeda et al., 2008

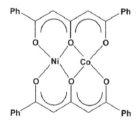

Fig. 8. $NiCo(DBA)_2$, DBA stands for Dibenzoylacetone.

5. Conclusions

Photocatalytic deposition methods have been shown to be of high potentiality for loading small-size dispersed metal oxides on powder or film semiconductors. This method also is a promising technique to obtain composite nanomaterials with the possibility to control the structural properties. Size, morphology and structure of the deposited oxides depend of the concentration of precursor and semiconductor, pH of the solution, light intensity and wavelength, illumination time and the type of sacrificial electron acceptor employed.

Although most work has been focused to the use of titanium dioxide as supporting material, other semiconductors have now been investigated (e.g. ZnO, WO$_3$, SnO$_2$, ZnS, GaO). So that, it is possible to design new advanced compositing materials by selecting the appropriate semiconductor and depositing pure or mixed oxides with specific applications in solar energy conversion, purification of water and air streams, metal corrosion and prevention, chemical synthesis and manufacturing, nanoelectronics, medicine, among others.

In addition, this method has the main advantage of not using high pressures and temperatures and in most cases the synthesis is carried out in aqueous solution. In spite of the photodeposition methods seem to be ideal for the synthesis of catalytic materials, to date, research reports have mainly focused in the photoreduction of noble metals. Therefore, the range of metal oxides deposited by a photoxidative or photoreductive routes has been limited. Finally, the oxidative deposition of metal oxides in semiconductors requires a deep investigation from fundamental to practical application. This is of crucial importance for understanding the mechanism of simple or mixed oxides formation (core-shell or alloys) during the irradiation step and the interfacial reactions of the process.

6. Acknowledgements

This work was supported by grants Nos. 153356 and 106891 from the National Council of Science and Technology (Conacyt, México).

7. References

Buono-Core G., Tejos M., Cabello G., Guzman N., Hill R, 2006, Photochemical deposition of NiCoO$_x$ thin films from Ni/Co heteronuclear triketonate complexes, *Materials Chemistry and Physics*, Vol. 96, (March 2006), pp. 98-102, ISSN 0254-0584.

Chan S., Barteau M., 2005, Preparation of highly uniform Ag/TiO$_2$ and Au/TiO$_2$ supported nanoparticle catalysts by photodeposition, *Langmuir*, Vol. 21, No. 12, (June 2005), pp. 5588-5595, ISSN 0743-7463.

Choi W., 2006, Pure and modified TiO$_2$ photocatalysts and their environmental applications, Catalysis surveys from Asia, Vol. 10, No. 1, (March 2006), pp. 16-28, ISSN 1571 1013.

Chong M., Jin B., Chow Ch., Saint Ch., 2010, Recent developments in photocatalytic water treatment technology: A review, *Water Research*, Vol. 44, No. 10 (May 2010), pp. 2997-3027, ISSN 0043-1354.

Clarkson T., Magos L., 2006, The toxicology of mercury and its chemical compounds, *Critical Reviews In Toxicology*, Vol. 36, pp. 609–662, ISSN:1040-8444.

Fagnoni M, Dondi D., Ravelli D., Albini A., 2007, Photocatalysis for the formation of C-C bond, *Chemical Reviews*, Vol. 107, No. 6, (June 2007), pp. 2725-2756, ISSN 0009-2665.

Flores S. O., Gutiérrez R., Rios-Bernÿ O., Valenzuela M., 2008, Simultaneous Cr(VI) reduction and naphthalene oxidation in aqueous solutions by UV/TiO2, *Materials Research Society Symposium Proceedings*, Vol. 1045, (2008), pp. 1045-V03-05 , ISSN 1946-4274.

Giocondi J., Rohrer G., 2003, Structure sensitivity of photochemical oxidation and reduction reactions on $SrTiO_3$ surfaces, *Journal of American Ceramic Society*, Vol. 86, No. 7, (July 2003), pp. 1182-1189, ISSN: 1551-2916.

Kamada K., Moriyasu A., 2011, Photo-excited electroless deposition of semiconducting oxide films and their electrocatalytic properties, *Journal of Materials Chemistry*, Vol. 21, (February 2011), pp. 4301-4306, ISSN 0959-9428.

Kraeutler B., Bard A.J., 1978, Heterogeneous photocatalytic preparation of supported catalysts. Photodeposition of platinum on titanium dioxide powder and other substrates, *Journal of the American Chemical Society*, Vol. 100, No. 13, (June 1978) pp. 4317-4318,

Li Y., Wu C.Y., 2007, Kinetic study for the photocatalytic oxidation of elemental mercury on a SiO_2-TiO_2 nanocomposite, *Environmental Engineering Science*, Vol. 24, No. 1, pp. 3-12.

Lin W. Wei Ch. Rajeshwar K., 1993, Photocatalytic reduction and immobilization of hexavalent chromiun at titanium dioxide in aqueous basic media, *Journal of The Electrochemical Society*, Vol. 140, No. 9, (September 1993) pp. 2477-2482, ISSN 1945-7111.

Litter M., 1999, Heterogeneous photocatalysis Transition metals ions in photocatalytic systems, *Applied Catalysis B: Environmental*, Vol. 23 No. 2-3, (November 1999), pp. 89-114, ISSN 0926-3373.

Liu K, Anderson M.A., 1996, Porous Nickel Oxide/Nickel films for electrochemical capacitors, *Journal of the Electrochemical Society*, (January 1996), Vol. 143, No. 1, pp. 124-130, ISSN 1945-7111.

Loganathan K., Bommusamy P., Muthaiahpillai P., Velayutham M., 2011, The syntheses, characterizations, and photocatalytic activities of silver, platinum, and gold doped TiO_2, *Environmental Engineering Research*, Vol. 16, No. 2 (June 2011), pp. 81-90, ISSN 1226-1025.

Maeda K., Teramura K., Lu D., Saito N., Inoue Y., Domen K., 2006, Noble metal/Cr_2O_3 core/shell nanoparticles as a cocatalyst for photocatalytic overall water splitting, *Angew. Chem. Int. Ed.*, Vol. 45, pp. 7806-7809.

Maeda K., Lu D., Teramura K., Domen K., 2008, Direct deposition of nanoparticulate rhodium–chromium mixed-oxides on a semiconductor powder by band-gap irradiation, *Journal of Materials Chemistry*, Vol. 18, (June 2008), pp. 3539-3542, ISSN 0959-9428.

Matsumoto Y., 2000, Electrochemical and photoelectrochemical processing for oxide films, *MRS Bulletin*, Vol.25, No. 9, (September 2000), pp. 47-50, ISSN 0025-5408.

Matsumoto Y., Ida, S., Inoue, T., 2008, Photodeposition of metal oxide at the TiO_x nanosheet to observe the photocatalytic active site. *The Journal of Physical Chemistry C*, Vol. 112, No. 31, pp. 11614-11616, ISSN 1932-7455.

Mills A., Duckmanton P.A., Reglinski J., 2010, A simple novel method for preparing an effective water oxidation catalyst, *Chemical Communications,* Vol. 46, pp. 2397-2398.

Murruni L., Leyva G., 2007, Photocatalytic removal of Pb(II) over TiO_2 and $Pt–TiO_2$ powders, *Catalysis Today*, Vol. 129, No. 1-2, (December2007), pp. 127-135, ISSN: 0920-5861.

Nishimura N., Tanikawa J., Fujii M., Kawahara T., Ino J., Akita T., Fujino T., Tada H., 2008, A green process for coupling manganese oxides with titanium (IV) dioxide, *Chemical Communications*, pp. 3564-3566.

Ohtani B, 2010, Photocatalysis A to Z- what we know and what we do not know in a scientific science, *Journal of Photochemistry and Photobiology C: Photochemistry Reviews*, (December 2010), Vol. 11, No. 4, pp. 157-178, ISSN 13895567.

Ohyama J., Yamamoto A., Teramura K., Shishido T., Tanaka T., 2011, Modification of metal nanoparticles with TiO_2 and metal-support interaction in photodeposition, *Catalysis*, Vol. 1, No. 3, (February 2011), pp. 187-192, ISSN 2155-5435.

Paz Y., 2010, Application of TiO2 for air treatment: Patent´s overview, Applied Catalysis B: Environmental, Vol. 99, No. 3-4, (September 2010) , pp. 448-460, ISSN 0926 3373.

Si S., Huang K., Wang X., Huang M., Chen H., 2002, Investigation of photoelectrochemical oxidation of Fe^{2+} ions on porous nanocrystalline TiO_2 electrodes using electrochemical quartz crystal microbalance, *Thin Solids Films*, Vol. 422, pp.205-210.

Steinmiller E., Choi K., 2009, Photochemical deposition of cobalt-based oxygen evolving catalyst on a semiconductor photoanode for solar oxygen production, *Proceedings of National Academy of Sciences*, Vol. 106, No.49, (December 2009), pp. 20633-20636, ISSN 0027-8424.

Snider G., Ariya P.,2010, Photocatalytic oxidation reaction of gaseous mercury over titanium dioxide nanoparticle surface, *Chemical Physics Letters*, Vol. 491, (May 2010) pp.23-28 and references therein.

Tak Y., Yong K., 2008, A novel heterostructure of Co_3O_4/ZnO nanowire array fabricated by photochemical coating method, *Journal of Physical Chemistry C*, Vol. 112, pp. 74-79.

Tanaka K. Harada K., Murata S., 1986, Photocatalytic deposition of metal ions onto TiO_2 powder, *Solar Energy*, Vol. 36, No 2, (March 1986), pp. 159-161, ISSN: 0038-092X.

Torres J, Cervera-March S., 1992, Kinetics of the photoassisted catalytic oxidation of Pb(II) in TiO_2 suspensions, *Chemical Engineering Science*, Vol. 47, No. 15-16, (October-November 1992), pp. 4107-4120, ISSN: 0009-2509.

Valenzuela M. A., Albiter E., Rios-Bernÿ O., Córdova I., Flores S. O., 2010, Photocatalytic reduction of organics compounds, *Journal of Advanced Oxidation Technology*, Vol. 13, No. 3, (July 2010), pp. 321-340, ISSN 1203-8407.

Wu T., Li Y., Chu M., 2003, *Handbook of Photochemistry and Photobiology Vol. I, Inorganic Photochemistry*, pp. 249-282, American Science Publisher, ISBN 1-58883-004-7,

3

Improved Photochemistry of TiO$_2$ Inverse Opals and some Examples

Fabrizio Sordello, Valter Maurino and Claudio Minero
Università degli Studi di Torino, Dipartimento di Chimica Analitica, Torino,
Italy

1. Introduction

TiO$_2$ inverse opals are porous TiO$_2$ structures in which the pore arrangement is well ordered in three dimensions. Frequently the pores are arranged in a fcc or hcp structure and each pore is connected to the twelve nearest neighbours. TiO$_2$ commonly occupies about 25% of the volume of the material, while the pores, which can be filled with gaseous or liquid solutions, account for the remaining 75% of the volume (Fig. 1).

The ordered arrangement of pores of the same size can be seen as a periodic modulation of the refractive index in the space, and therefore TiO$_2$ inverse opals are by definition photonic crystals (John 1987; Yablonovitch 1987). Photonic crystals are very useful in controlling the propagation of light, and they can represent for photonics the same improvement semiconductors represented in electronics. Hence properly designed TiO$_2$ inverse opals find application in solar energy recovery (Nishimura et al. 2003; Mihi et al. 2008; Chutinan et al. 2009) and photocatalysis (J. I. L. Chen et al. 2006; Y. Li et al. 2006; Ren et al. 2006; Srinivasan & White 2007; J. I. L. Chen et al. 2008; Sordello et al. 2011a).

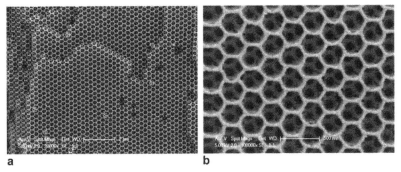

a   **b**

Fig. 1. SEM micrographs of TiO$_2$ inverse opals at different magnification: (a) 20 000x magnification, (b) 100 000x magnification (Reprinted from Waterhouse & Waterland 2007, Copyright (2007), with permission from Elsevier)

This chapter reviews the literature to give a complete picture of the state of the art of the photochemistry on TiO$_2$ inverse opals and outlines the more promising perspectives of the field in the near future.

2. The context: Light driven processes on TiO$_2$

TiO$_2$ is used in heterogeneous photocatalysis as photocatalyst in water photosplitting and hydrogen production, in solar cells for the production of electricity, and in other applications that do not require light, such as lithium ion batteries, bone implants and gate insulator for MOSFETs. When TiO$_2$ is irradiated with light with energy higher than the band gap an electron-hole couple is generated. The charge carriers can separate and migrate towards the surface where they can be trapped or react with solution species, as it can be seen in Fig. 2 (Diebold 2003). Reactive oxygen species are formed, the degradation of organic molecules and pollutants occurs, and the complete mineralization to CO$_2$, H$_2$O and inorganic ions has been reported (Pelizzetti et al. 1989). From the environmental point of view this process can be very useful and effective, since the harmful pollutant is not displaced into another phase, but ultimately decomposed to non harmful inorganic compounds. In the presence of a metallic cocatalyst, platinum for example, and in the absence of oxygen, the photogenerated electrons can reduce H$_3$O$^+$ to hydrogen, and in the absence of an effective hole scavenger, the photogenerated holes can oxidize water to oxygen, leading to water photosplitting (Ekambaram 2008; Yun et al. 2011). Recently, research focused its attention on the photocatalytic productions of value added chemicals starting from glycerol (Maurino et al. 2008) or directly from CO$_2$ in artificial photosynthesis (Benniston & Harriman 2008; Roy et al. 2010).

The technological and commercial affirmation of these light driven processes is delayed because (Fujishima et al. 2008; Gaya & Abdullah 2008):

1. TiO$_2$ is an indirect band gap semiconductor, and therefore its light absorption is limited.

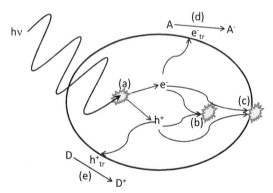

Fig. 2. Processes occurring on TiO$_2$ under UV irradiation: (a) light absorption and charge carriers formation, (b) electron-hole recombination at bulk trapping sites with release of heat, (c) electron-hole recombination at surface trapping sites, (d) trapped electron reacts with acceptor (e) trapped hole reacts with donor.

2. TiO$_2$ is a wide band gap semiconductor and absorbs only the UV fraction of the solar spectrum.
3. TiO$_2$ has a quite high refractive index and light absorption is limited also by reflection.
4. The efficiencies of the above mentioned processes are quite low, because the charge carriers recombination is fast.

TiO$_2$ inverse opals can especially improve the light absorption, allowing also a fast mass transfer of solution species due to the large-pore structure. This approach can be combined with other strategies, new or already employed to improve the efficiency of the photocatalytic process on TiO$_2$ surfaces, for example synthesizing doped or dye sensitized TiO$_2$ inverse opals, or realizing structures with controlled exposed surfaces. In this case the achievable structures are limited only by the creativity of the researchers and the synthetic procedures, which can become very complicated and difficult to implement.

In the next section the origin of the better absorption of light of TiO$_2$ inverse opals will be discussed.

3. Photonic band gap and slow photons

As we mentioned in section 1 TiO$_2$ inverse opals are photonic crystals, that is materials characterized by a periodic modulation in the space of the refractive index. The variation of the dielectric constant can be periodic in one dimension, two dimensions or three dimensions. Inverse opals are three dimensional photonic crystals, but for simplicity in the following we will consider the interaction of light with a monodimensional photonic crystal.

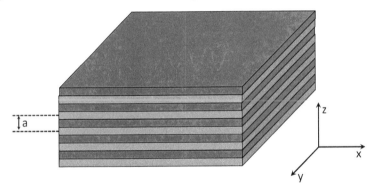

Fig. 3. The multilayer film is a monodimensional photonic crystal. Monodimensional means that the dielectric constant varies only along one direction (z). The system is composed by alternating layers of different materials with different refractive indexes and with spatial period a. Every layer is uniform and extends to infinity along the xy plane. Also the periodicity in the z direction extends to infinity.

A monodimensional photonic crystal is a multilayer film formed by alternating layers of constant thickness, with different refractive index and constant spacing among them (Fig. 3). Referring to Fig. 3 it can be noticed that light propagating in the z direction encounters on its path several interfaces between the two dielectrics. At each interface light is reflected and refracted following the Snell law. If the wavelength of light propagating along the z direction matches perfectly the periodicity of the one-dimensional photonic crystal, the reflected waves will be in phase, and as a consequence i) light will be reflected by the photonic crystal and ii) its propagation inside the material will be forbidden (Yablonovitch 2001). We say that the wavelength (or the frequency) of the incident light falls inside the photonic band gap (Fig. 4).

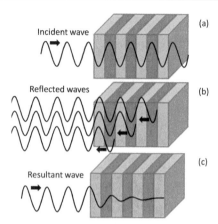

Fig. 4. Schematic representation of an electromagnetic wave impinging a photonic band gap material (a), partial reflection occurs at every interface and, since the frequency of the incident wave falls inside the photonic band gap the reflected waves are all in phase (b); as a result the light cannot travel through the material (c) (adapted from Yablonovitch 2001)

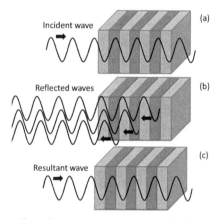

Fig. 5. Schematic representation of an electromagnetic wave with frequency outside the photonic band gap propagating into a photonic crystal (a), in this case the reflected waves are out of phase (b), and the light propagates inside the material only slightly attenuated (c) (adapted from Yablonovitch 2001)

On the contrary, if the wavelength of the incident light does not match the periodicity of the photonic crystal lattice, the waves reflected at each interface will be out of phase, and they will cancel out each other (Fig. 5). As a result the light will be able to propagate inside the material. In this case the wavelength (or the frequency) of the incident light is said to fall outside the photonic band gap.

This approach is useful from a qualitative point of view and gives a general idea of physical phenomena involved, but it cannot be extended to the bidimensional and tridimensional cases. Moreover, it lacks of quantitative understanding of the phenomenon, impeding, for example, the evaluation of the magnitude of the photonic band gap.

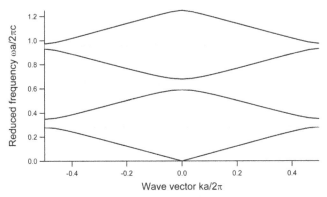

Fig. 6. The photonic band structure of the multilayer film depicted in Fig. 3 calculated with a freely available software (S. G. Johnson & Joannopoulos 2001). The dielectric constants of the layers are ε=5 and ε=2. The photonic band gap frequencies are highlighted in yellow.

A more rigorous approach starts from the treatment of the electromagnetic problem in mixed dielectric media, where the dielectric constant ε(r) becomes a function of the spatial coordinate r. For this case the Maxwell equations have the form (Joannopoulos et al. 2008):

$$\nabla \cdot \mathbf{H}(\mathbf{r},t) = 0 \tag{1}$$

$$\nabla \cdot [\, \varepsilon(\mathbf{r})\, \mathbf{E}(\mathbf{r},t)] = 0 \tag{2}$$

$$\nabla \times \mathbf{H}(\mathbf{r},t) - \varepsilon_0 \varepsilon(\mathbf{r})\, \partial \mathbf{E}(\mathbf{r},t)/\partial t = 0 \tag{3}$$

$$\nabla \times \mathbf{E}(\mathbf{r},t) + \mu_0\, \partial \mathbf{H}(\mathbf{r},t)/\partial t = 0 \tag{4}$$

where E and H are the macroscopic electric and magnetic fields, $\varepsilon_0 \approx 8.854 \times 10^{-12}$ F/m is the vacuum permittivity, $\mu_0 = 4\pi \times 10^{-7}$ H/m is the vacuum permeability and ε(r) is the scalar dielectric function. It can be demonstrated that Maxwell equations can be rearranged to yield equation 5:

$$\nabla \times \{[1/\, \varepsilon(\mathbf{r})]\, \nabla \times \mathbf{H}(\mathbf{r})\} = (\omega/c)^2\, \mathbf{H}(\mathbf{r}) \tag{5}$$

where the electromagnetic problem takes the form of an eigenvalue problem. It can be demonstrated that the operator working on the magnetic field is linear and hermitian. In a mixed periodic medium the eigenfuctions that satisfy equation 5 are in the form of Bloch waves in which the expression of a plane wave is multiplied by a periodic function that accounts for the periodicity of ε(r) in the photonic crystal:

$$\mathbf{H}(\mathbf{r}) = e^{i\mathbf{k}\cdot\mathbf{r}}\, \mathbf{u}(\mathbf{r}) \tag{6}$$

where u(r) is a periodic function of the type u(r) = u(r+R) for every lattice vector R (Joannopoulos et al. 2008). In the case of the multilayer film in Fig. 3 the periodic function u depends only on the z coordinate.

Introducing the Bloch waves of equation 6 in equation 5 the eigenproblem can be solved and the photonic band structure of the photonic crystal of interest can be studied. If in the photonic band diagram there are frequencies for which there are no photonic modes allowed for every wavevector k, a photonic band gap is present (Fig. 6). At those frequencies light cannot propagate inside the material. In such cases the photonic crystals can find application in lasing cavity, optical filter and dielectric mirrors.

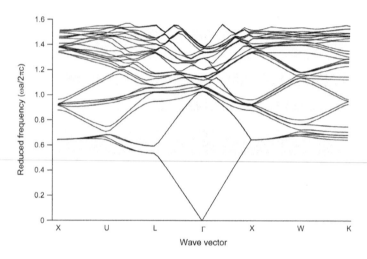

Fig. 7. Photonic band diagram of a TiO$_2$ inverse opal with the pores filled with water. The frequencies associated with light with low group velocity are highlighted. The band structure has been calculated with a freely available software (S. G. Johnson & Joannopoulos 2001) assuming for the dielectric constants of TiO$_2$ and water the values 4.4 and 1.7, respectively.

If equation 5 is solved for the TiO$_2$ inverse opal (Fig. 7) no photonic band gaps are present. To have a photonic band gap in the inverse opal structure the dielectric contrast (the difference between the dielectric constants of the two media) has to be at least 9, whereas TiO$_2$ anatase has a dielectric constant higher than ten only for photon energies above 3.8 – 4.0 eV (310 – 325 nm) (Jellison et al. 2003). Therefore in the visible and UVA TiO$_2$ will not have a complete photonic band gap. Nevertheless, at certain frequencies light will not be able to propagate in some direction (Fig. 7). For example in the Γ-L direction there is a pseudo photonic gap that forbids the propagation of light at a value of the reduced frequency around 0.55. This feature is not so unsuitable for photosynthetic or photocatalytic application, since in those cases light has to propagate inside the catalyst to be absorbed and create charge carriers.

The photonic band diagram of TiO$_2$ inverse opal shows that for some photonic bands the behaviour of the frequency as a function of the wavevector presents an almost flat trend. As the group velocity of light $\mathbf{v_g}$ is defined as:

$$\mathbf{v_g(k)} = \nabla_k \omega \qquad (7)$$

from equation 7 we can argue that a flat trend of a photonic mode in the frequency ω vs. wavevector **k** plot is indicative of low group velocity. Looking at Fig. 7, we can observe that almost flat photonic bands are present for different frequency ranges. At those frequencies light will be able to travel inside the TiO₂ inverse opal, but its group velocity will be strongly reduced. Hence the light interaction with the material will be incremented.

When light with low group velocity (or slow light or slow photons) can be exploited, the optical absorption of the material can be improved as if the optical path inside the material would be lengthened. An elegant experimental demonstration of this phenomenon is the change of the absorbance spectrum of an adsorbed dye on different TiO₂ inverse opals (Fig. 8).

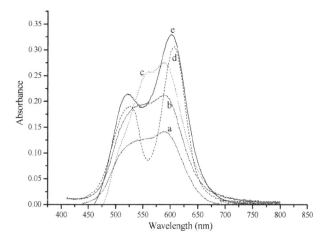

Fig. 8. Absorption spectra of crystal violet adsorbed on different TiO₂ films: (a) reference TiO₂ flat film, (b) and (c) TiO₂ inverse opals slowing photons outside the 450-650 nm range, (d) and (e) TiO₂ inverse opals slowing photons in the 600-650 nm range Reprinted with permission from Y. Li et al. 2006 Copyright (2006) American Chemical Society.

When the crystal violet dye is adsorbed on a TiO₂ inverse opal, which can slow down the photons possibly absorbed by the dye, its absorbance is increased with respect to flat TiO₂ or TiO₂ inverse opals with not properly tuned periodicity (Y. Li et al. 2006). Absorption spectra reported in Fig. 8 show that also in the case of TiO₂ inverse opals with not properly tuned photonic band gap (slow photons outside the 450-650 nm range) there is a higher absorption that can be explained in terms of porosity and larger amount of adsorbed dye. In the case of TiO₂ inverse opals slowing photons in the 600-650 nm range, higher absorption cannot be explained only in terms of porosity and higher amount of adsorbed dye, because not only the intensity of the spectra is different from the reference film, but also because the shape of the spectra changes, as the absorption at 600-625 nm becomes predominant over that at 500 nm.

This important feature, together with porosity and indeed high surface area, makes TiO₂ inverse opals very good materials for application in semiconductor photocatalysis, solar energy recovery (dye sensitized solar cells) and artificial photosynthesis (Ren & Valsaraj 2009). Owing to this, they have drawn and they are drawing a lot of research into this field.

4. Synthesis

TiO$_2$ inverse opals can be prepared in several ways and, probably, new synthetic routes will be discovered in the next years. Different approaches are possible because three dimensional ordered porous structures can be obtained exploiting many different physical principles. The invention of new methods is only limited by physics and chemistry and by the creativity of the researchers. An ideal method would be fast, requiring only standard procedures and instruments, it would be cheap and it should be able to produce homogeneous TiO$_2$ inverse opals over large surfaces with the possibility to control the thickness of the synthesized material. A good method would be also able to produce TiO$_2$ inverse opals with few defects both at microscopic (vacancies, dislocations, grain boundaries, stacking faults, ...) and macroscopic level (cracks, thickness inhomogeneity).

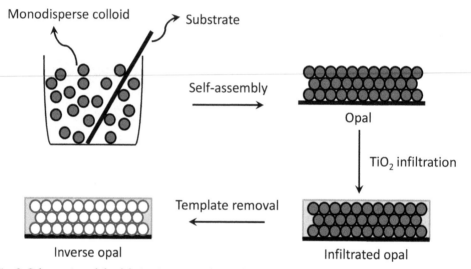

Fig. 9. Schematics of the fabrication procedure of a TiO$_2$ inverse opal.
(adapted from L. Liu et al. 2011)

In general TiO$_2$ inverse opals are synthesized in a two steps procedure (Fig. 9). Firstly, a well monodisperse colloidal suspension of SiO$_2$ or polymer spheres has to be prepared. The monodisperse colloid is then conveniently deposited onto a clean substrate to form a colloidal crystal, that is a solid characterized by an ordered disposition of silica or polymer spheres in three dimensions. A colloidal crystal can be described also as a superlattice of closely packed colloidal particles. If the colloidal particles are spherical the resulting colloidal crystal is called opal or synthetic opal, in analogy with natural opals characterized by an ordered arrangement of silica spheres in three dimensions. Most of the time colloidal crystals obtained in this way are synthetic opals and can find application as sensors (Endo et al. 2007; Shi et al. 2008) and as model systems in the study of crystals, phase transitions (Bosma et al. 2002) and the interaction of photonic crystals with light (Miguez et al. 2004; Pavarini et al. 2005; M. Ishii et al. 2007). Beyond these applications, colloidal crystals serve as sacrificial templates for the synthesis of TiO$_2$ inverse opals. In this case the quality of the colloidal-crystal template is very important, because every defect in the structure will be replicated in the TiO$_2$ inverse opal. For the TiO$_2$ inverse opal preparation a TiO$_2$ precursor solution (titanium(IV) isopropoxide, titanium(IV) butoxide, other titanium(IV) species) or a dispersion of TiO$_2$ nanoparticles is used to fill the voids of the synthetic opal, a process normally called infiltration. Once the infiltration process is completed and TiO$_2$ backbone is formed, the template is removed by etching with NaOH or HF in the case of silica templates, with toluene in the case of polystyrene templates, or by calcination if a cross linked polymeric template has been used.

4.1 Preparation of monodisperse colloids

In the literature many methods to synthesize monodisperse colloids are described. For the successive colloidal crystal preparation the most suitable method to obtain monodisperse polymeric colloids is the emulsion polymerization without emulsifier in water. To synthesize monodisperse polystyrene colloids the emulsion water-styrene is heated and vigorously stirred, the addition of the initiator makes the reaction start, and the relative amounts of monomer/initiator/ionic strength affect the size of the final polymer spheres (Goodwin et al. 1974). It is also possible to synthesize positively charged polystyrene colloids (Reese et al. 2000) or functionalized colloids (X. Chen et al. 2002) to ease the colloidal crystal formation and the infiltration of TiO$_2$. The preparation of monodisperse polymethylmethacrylate colloids is very similar (Waterhouse & Waterland 2007), and examples of synthesis in non polar solvents are also reported (Klein et al. 2003).

Monodisperse silica spheres can be synthesized in ethanol with determined amount of water in the presence of ammonia as shape controller (Jiang et al. 1999).

4.2 Colloidal crystal synthesis

Many different strategies to obtain colloidal crystals are available depending on the final goal, because they can be obtained as powders, thick films, and also thin films containing one, two or at least three layers.

a. Direct centrifugation of the monodisperse colloid leads to the formation of a colloidal crystal. Although the method is simple, it is not suitable if a film of controlled thickness has to be obtained (Wijnhoven & Vos 1998; Waterhouse & Waterland 2007). Another drawback is the slowness, if small colloidal particles (size < 150 nm) are involved, because more than one hour of centrifugation is needed (Sordello et al. 2011a).

b. A related method is the sedimentation. A drop of the monodisperse colloid is deposited on a clean substrate, where the drop can broaden due to the hydrophilicity of the substrate with contact angle close to zero. The evaporation of the solvent leads to the formation of the colloidal crystal (Denkov et al. 1993). To avoid thickness non uniformity the crystallization can be carried out under silicone liquid (Fudouzi 2004). Since in most cases the evaporation of water is necessary for the synthesis of the colloidal crystal, relative humidity plays an important role (Liau & Huang 2008), and the deposition is quite slow. To accelerate the process ethanol can be used instead of water (Shin et al. 2011).

c. The most popular method used to synthesize colloidal crystals is based on the vertical, rather than horizontal, position of the substrate on which the monodisperse colloid is deposited. When the substrate is dipped into the colloidal dispersion of monodisperse spheres the particles self-assembly in an opaline structure in the meniscus region (Fig.

10). With the evaporation of the solution the meniscus moves downwards and the opal film grows in the same direction (Jiang et al. 1999; Z. Z. Gu et al. 2002; Norris et al. 2004; Shimmin et al. 2006). The method allows the formation of well ordered opals, but the process is slow and the thickness of the film is not uniform, because the colloid concentrates during the process. Moreover, it has been reported that in the vertical deposition method the thickness of the film has a sinusoidal trend in the length scale of the order of 100 μm. The oscillations are more pronounced with increasing particle concentration and with decreasing temperature (Lozano & Miguez 2007).

d. In the dip-drawing technique (Z. Liu et al. 2006) the colloidal crystal is formed by the downwards movement of the meniscus. The substrate is immersed vertical in the colloidal suspension and the meniscus is moved downwards withdrawing the suspension by means of a peristaltic pump. The rapidity of the method can be tuned, but if suspension withdrawal is too fast the quality of the resulting colloidal crystal is poor.

e. Polymer or silica spheres can self assembly in ordered structures in the liquid phase (Reese et al. 2000) forming single crystals over large areas (cm²) by simply applying a shear flow if the colloidal particles are charged and the ionic strength of the medium is sufficiently low (Amos et al. 2000; Sawada et al. 2001). This method is fast, allows the formation of large single crystals without grain boundaries and, if colloidal particles with opposed charge are employed, colloidal crystals with particle packing different from the usual fcc or hcp geometry are obtained (Leunissen et al. 2005; Shevchenko et al. 2006).

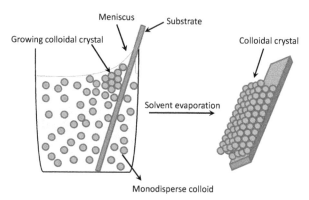

Fig. 10. Schematic representation of a colloidal crystal formed in the meniscus region by slow evaporation of the suspension (adapted from Norris et al. 2004).

The shear flow method can also produce colloidal crystals without cracks over large domains. Normally, once the colloidal crystal is deposited the drying process causes the shrinking of the colloidal particles that produce macroscopic fractures in the structure (Sun et al. 2011). To avoid the formation of these imperfections core-shell spheres can be used. The colloidal crystal is assembled under high external pressure, and during the drying the soft shells will expand, compensating the volume shrinkage of the whole structure (Ruhl et al. 2004; Pursiainen et al. 2008). A further possibility is the use of a photosensitive monomer dissolved in the slurry. The irradiation after colloidal crystal formation will freeze the structure in a polymeric gel that will prevent any shrinking during the drying process (Kanai & Sawada 2009). These procedures allow the synthesis of almost perfect colloidal crystals over large areas. Nevertheless the subsequent TiO₂ infiltration will be considerably hindered.

f.   Colloidal crystals are also produced by confinement between two flat surfaces (Fig. 11). The colloidal dispersion of monodisperse spheres can enter the cell because of capillarity forces (M. Ishii et al. 2005; X. Chen et al. 2008) or by means of a hole in the cell (Park et al. 1998; Lu et al. 2001). Evaporation or removal of the solvent with a gas stream leaves behind large area colloidal crystals slightly fractured.

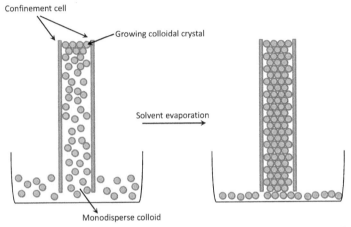

Fig. 11. Schematic representation of the formation of a colloidal crystal in a confinement cell (adapted from X. Chen et al. 2008)

g.   Spin coating can be a valuable and rapid method to produce thin film colloidal crystals. Even if many grain boundaries are produced, the distinctive feature of the method is the ability to produce very thin films. The synthesis of monolayered colloidal crystals has been demonstrated (Mihi et al. 2006).

h.   Finally, polymeric photonic crystal can also be prepared by mixing immiscible polymers (C. W. Wang & Moffitt 2005) and by laser holographic lithography (Moon et al. 2006; Y. Xu et al. 2008; Lin et al. 2009; Miyake et al. 2009).

4.3 Infiltration and TiO₂ inverse opal synthesis

The infiltration in the sacrificial template is the crucial step in the TiO₂ inverse opal synthesis, because this phase is responsible for the major production of defects and macroscopic imperfections. If the infiltrated solution or colloid cannot form a stable network before the template is removed, the structure will collapse with the destruction of the three dimensional lattice. Nowadays several infiltration methods exist and in the following we will give the reader a general picture of the available techniques.

The sol-gel infiltration is probably the most popular and widespread method, because it is simple and low cost. Nevertheless, it has to be implemented with great care especially if TiO₂ precursors with large tendency to hydrolyze rapidly are used, because the amount of deposited TiO$_2$ is difficult to control (Wei et al. 2011). In some cases the deposition of an extra layer of bulk TiO$_2$ over the inverse opal can be turned into an advantage with the lift-off/turn-over method (Fig. 12). Depending on the application both a self standing TiO$_2$ inverse opal or a Bragg mirror behind a TiO$_2$ active layer can be created (Galusha et al. 2008). To avoid dealing with violent hydrolysis it is possible to introduce TiO$_2$ in the template in the form of nanoparticle already hydrolyzed (Yip et al. 2008; Shin et al. 2011). In this case the difficulty is the preparation of sufficiently small particles capable to penetrate into the pores of the opal template, because the resulting infiltration will be poor if the nanoparticles are too big, and the inverse opal structure will collapse after template removal.

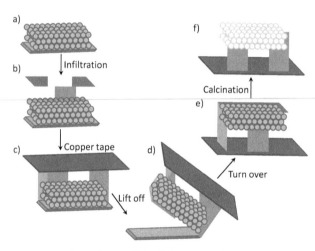

Fig. 12. Schematic representation of the "lift-off/turn-over" technique. (a) polystyrene opal template. (b) Infiltrated opal with TiO$_2$. (c) Infiltrated opal with adhesive copper tape placed on top. (d) Lift off of the infiltrated opal from the substrate. (e) The infiltrated opal is turned over so that the flat opal-terminated surface is on top. (f) Calcined TiO$_2$ inverse opal with porous surface (adapted from Galusha et al. 2008).

Electrodeposition is a more complicated method requiring that the sacrificial template is deposited onto a conductive substrate, but it allows a better control of Ti^{4+} hydrolysis and a superior filling of the pores of the template. The potential of the conductive glass that supports the template is brought to negative potential to reduce nitrate present in solution according to the reaction:

$$NO_3^- + 6\ H_2O + 8\ e^- \rightarrow NH_3 + 9\ OH^- \tag{8}$$

The OH$^-$ produced causes a local increase of pH and the precipitation of TiO^{2+} from the precursor solution. In this way the TiO$_2$ deposition is compact and finely tunable varying the concentration of NO$_3^-$, the applied current and the deposition time (Y. Xu et al. 2008). A similar technique is electrophoresis. The method can infiltrate a TiO$_2$ colloid constituted by charged particles, and according to Gu et al. (Z.-Z. Gu et al. 2001), the deposition can be fast and uniform.

Spectacular structures (Fig. 13) can be obtained with infiltration by atomic layer deposition. The technique allows a fine control of the amount of deposited TiO$_2$, the filling of the pores of the template is optimal and, as a consequence, the synthesized TiO$_2$ inverse opals are very resistant. The drawbacks are the slowness and the cost of the needed equipment, which cannot be considered a standard facility of every laboratory (King et al. 2005; L. Liu et al. 2011).

Infiltration can also be carried out by spin-coating, the method is fast and the produced TiO$_2$ inverse opals are regular and characterized by smooth surfaces (Matsushita et al. 2007).

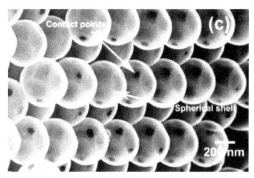

Fig. 13. SEM micrograph of the fractured surface of a TiO$_2$ inverse opal produced by atomic layer deposition (Copyright (2005) Wiley. Used with permission from King at al. 2005)

TiO$_2$ inverse opals can be obtained also in only one synthetic step, in which the silica or polymeric colloid, that will build the colloidal crystal template, is mixed together with a TiO$_2$ colloid that will occupy the interstices of the colloidal crystal structure (Meng et al. 2002). For this reason TiO$_2$ particles have to be one or two orders of magnitude smaller than the silica or polymer particles. The presence of smaller TiO$_2$ particles forces the larger particles to self assembly in close packing. In this way the volume available for the diffusion of the small particles increases, with an overall gain in entropy for the entire system (Yodh et al. 2001).

Beyond the classic filling of TiO$_2$ inverse opal, in which TiO$_2$ occupies the residual volume of the former opal template, a structure also called shell structure or residual volume structure, there are synthetic procedures that can build TiO$_2$ inverse opals with a skeleton structure (Dong et al. 2003) or with a fractal distribution of the pores (Ramiro-Manzano et al. 2007).

5. Improved photochemistry on TiO$_2$ inverse opals

Although the unique properties of TiO$_2$ inverse opals and their possible applications for improved photochemistry are many, surprisingly the first demonstration of better use of light on TiO$_2$ inverse opals dates back only in 2006 (Ren et al. 2006). Before they received attention as back reflector in dye sensitized solar cells (Nishimura et al. 2003; Halaoui et al. 2005; Somani et al. 2005), or in fundamental studies (N. P. Johnson et al. 2001; Schroden et al. 2002; Dong et al. 2003). In that seminal work Ren et al. demonstrated that TiO$_2$ inverse opal with photonic pseudo gap in the UV is more efficient than P25 TiO$_2$ in the photodegradation of 1,2-dichlorobenzene in the gas phase (Fig. 14). They also found that the rate constant for the degradation of the pollutant is proportional to the radiant flux intensity even at intensities higher than 25 mW cm^{-2}, whereas usually for TiO$_2$ the rate constant is proportional to the square root of the radiant flux intensity (Minero 1999).

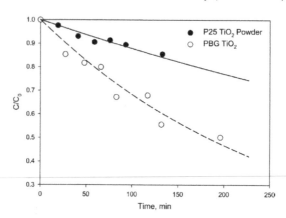

Fig. 14. Normalized concentrations of 1,2-dichlorobenzene as a function of the irradiation time in the presence of P25 TiO$_2$ powder (closed circles) and TiO$_2$ inverse opal (PBG TiO$_2$, open circles) Reprinted with permission from Ren et al. 2006 Copyright (2006) American Chemical Society.

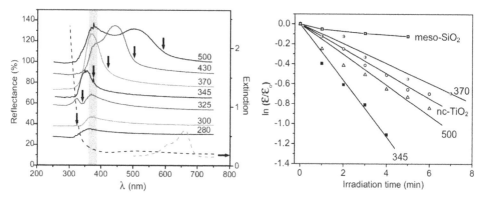

Fig. 15. Left: Reflectance spectra of TiO$_2$ inverse opals with photonic pseudo gaps located at 500, 430, 370, 345, 325, 300, and 280 nm (solid lines), and extinction spectra of nanocrystalline TiO$_2$ (black dashed line) and methylene blue (gray dashed line). For clarity the spectra have been displaced in the vertical axis. The highlighted region indicates the wavelengths used during irradiation. Right: Logarithmic plot of the photodegradation of methylene blue showing the first-order decay rate for nanocrystalline TiO$_2$ (nc-TiO$_2$) and TiO$_2$ inverse opals with photonic pseudo gaps at 345, 370, and 500 nm. Mesoporous SiO$_2$ (meso-SiO$_2$) was used as the blank (Copyright (2006) Wiley. Used with permission from J.I.L. Chen et al. 2006)

To elucidate the reason of this behaviour Chen et al. (J. I. L. Chen et al. 2006) synthesized TiO$_2$ inverse opals with different pore size, and hence with different position of the photonic pseudo gaps (Fig. 15). The authors observed that with narrow spectrum irradiation (370 ± 10 nm) at the red edge of the photonic pseudo gap the degradation of methylene blue was significantly accelerated with respect to nanocrystalline TiO$_2$, whereas when irradiation was carried out at wavelengths of the photonic pseudo gap the degradation was even slower than in the presence of the reference nanocrystalline TiO$_2$ photocatalyst (Fig. 15). These evidences suggest that the better activity arises from the exploitation of slow photons, which is maximized when the wavelength used for the irradiation falls at the red edge of the photonic pseudo gap, whereas porosity and the improved mass transfer of the species cannot explain such important variations in activity as a consequence of the small difference in the pore sizes used.

The same authors studied also the effect of the disorder on the activity of TiO$_2$ inverse opals (J. I. L. Chen et al. 2007). They found that such systems can tolerate a certain disorder, but the addition in the polymer template of up to 40% of particles with a different diameter (up to 20% different, Fig. 16) leads to a significant loss in activity.

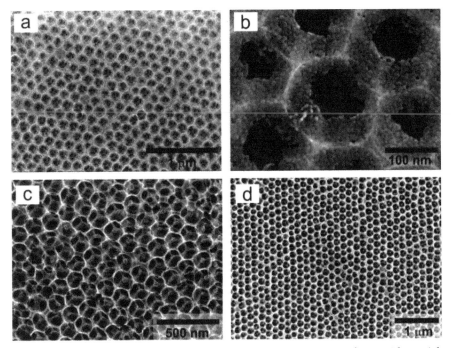

Fig. 16. SEM micrographs of TiO$_2$ inverse opals obtained from PS templates with particle dimensions (1-x)150-x180 nm, with x=0.13 (a and b), 0.37 (c), and 0.57 (d) Reprinted with permission from J.I.L. Chen et al. 2007 Copyright (2007) American Chemical Society.

Nevertheless, partially disordered inverse opals keep an enhancement factor of 1.6, calculated as the activity ratio between TiO$_2$ inverse opal and nanocrystalline TiO$_2$, while the well monodispersed material can attain an enhancement factor of 2.3. This is an important result because it shows that the light scattering is not crucial in increasing the efficiency of these materials, and that the unavoidable imperfections in the periodic structure do not prevent the practical applications in environmental remediation or water purification.

An improved degradation of methylene blue on TiO_2 inverse opals under UV irradiation was also observed (Srinivasan & White 2007), but in this case the better efficiency was explained in terms of better diffusion of the species due to the porosity of the material and to the large surface area. This interpretation was supported also by Chen & Ozin in a 2009 paper (J. I. L. Chen & Ozin 2009), where they partially revised the conclusions of their previous works, limiting the effectiveness of slow light only in the case of TiO_2 inverse opals with high fill factors.

To clarify the relative importance of slow light, light scattering and improved mass transfer due to the porosity Sordello et al. performed the photocatalytic degradation of phenol at two different wavelengths on TiO_2 inverse opal and TiO_2 disordered macroporous powders (Sordello et al. 2011a). The wavelengths used were 365 nm, where, for the TiO_2 inverse opal employed, the slow photon effect was maximized, and 254 nm, where, on the contrary, the slow light was negligible. At 365 nm the inverse opal is four times more active than the disordered macroporous structure. The key experiment showed that the pristine inverse opal powder is four times more active than the inverse opal crushed in a mortar, which has lost its periodicity in three dimensions. These differences vanish irradiating at 254 nm, as at that wavelength the three powders show almost the same activity. These evidences suggest that slow light plays an important role in increasing the absorption of light of TiO_2 inverse opals and hence in improving their photoactivity. The photoelectrochemical study of TiO_2 inverse opals confirmed the better light absorption of these materials when slow photons are involved. Furthermore, it was evidenced a faster electron transfer to the oxygen present in solution with respect to disordered macroporous TiO_2 (Sordello et al. 2011b).

The better absorption of light, the porosity and the high surface area derive from the structuration of these materials and can be coupled with other physical and chemical modifications to boost efficiency of a great variety of photoreactions. A frequent modification is the addition of metallic platinum to improve the separation of the photogenerated charge carriers and reduce recombination (Kraeutler & Bard 1978). The addition of platinum to TiO_2 inverse opals leads to a significant improvement of the photoactivity (J. I. L. Chen et al. 2008). The activity ratio between platinised inverse opal and nanocrystalline TiO_2 is four, whereas a lower value (around 2.5) should be expected considering the activity ratios of TiO_2 inverse opal (1.7) and platinised nanocrystalline TiO_2 (1.8) alone, evidencing that there is a cooperation between slow light and platinisation. In summary the total effect is not merely the sum of the two contributions (J. I. L. Chen et al. 2008). A different approach consists in synthesizing the metallic platinum inverse opal first, and coating it with TiO_2 in a second step (H. Chen et al. 2010). The recombination of photogenerated charge carriers is effectively reduced by the Schottky junction Pt/TiO_2 as the platinised samples show higher photocurrents and faster photodegradation rate for aqueous phenol. Moreover, Pt/TiO_2 inverse opal with properly tuned photonic pseudo gap has even higher photocurrents and photocatalytic activity with respect to inverse opals that cannot exploit slow light because the irradiation wavelength does not correspond to photonic modes with slow group velocity. It is also interesting to note that without the chance to harvest slow light Pt/TiO₂ inverse opals have the same characteristics of the macroporous disordered Pt/TiO₂ material (H. Chen et al. 2010).

An alternative strategy used to improve photochemistry on TiO₂ is the doping with transition metals, in order to extend the absorption of light to the visible and to allow practical applications with solar light. Wang et al. (C. Wang et al. 2006) were successful in synthesizing TiO₂, ZrO₂, Ta₂O₅ and Zr or Ta doped TiO₂ inverse opals which were tested in the photocatalytic degradation of 4-nitrophenol and Rhodamine B in aqueous solution. They demonstrated the red shift of the absorption edge of doped TiO₂, and measured higher photoactivity for the doped TiO₂ inverse opal samples. The better activity was attributed to the improved absorption of light due to the doping and to the porosity of the structures constituted by interconnected macropores and mesopores that allow faster migration of photogenerated electrons and holes.

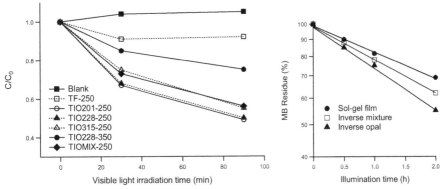

Fig. 17. Left: Methylene blue concentration as a function of the irradiation time for N,F-doped TiO₂ inverse opals with different pore dimensions (e.g. TIO201-250 refers to TiO₂ inverse opal with macropore radius of 201 nm and calcined at 250°C), for a N,F doped TiO₂ macroporous disordered sample (TIOMIX-250), and for N,F doped TiO₂ sol–gel thin film Reprinted with permission from J.A. Xu et al. 2010 Copyright (2010) American Chemical Society. Right: Percent residual methylene blue as a function of the irradiation time (>400 nm) for nitrogen-doped TiO₂ inverse opal (macropore radius ≈ 280 nm), nitrogen-doped TiO₂ macroporous disordered structure (inverse mixture), and nitrogen-doped TiO₂ sol–gel thin film (Copyright (2008) Wiley. Used with permission from Q. Li & Shang 2008)

To extend the absorption of TiO₂ into the visible doping with nitrogen, carbon, sulphur or fluorine is usually adopted. In this case the absorption in the visible is due to colour centres and not to a narrowing of valence and conduction band (Serpone 2006). The doping with nitrogen (Q. Li & Shang 2008) and with nitrogen and fluorine (J. A. Xu et al. 2010) of TiO₂ inverse opal led to an improved photocatalytic degradation of methylene blue in water with visible light, even if in the case of nitrogen and fluorine codoping the activity of the TiO₂ inverse opals is only slightly higher than in the case of a macroporous disordered TiO₂. This is probably due to a non optimal choice of the photonic band gap position, although in the case of nitrogen doping the improvement from disordered to ordered structure is significant (Fig. 17). The photodegradation of methylene blue with visible light alone is not sufficient to demonstrate that doped TiO₂ can effectively absorb visible photons with the consequent production of holes and electrons that can respectively react with the organic dye and molecular oxygen. According to Zhao et al. (Zhao et al. 2002) the disappearance of the dye is possible under visible light irradiation, without the need of photon absorption by TiO_2. In this case a dye sensitization occurs due to the electron injection from the excited dye (S^*) to the TiO_2 conduction band (equations 9 and 10):

$$S + h\nu \rightarrow S^* \tag{9}$$

$$S^* \rightarrow e^-_{CB} + S^+ \tag{10}$$

The oxidized dye S^+ can directly react with oxygen:

$$S^+ + O_2 \rightarrow SO_2^+ \tag{11}$$

whereas the conduction band electron can react with the dissolved oxygen to yield several reactive oxygen species (Minero & Vione 2006 and references therein):

$$O_2 + e^-_{CB} \rightarrow O_2^{\cdot-} \tag{12}$$

$$O_2^{\cdot-} + H^+ \rightarrow HO_2^{\cdot} \tag{13}$$

$$2\, HO_2^{\cdot} \rightarrow O_2 + H_2O_2 \tag{14}$$

$$O_2^{\cdot-} + HO_2^{\cdot} \rightarrow O_2 + HO_2^- \tag{15}$$

$$H_2O_2 + e^-_{CB} \rightarrow HO^{\cdot} + HO^- \tag{16}$$

$$H_2O_2 + HO^{\cdot} \rightarrow HO_2^{\cdot} + H_2O \tag{17}$$

These species can react with the adsorbed dye and carry out its degradation up to CO_2 and water. In the case of dye molecules adsorbed on the internal surfaces of a TiO_2 inverse opal with properly tuned photonic band gap, dye absorption is enhanced and its degradation improved. This mechanism has been hypothesized to account for the degradation of crystal violet on undoped TiO_2 inverse opal under visible irradiation (Y. Li et al. 2006). The role of the inverse opal is to slow down the photons absorbed by the dye to increase the interaction of light with organic compounds.

TiO_2 inverse opals were also coupled to metallic copper in the photoreduction of CO_2 to methanol in the presence of water vapour and UV light (Ren & Valsaraj 2009). With respect to nanocrystalline TiO_2, on inverse opals the reaction proceeded also at lower light intensities, with a rate dependence on light intensity $I^{0.74}$ with respect to nanocrystalline TiO_2 that showed a dependence $I^{0.20}$.

6. Conclusion

TiO_2 inverse opals and their role in improving the efficiency of photochemical reactions have been presented. The physical origin of the photonic band gap and of slow light has been discussed together with the synthetic routes to obtain such structures. Some applications and practical examples of improved photochemistry on TiO_2 inverse opals have been reviewed, demonstrating how such materials can help photocatalysis to be competitive in solar energy recovery, environmental remediation, water purification and also in the synthesis of chemicals. The possibility of better light absorption of inverse opals is very promising to improve the efficiency of light driven reactions for a great variety of implementations, provided that knowledge and competences are transferred among different fields. This is a necessary condition to produce complex systems that take advantage of cooperative effects.

7. References

Amos, R. M.; Rarity, J. G.; Tapster, P. R.; Shepherd, T. J. & Kitson, S. C. (2000). Fabrication of large-area face-centered-cubic hard-sphere colloidal crystals by shear alignment. *Physical Review E*, Vol. 61, No. 3, pp. 2929, ISSN:

Benniston, A. C. & Harriman, A. (2008). Artificial photosynthesis. *Materials Today*, Vol. 11, No. 12, pp. 26-34, ISSN: 1369-7021

Bosma, G.; Pathmamanoharan, C.; de Hoog, E. H. A.; Kegel, W. K.; van Blaaderen, A. & Lekkerkerker, H. N. W. (2002). Preparation of Monodisperse, Fluorescent PMMA-Latex Colloids by Dispersion Polymerization. *Journal of Colloid and Interface Science*, Vol. 245, No. 2, pp. 292-300, ISSN:

Chen, H.; Chen, S.; Quan, X. & Zhang, Y. (2010). Structuring a TiO2-Based Photonic Crystal Photocatalyst with Schottky Junction for Efficient Photocatalysis. *Environmental Science & Technology*, Vol. 44, No. 1, pp. 451-455, ISSN: 0013-936X

Chen, J. I. L.; Loso, E.; Ebrahim, N. & Ozin, G. A. (2008). Synergy of slow photon and chemically amplified photochemistry in platinum nanocluster-loaded inverse titania opals. *Journal of the American Chemical Society*, Vol. 130, No. 16, (Apr), pp. 5420-5421, ISSN: 0002-7863

Chen, J. I. L. & Ozin, G. A. (2009). Heterogeneous photocatalysis with inverse titania opals: probing structural and photonic effects. *Journal of Materials Chemistry*, Vol. 19, No. 18, pp. 2675-2678, ISSN: 0959-9428

Chen, J. I. L.; von Freymann, G.; Kitaev, V. & Ozin, G. A. (2007). Effect of Disorder on the Optically Amplified Photocatalytic Efficiency of Titania Inverse Opals. *Journal of the American Chemical Society*, Vol. 129, No. 5, pp. 1196-1202, ISSN:

Chen, J. I. L.; vonFreymann, G.; Choi, S. Y.; Kitaev, V. & Ozin, G. A. (2006). Amplified Photochemistry with Slow Photons. *Advanced Materials*, Vol. 18, No. 14, pp. 1915-1919, ISSN: 1521-4095

Chen, X.; Cui, Z.; Chen, Z.; Zhang, K.; Lu, G.; Zhang, G. & Yang, B. (2002). The synthesis and characterizations of monodisperse cross-linked polymer microspheres with carboxyl on the surface. *Polymer*, Vol. 43, No. 15, pp. 4147-4152, ISSN: 0032-3861

Chen, X.; Sun, Z.; Chen, Z.; Shang, W.; Zhang, K. & Yang, B. (2008). Alternative preparation and morphologies of self-assembled colloidal crystals via combining capillarity and vertical deposition between two desired substrates. *Colloids and Surfaces A: Physicochemical and Engineering Aspects*, Vol. 315, No. 1-3, pp. 89-97, ISSN: 0927-7757

Chutinan, A.; Kherani, N. P. & Zukotynski, S. (2009). High-efficiency photonic crystal solar cell architecture. *Optics Express*, Vol. 17, No. 11, (May), pp. 8871-8878, ISSN: 1094-4087

Denkov, N. D.; Velev, O. D.; Kralchevsky, P. A.; Ivanov, I. B.; Yoshimura, H. & Nagayama, K. (1993). Two-dimensional crystallization. *Nature*, Vol. 361, No. 6407, (07 January 1993), pp. 26, ISSN:

Diebold, U. (2003). The surface science of titanium dioxide. *Surface Science Reports*, Vol. 48, No. 5-8, pp. 53-229, ISSN: 0167-5729

Dong, W. T.; Bongard, H. J. & Marlow, F. (2003). New type of inverse opals: Titania with skeleton structure. *Chemistry of Materials*, Vol. 15, No. 2, (Jan), pp. 568-574, ISSN: 0897-4756

Ekambaram, S. (2008). Photoproduction of clean H2 or O2 from water using oxide semiconductors in presence of sacrificial reagent. *Journal of Alloys and Compounds*, Vol. 448, No. 1-2, pp. 238-245, ISSN: 0925-8388

Endo, T.; Yanagida, Y. & Hatsuzawa, T. (2007). Colorimetric detection of volatile organic compounds using a colloidal crystal-based chemical sensor for environmental applications. *Sensors and Actuators B-Chemical*, Vol. 125, No. 2, (Aug), pp. 589-595, ISSN: 0925-4005

Fudouzi, H. (2004). Fabricating high-quality opal films with uniform structure over a large area. *Journal of Colloid and Interface Science*, Vol. 275, No. 1, pp. 277-283, ISSN: 0021-9797

Fujishima, A.; Zhang, X. & Tryk, D. A. (2008). TiO2 photocatalysis and related surface phenomena. *Surface Science Reports*, Vol. 63, No. 12, pp. 515-582, ISSN:

Galusha, J. W.; Tsung, C. K.; Stucky, G. D. & Bartl, M. H. (2008). Optimizing sol-gel infiltration and processing methods for the fabrication of high-quality planar Titania inverse opals. *Chemistry of Materials*, Vol. 20, No. 15, (Aug), pp. 4925-4930, ISSN: 0897-4756

Gaya, U. I. & Abdullah, A. H. (2008). Heterogeneous photocatalytic degradation of organic contaminants over titanium dioxide: A review of fundamentals, progress and problems. *Journal of Photochemistry and Photobiology C: Photochemistry Reviews*, Vol. 9, No. 1, pp. 1-12, ISSN: 1389-5567

Goodwin, J. W.; Hearn, J.; Ho, C. C. & Ottewill, R. H. (1974). Studies on the preparation and characterisation of monodisperse polystyrene laticee. *Colloid and Polymer Science*, Vol. 252, No. 6, pp. 464-471, ISSN: 0303-402X

Gu, Z.-Z.; Hayami, S.; Kubo, S.; Meng, Q.-B.; Einaga, Y.; Tryk, D. A.; Fujishima, A. & Sato, O. (2001). Fabrication of Structured Porous Film by Electrophoresis. *Journal of the American Chemical Society*, Vol. 123, No. 1, pp. 175-176, ISSN: 0002-7863

Gu, Z. Z.; Fujishima, A. & Sato, O. (2002). Fabrication of high-quality opal films with controllable thickness. *Chemistry of Materials*, Vol. 14, No. 2, (Feb), pp. 760-765, ISSN: 0897-4756

Halaoui, L. I.; Abrams, N. M. & Mallouk, T. E. (2005). Increasing the conversion efficiency of dye-sensitized TiO2 photoelectrochemical cells by coupling to photonic crystals. *Journal of Physical Chemistry B*, Vol. 109, No. 13, (Apr), pp. 6334-6342, ISSN: 1520-6106

Ishii, M.; Harada, M.; Tsukigase, A. & Nakamura, H. (2007). Three-dimensional structure analysis of opaline photonic crystals by angle-resolved reflection spectroscopy. *Journal of Optics a-Pure and Applied Optics*, Vol. 9, No. 9, pp. S372-S376, ISSN: 1464-4258

Ishii, M.; Nakamura, H.; Nakano, H.; Tsukigase, A. & Harada, M. (2005). Large-Domain Colloidal Crystal Films Fabricated Using a Fluidic Cell. *Langmuir*, Vol. 21, No. 12, pp. 5367-5371, ISSN:

Jellison, G. E.; Boatner, L. A.; Budai, J. D.; Jeong, B. S. & Norton, D. P. (2003). Spectroscopic ellipsometry of thin film and bulk anatase (TiO2). *Journal of Applied Physics*, Vol. 93, No. 12, pp. 9537-9541, ISSN: 0021-8979

Jiang, P.; Bertone, J. F.; Hwang, K. S. & Colvin, V. L. (1999). Single-crystal colloidal multilayers of controlled thickness. *Chemistry of Materials*, Vol. 11, No. 8, (Aug), pp. 2132-2140, ISSN: 0897-4756

Joannopoulos, J. D.; Johnson, S. G.; Winn, J. N. & Meade, R. D. (2008). Photonic Crystals Molding the Flow of Light (second), Princeton University Press, 978-0-691-12456-8, Princeton

John, S. (1987). Strong localization of photons in certain disordered dielectric superlattices. *Physical Review Letters*, Vol. 58, No. 23, pp. 2486, ISSN:

Johnson, N. P.; McComb, D. W.; Richel, A.; Treble, B. M. & De la Rue, R. M. (2001). Synthesis and optical properties of opal and inverse opal photonic crystals. *Synthetic Metals*, Vol. 116, No. 1-3, (Jan), pp. 469-473, ISSN: 0379-6779

Johnson, S. G. & Joannopoulos, J. D. (2001). Block-iterative frequency-domain methods for Maxwell's equations in a planewave basis. *Optics Express*, Vol. 8, No. 3, pp. 173-190, ISSN: 1094-4087

Kanai, T. & Sawada, T. (2009). New Route to Produce Dry Colloidal Crystals without Cracks. *Langmuir*, Vol. 25, No. 23, pp. 13315-13317, ISSN:

King, J. S.; Graugnard, E. & Summers, C. J. (2005). TiO2 Inverse Opals Fabricated Using Low-Temperature Atomic Layer Deposition. *Advanced Materials*, Vol. 17, No. 8, pp. 1010-1013, ISSN: 1521-4095

Klein, S. M.; Manoharan, V. N.; Pine, D. J. & Lange, F. F. (2003). Preparation of monodisperse PMMA microspheres in nonpolar solvents by dispersion polymerization with a macromonomeric stabilizer. *Colloid and Polymer Science*, Vol. 282, No. 1, (Dec), pp. 7-13, ISSN: 0303-402X

Kraeutler, B. & Bard, A. J. (1978). Heterogeneous photocatalytic synthesis of methane from acetic acid - new Kolbe reaction pathway. *Journal of the American Chemical Society*, Vol. 100, No. 7, (1978/03/01), pp. 2239-2240, ISSN: 0002-7863

Leunissen, M. E.; Christova, C. G.; Hynninen, A.-P.; Royall, C. P.; Campbell, A. I.; Imhof, A.; Dijkstra, M.; van Roij, R. & van Blaaderen, A. (2005). Ionic colloidal crystals of oppositely charged particles. *Nature*, Vol. 437, No. 7056, pp. 235-240, ISSN:

Li, Q. & Shang, J. K. (2008). Inverse opal structure of nitrogen-doped titanium oxide with enhanced visible-light photocatalytic activity. *Journal of the American Ceramic Society*, Vol. 91, No. 2, (Feb), pp. 660-663, ISSN: 0002-7820

Li, Y.; Kunitake, T. & Fujikawa, S. (2006). Efficient Fabrication and Enhanced Photocatalytic Activities of 3D-Ordered Films of Titania Hollow Spheres. *Journal of Physical Chemistry B*, Vol. 110, No. 26, pp. 13000-13004, ISSN:

Liau, L. C. K. & Huang, Y. K. (2008). Process optimization of sedimentation self-assembly of opal photonic crystals under relative humidity-controlled environments. *Expert Systems with Applications*, Vol. 35, No. 3, (Oct), pp. 887-893, ISSN: 0957-4174

Lin, Y. K.; Harb, A.; Lozano, K.; Xu, D. & Chen, K. P. (2009). Five beam holographic lithography for simultaneous fabrication of three dimensional photonic crystal templates and line defects using phase tunable diffractive optical element. *Optics Express*, Vol. 17, No. 19, pp. 16625-16631, ISSN: 1094-4087

Liu, L.; Karuturi, S. K.; Su, L. T. & Tok, A. I. Y. (2011). TiO2 inverse-opal electrode fabricated by atomic layer deposition for dye-sensitized solar cell applications. *Energy & Environmental Science*, Vol. 4, No. 1, pp. 209-215, ISSN: 1754-5692

Liu, Z.; Ya, J.; Xin, Y.; Ma, J. & Zhou, C. (2006). Assembly of polystyrene colloidal crystal templates by a dip-drawing method. *Journal of Crystal Growth*, Vol. 297, No. 1, pp. 223-227, ISSN:

Lozano, G. & Miguez, H. (2007). Growth Dynamics of Self-Assembled Colloidal Crystal Thin Films. *Langmuir*, Vol. 23, No. 20, pp. 9933-9938, ISSN:

Lu, Y.; Yin, Y.; Gates, B. & Xia, Y. (2001). Growth of Large Crystals of Monodispersed Spherical Colloids in Fluidic Cells Fabricated Using Non-photolithographic Methods. *Langmuir*, Vol. 17, No. 20, pp. 6344-6350, ISSN:

Matsushita, S.; Fujikawa, S.; Onoue, S.; Kunitake, T. & Shimomura, M. (2007). Rapid fabrication of a smooth hollow-spheres array. *Bulletin of the Chemical Society of Japan*, Vol. 80, No. 6, pp. 1226-1228, ISSN: 0009-2673

Maurino, V.; Bedini, A.; Minella, M.; Rubertelli, F.; Pelizzetti, E. & Minero, C. (2008). Glycerol transformation through photocatalysis: A possible route to value added chemicals. *Journal of Advanced Oxidation Technologies*, Vol. 11, No. 2, (Jul), pp. 184-192, ISSN: 1203-8407

Meng, Q. B.; Fu, C. H.; Einaga, Y.; Gu, Z. Z.; Fujishima, A. & Sato, O. (2002). Assembly of Highly Ordered Three-Dimensional Porous Structure with Nanocrystalline TiO2 Semiconductors. *Chemistry of Materials*, Vol. 14, No. 1, pp. 83-88, ISSN: 0897-4756

Miguez, H.; Kitaev, V. & Ozin, G. A. (2004). Band spectroscopy of colloidal photonic crystal films. *Applied Physics Letters*, Vol. 84, No. 8, pp. 1239-1241, ISSN: 0003-6951

Mihi, A.; Calvo, M. E.; Anta, J. A. & Miguez, H. (2008). Spectral response of opal-based dye-sensitized solar cells. *Journal of Physical Chemistry C*, Vol. 112, No. 1, (Jan), pp. 13-17, ISSN: 1932-7447

Mihi, A.; Ocana, M. & Miguez, H. (2006). Oriented colloidal-crystal thin films by spin-coating microspheres dispersed in volatile media. *Advanced Materials*, Vol. 18, No. 17, (Sep), pp. 2244-2249, ISSN: 0935-9648

Minero, C. (1999). Kinetic analysis of photoinduced reactions at the water semiconductor interface. *Catalysis Today*, Vol. 54, No. 2-3, pp. 205-216, ISSN:

Minero, C. & Vione, D. (2006). A quantitative evalution of the photocatalytic performance of TiO2 slurries. *Applied Catalysis B: Environmental*, Vol. 67, No. 3-4, pp. 257-269, ISSN: 0926-3373

Miyake, M.; Chen, Y. C.; Braun, P. V. & Wiltzius, P. (2009). Fabrication of Three-Dimensional Photonic Crystals Using Multibeam Interference Lithography and Electrodeposition. *Advanced Materials*, Vol. 21, No. 29, pp. 3012-+, ISSN: 0935-9648

Moon, J. H.; Yang, S.; Dong, W.; Perry, J. W.; Adibi, A. & Yang, S.-M. (2006). Core-shell diamond-like silicon photonic crystals from 3D polymer templates created by holographic lithography. *Optics Express*, Vol. 14, No. 13, pp. 6297-6302, ISSN:

Nishimura, S.; Abrams, N.; Lewis, B. A.; Halaoui, L. I.; Mallouk, T. E.; Benkstein, K. D.; van de Lagemaat, J. & Frank, A. J. (2003). Standing wave enhancement of red absorbance and photocurrent in dye-sensitized titanium dioxide photoelectrodes coupled to photonic crystals. *Journal of the American Chemical Society*, Vol. 125, No. 20, (May), pp. 6306-6310, ISSN: 0002-7863

Norris, D. J.; Arlinghaus, E. G.; Meng, L.; Heiny, R. & Scriven, L. E. (2004). Opaline Photonic Crystals: How Does Self-Assembly Work? *Advanced Materials*, Vol. 16, No. 16, pp. 1393-1399, ISSN: 1521-4095

Park, S. H.; Qin, D. & Xia, Y. (1998). Crystallization of Mesoscale Particles over Large Areas. *Advanced Materials*, Vol. 10, No. 13, pp. 1028-1032, ISSN: 1521-4095

Pavarini, E.; Andreani, L. C.; Soci, C.; Galli, M.; Marabelli, F. & Comoretto, D. (2005). Band structure and optical properties of opal photonic crystals. *Physical Review B*, Vol. 72, No. 4, pp. 045102, ISSN: 1098-0121

Pelizzetti, E.; Minero, C.; Maurino, V.; Sclafani, A.; Hidaka, H. & Serpone, N. (1989). Photocatalytic degradation of nonylphenol ethoxylated surfactants. *Environmental Science & Technology*, Vol. 23, No. 11, pp. 1380-1385, ISSN:

Pursiainen, O. L. J.; Baumberg, J. J.; Winkler, H.; Viel, B.; Spahn, P. & Ruhl, T. (2008). Shear-Induced Organization in Flexible Polymer Opals. *Advanced Materials*, Vol. 20, No. 8, pp. 1484-1487, ISSN: 1521-4095

Ramiro-Manzano, F.; Atienzar, P.; Rodriguez, I.; Meseguer, F.; Garcia, H. & Corma, A. (2007). Apollony photonic sponge based photoelectrochemical solar cells. *Chemical Communications*, Vol. No. 3, pp. 242-244, ISSN: 1359-7345

Reese, C. E.; Guerrero, C. D.; Weissman, J. M.; Lee, K. & Asher, S. A. (2000). Synthesis of highly charged, monodisperse polystyrene colloidal particles for the fabrication of photonic crystals. *Journal of Colloid and Interface Science*, Vol. 232, No. 1, (Dec), pp. 76-80, ISSN: 0021-9797

Ren, M. M.; Ravikrishna, R. & Valsaraj, K. T. (2006). Photocatalytic degradation of gaseous organic species on photonic band-gap titania. *Environmental Science and Technology*, Vol. 40, No. 22, (Nov), pp. 7029-7033, ISSN: 0013-936X

Ren, M. M. & Valsaraj, K. (2009). Inverse Opal Titania on Optical Fiber for the Photoreduction of CO2 to CH3OH. *International Journal of Chemical Reactor Engineering*, Vol. 7, No. pp. A90, ISSN: 1542-6580

Roy, S. C.; Varghese, O. K.; Paulose, M. & Grimes, C. A. (2010). Toward Solar Fuels: Photocatalytic Conversion of Carbon Dioxide to Hydrocarbons. *ACS Nano*, Vol. 4, No. 3, (Mar), pp. 1259-1278, ISSN: 1936-0851

Ruhl, T.; Spahn, P.; Winkler, H. & Hellmann, G. P. (2004). Large Area Monodomain Order in Colloidal Crystals. *Macromolecular Chemistry and Physics*, Vol. 205, No. 10, pp. 1385-1393, ISSN: 1521-3935

Sawada, T.; Suzuki, Y.; Toyotama, A. & Iyi, N. (2001). Quick Fabrication of Gigantic Single-Crystalline Colloidal Crystals <br> for Photonic Crystal Applications. *Japanese Journal of Applied Physics*, Vol. 40, No. Copyright (C) 2001 The Japan Society of Applied Physics, pp. L1226-L1228, ISSN:

Schroden, R. C.; Al-Daous, M.; Blanford, C. F. & Stein, A. (2002). Optical properties of inverse opal photonic crystals. *Chemistry of Materials*, Vol. 14, No. 8, (Aug), pp. 3305-3315, ISSN: 0897-4756

Serpone, N. (2006). Is the Band Gap of Pristine TiO2 Narrowed by Anion- and Cation-Doping of Titanium Dioxide in Second-Generation Photocatalysts? *The Journal of Physical Chemistry B*, Vol. 110, No. 48, (2006/12/01), pp. 24287-24293, ISSN: 1520-6106

Shevchenko, E. V.; Talapin, D. V.; Kotov, N. A.; O'Brien, S. & Murray, C. B. (2006). Structural diversity in binary nanoparticle superlattices. *Nature*, Vol. 439, No. 7072, pp. 55-59, ISSN:

Shi, J.; Hsiao, V. K. S.; Walker, T. R. & Huang, T. J. (2008). Humidity sensing based on nanoporous polymeric photonic crystals. *Sensors and Actuators B-Chemical*, Vol. 129, No. 1, (Jan), pp. 391-396, ISSN: 0925-4005

Shimmin, R. G.; DiMauro, A. J. & Braun, P. V. (2006). Slow Vertical Deposition of Colloidal Crystals: A Langmuir-Blodgett Process? *Langmuir*, Vol. 22, No. 15, pp. 6507-6513, ISSN: 0743-7463

Shin, J.-H.; Kang, J.-H.; Jin, W.-M.; Park, J. H.; Cho, Y.-S. & Moon, J. H. (2011). Facile Synthesis of TiO2 Inverse Opal Electrodes for Dye-Sensitized Solar Cells. *Langmuir*, Vol. 27, No. 2, pp. 856-860, ISSN: 0743-7463

Somani, P. R.; Dionigi, C.; Murgia, M.; Palles, D.; Nozar, P. & Ruani, G. (2005). Solid-state dye PV cells using inverse opal TiO2 films. *Solar Energy Materials and Solar Cells*, Vol. 87, No. 1-4, (May), pp. 513-519, ISSN: 0927-0248

Sordello, F.; Duca, C.; Maurino, V. & Minero, C. (2011a). Photocatalytic metamaterials: TiO2 inverse opals. *Chemical Communications*, Vol. 47, No. 21, pp. 6147-6149, ISSN: 1359-7345

Sordello, F.; Maurino, V. & Minero, C. (2011b). Photoelectrochemical study of TiO2 inverse opals. *Journal of Materials Chemistry*, Vol. 21, No. 47, pp. 19144 - 19152, ISSN: 0959-9428

Srinivasan, M. & White, T. (2007). Degradation of methylene blue by three-dimensionally ordered macroporous titania. *Environmental Science and Technology*, Vol. 41, No. 12, (Jun), pp. 4405-4409, ISSN: 0013-936X

Sun, W.; Jia, F.; Sun, Z.; Zhang, J.; Li, Y.; Zhang, X. & Yang, B. (2011). Manipulation of Cracks in Three-Dimensional Colloidal Crystal Films via Recognition of Surface Energy Patterns: An Approach to Regulating Crack Patterns and Shaping Microcrystals. *Langmuir*, Vol. 27, No. 13, pp. 8018-8026, ISSN: 0743-7463

Wang, C.; Geng, A.; Guo, Y.; Jiang, S.; Qu, X. & Li, L. (2006). A novel preparation of three-dimensionally ordered macroporous M/Ti (M = Zr or Ta) mixed oxide nanoparticles with enhanced photocatalytic activity. *Journal of Colloid and Interface Science*, Vol. 301, No. 1, pp. 236-247, ISSN: 0021-9797

Wang, C. W. & Moffitt, M. G. (2005). Nonlithographic Hierarchical Patterning of Semiconducting Nanoparticles via Polymer/Polymer Phase Separation. *Chemistry of Materials*, Vol. 17, No. 15, pp. 3871-3878, ISSN: 0897-4756

Waterhouse, G. I. N. & Waterland, M. R. (2007). Opal and inverse opal photonic crystals: Fabrication and characterization. *Polyhedron*, Vol. 26, No. 2, (Jan), pp. 356-368, ISSN: 0277-5387

Wei, N. N.; Han, T.; Deng, G. Z.; Li, J. L. & Du, J. Y. (2011). Synthesis and characterizations of three-dimensional ordered gold- nanoparticle-doped titanium dioxide photonic crystals. *Thin Solid Films*, Vol. 519, No. 8, pp. 2409-2414, ISSN: 0040-6090

Wijnhoven, J. & Vos, W. L. (1998). Preparation of photonic crystals made of air spheres in titania. *Science*, Vol. 281, No. 5378, (Aug), pp. 802-804, ISSN: 0036-8075

Xu, J. A.; Yang, B. F.; Wu, M.; Fu, Z. P.; Lv, Y. & Zhao, Y. X. (2010). Novel N-F-Codoped TiO2 Inverse Opal with a Hierarchical Meso-/Macroporous Structure: Synthesis, Characterization, and Photocatalysis. *Journal of Physical Chemistry C*, Vol. 114, No. 36, pp. 15251-15259, ISSN: 1932-7447

Xu, Y.; Zhu, X.; Dan, Y.; Moon, J. H.; Chen, V. W.; Johnson, A. T.; Perry, J. W. & Yang, S. (2008). Electrodeposition of Three-Dimensional Titania Photonic Crystals from Holographically Patterned Microporous Polymer Templates. *Chemistry of Materials*, Vol. 20, No. 5, pp. 1816-1823, ISSN: 0897-4756

Yablonovitch, E. (1987). Inhibited Spontaneous Emission in Solid-State Physics and Electronics. *Physical Review Letters*, Vol. 58, No. 20, pp. 2059-2062, ISSN:

Yablonovitch, E. (2001). Photonic crystals: Semiconductors of light. *Scientific American*, Vol. 285, No. 6, (Dec), pp. 46-50, ISSN: 0036-8733

Yip, C. H.; Chiang, Y. M. & Wong, C. C. (2008). Dielectric band edge enhancement of energy conversion efficiency in photonic crystal dye-sensitized solar cell. *Journal of Physical Chemistry C*, Vol. 112, No. 23, (Jun), pp. 8735-8740, ISSN: 1932-7447

Yodh, A. G.; Lin, K.-H.; Crocker, J. C.; Dinsmore, A. D.; Verma, R. & Kaplan, P. D. (2001). Entropically Driven Self-Assembly and Interaction in Suspension. *Philosophical Transactions of the Royal Society of London, Series A: Physical Sciences and Engineering*, Vol. 359, No. 1782, pp. 921-937, ISSN:

Yun, H. J.; Lee, H.; Joo, J. B.; Kim, N. D. & Yi, J. (2011). Effect of TiO2 Nanoparticle Shape on Hydrogen Evolution via Water Splitting. *Journal of Nanoscience and Nanotechnology*, Vol. 11, No. 2, (Feb), pp. 1688-1691, ISSN: 1533-4880

Zhao, W.; Chen, C.; Li, X.; Zhao, J.; Hidaka, H. & Serpone, N. (2002). Photodegradation of Sulforhodamine-B Dye in Platinized Titania Dispersions under Visible Light Irradiation: Influence of Platinum as a Functional Co-catalyst. *The Journal of Physical Chemistry B*, Vol. 106, No. 19, (2002/05/01), pp. 5022-5028, ISSN: 1520-6106

4

Photoisomerization of Norbornadiene to Quadricyclane Using Ti-Containing Photocatalysts

Ji-Jun Zou, Lun Pan, Xiangwen Zhang and Li Wang

Key Laboratory for Green Chemical Technology of Ministry of Education,
School of Chemical Engineering and Technology,
Tianjin University
P.R. China

1. Introduction

Photoisomerization, an important aspect of photochemistry, is molecular behavior in which the structural change between isomers is caused by photoexcitation. Photoisomerization is already applied or has potential in many fields, such as the synthesis of compounds that can not be obtained by other methods, pigments in digital data storage and recording, solar energy harvesting, and nanoscale devices and materials with photo-modulable properties.

Conformation transformation, especially the *trans-cis* photoisomerization of alkenes, see Scheme 1, is the most studied photoisomerization (Waldeck, 1991; Dou & Allen, 2003; Quenneville & Martínez, 2003; Minezawa & Gordon, 2011). Stilbene is a prototypical molecule that has been extensively investigated by both experimental and theoretical approaches. The primary mechanism of isomerization is through the excited singlet state starting from either the *cis* or the *trans* geometry. After photoexcitation, the molecule can overcome a small activation barrier and twist about its central C=C bond to form a twisted intermediate. This intermediate then decays with equal probability to either ground state *cis*-stilbene or ground state *trans*-stilbene. Similarly, the torsion around N=N bond also induces photoisomerization (Ciminelli et al., 2004; Mita et al., 1989), with azobenzene as the prototype. Moreover, compounds with photoisomerizable core have been designed for some special purposes. For example, highly branched dendrimers containing azobenzene core can be excited and converted to isomers by infrared irradiation, which represents a strategy for harvesting low-energy photons via chemical transformation (Jiang & Aida, 1997).

Geometric isomerization is another important type of photoisomerization that involves bond cleavage and creation in alkenes, see Scheme 2. One typical transformation is intramolecular cycloaddition such as [2+2] and [2+3] cycloadditions (Xu et al., 2009; Filley et al., 2001; Lu et al., 2011; Somekawa et al., 2009), which is very attractive in synthetic applications. In addition, the cycloaddition may produce strained and energy-rich products, which has received attention as a way to store solar energy.

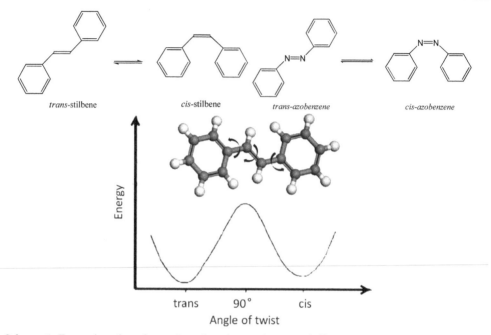

trans-stilbene   *cis*-stilbene   *trans-azobenzene*   *cis-azobenzene*

Scheme 1. Examples of conformation photoisomerization of alkenes along with the prototype surface diagram of stilbene isomerization.

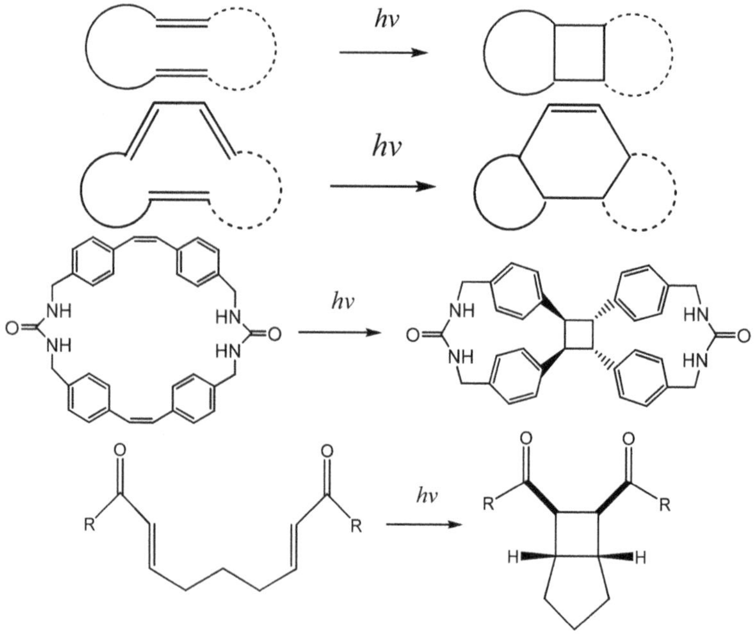

Scheme 2. Examples of geometric photoisomerization of alkenes

Generally, photoisomerization is sensitized by homogenous organics and/or metal complexes. However, solid semiconductors and even zeolites have been found to effective for these photo-induced processes. For example, CdS has been extensively studied for the *trans-cis* transformation of alkenes (Gao et al., 1998; Yanagida et al., 1986; Al-Ekabi & Mayo, 1985). Unfortunately, the instability of CdS under irradiation is a big problem for application.

2. Photosensitized isomerization of norbornadiene

Photoisomerization of norbornadiene (NBD) to quadricyclane (QC) is typical intramolecular [2+2] cycloaddition. It continues to be an interesting field as potential way for storage and conversion for solar energy (Hammond et al., 1964; Bren' et al., 1991; Dubonosov et al., 2002). The photoisomerization of NBD results in metastable structure that contains highly strained cyclobutane and two cyclopropane fragments. When one mole of NBD is transformed to QC, 89 kJ of solar energy could be stored in form of strain energy. Under some catalytic conditions, the inverse QC→NBD transformation occurs easily, accompanied with considerable thermal effect (ΔH=-89 kJ/mol). This represents an idea cycle for energy conversion and storage, see Scheme 3.

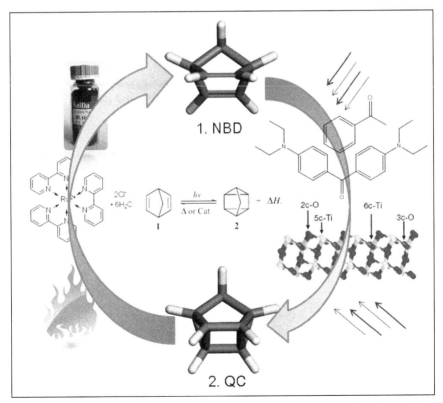

Scheme 3. Solar energy harvesting cycle based on photoisomerization of norbornadiene.

Recently, QC has been identified as a very promising high-energy compound as replacement for, or additive to, current hydrocarbon-based rocket propellants, because the extraordinary high strain energy offers a very high specific impulse (Kokan et al., 2009; Striebich & Lawrence, 2003). It is reported that QC-based fuels provide more propulsion than most of the hydrocarbon fuels like rocket propellant RP-1. QC is also designed for satellite propulsion system to replace highly toxic fuels like hydrazine and dinitrogen tetroxide. Moreover, QC is thermally and chemically stable, which means that it can be easily stored and transported like other hydrocarbon fuels.

The quantum yield of pure NBD photoisomerization is extremely low because the absorption edge of NBD is less than 300nm. Many efforts have been done to drive this photoisomerization using longer light and improve the quantum yield, which can be categorized into three directions: use of sensitizer, modification of NBD molecule and use of NBD-containing compounds. Dubonosov et al already presented two comprehensive reviews on the photoisomerization of NBD and its derivatives in 1991 and 2001 (Bren' et al., 1991; Dubonosov et al., 2002). This chapter focuses on the synthesis of QC from NBD, so only a brief summary is given to the direct photoisomerization of NBD, i.e. the first direction. The photosensitized isomerization of NBD occurs via triplet, so many carbonyl compounds like acetophenone, benzophenone and Michlers' ketone were used as triplet sensitizers. Actually, a recent patent claimed a solution phase photoisomerization process of NBD based on substituted Michlers' ketone (Cahill & Steppel, 2004). However, since the energy of the triplet state of NBD (^{3}NBD) is very high (~257 kJ/mol), only small amount of sensitizers are qualified. Then, metal complexes and derivatives of carbonyl compounds were studied. In this case, the isomerization proceeds through the formation of sensitizer-NBD complexes in electron-excited states, with or without the formation of ^{3}NBD.

However, the photosensitized reaction suffers from many drawbacks. First, homogenous reaction brings some difficulties in product purification and sensitizer recycling. Second, sensitizer tends to decompose under UV irradiation and induces some side-reactions like polymerization of NBD. In fact, in the past decade, work on the direct photoisomerization of NBD is very scare, and only some NBD derivatives were synthesized to prepare photo-responsive materials (Chen et al., 2007; Vlaar et al., 2001).

Heterogeneous semiconductors are extensively used in photocatalytic processes such as degradation of pollutants, hydrogen generation, and solar cell. They are also attractive for photoisomerization when considering the easy purification of product and reuse of catalyst. In fact, zeolites and semiconductors were already found to be active for the photoisomerization of NBD. In a brief communication, Lahiry and Haldar firstly reported that NBD can be isomerized over semiconductors like ZnO, ZnS and CdS (Lahiry & Haldar, 1986). Then Gandi et al. reported that Y-zeolites exchanged with K+, Cs+ and Tl+ ions can sensitize the intramolecular addition of some dienes like NBD and afford the corresponding triplet products through heavy atom effect (Ghandi, 2006). In this case the reactant is pre-adsorbed in the micropores. Similarly, Gu and Liu compared La-, Cs-, Zn- and K-exchanged Y zeolites for the photoisomerization of NBD in liquid phase, and found LaY shows relatively high activity (Gu & Liu, 2008). They postulated that the heavy atom effect and Brönsted acid account for the result.

3. Photoisomerization of NBD over Ti-containing photocatalysts

Among the photocatalysts studied, TiO_2 is the most widely used material owing to its low-cost, non-toxicity, chemical and biological inertness, and photostability. Previous literatures already hint that TiO_2 can facilitate the photoisomerization of NBD. Although the activity of TiO_2 is relatively low due to the low optical absorbance and high charge–hole recombination rate, many methods such as doping with metal and nonmetal atoms and preparation of highly dispersed Ti-O species have been established to overcome this problem.

Recently, we focused on the photocatalytic isomerization of NBD using Ti-containing materials including metal-doped TiO_2 (Pan et al., 2010; Zou et al., 2008a), Ti-containing MCM-41 molecule sieves (Zou et al., 2008b) and metal-incorporated Ti-MCM-41 (Zou et al., 2010). These photocatalysts do show improved activity compared with pure TiO_2, suggesting that the photocatalysts used in environmental photocatalysis can be applied in the photoisomerization. In the following sections, a mini review of our work will be given, with the aim to show a new and promising way for photoisomerization.

3.1 Synthesis of materials and evaluation of activity

Three kinds of photocatalysts, including metal doped TiO_2 (M-TiO_2), Ti-substituted (Ti-MCM-41) and Ti-grafted MCM-41(TiO_2-MCM-41), and metal incorporated Ti-MCM-41 (M-Ti-MCM-41) were studied. M-TiO_2 materials were synthesized using sol-gel method with tetrabutyl titanate, $VO(SO_4)$, $Fe(SO_4)_3$, $Cu(NO_3)_2$, $Cr(NO_3)_3$, $Ce(NO_3)_3$ and $ZnSO_4$ as the metal resources (Pan et al., 2010; Zou et al., 2008a). Ti-MCM-41 and M-Ti-MCM-41 materials were synthesized via hydrothermal method using cetyltrimethyl ammonium bromide and tetrathyorthosilicate as the structure director and Si resource, respectively (Zou et al., 2008b, 2010), and TiO_2-MCM-41 materials were prepared through chemical grafting (Zou et al., 2008b). All the prepared materials were calcined at 500°C for 3 or 5 hours. The abbreviation of materials was suffixed with a symbol x in parentheses to describe the original molar Ti/M or Si/M ratio in starting synthetic mixtures.

The photoisomerization reaction was conducted under UV irradiation in closed quartz reactor with magnetic stirring (Pan et al., 2010; Zou et al., 2008a, 2008b, 2010). For M-TiO_2(M=V, Fe, Cu, Ce and Cr), a quartz chamber was irradiated vertically by a 300 W high-pressure xenon lamp located on the upper position. The wavelength was limited in the range of 220-420 nm by an optical filter and dimethyl sulfoxide was used as the solvent. For M-TiO_2(M=Zn) and Ti-contaning MCM-41 materials, a cylindrical quartz vessel was irradiated by a 400 W high pressure mercury lamp positioned inside the vessel. In this case the wavelength was not controlled and p-xylene was used as the solvent. The composition of the resulted mixture was determined by a gas chromatograph equipped with BP-1 capillary column and flame ionization detector. The rate constant k for each photocatalyst was calculated via kinetics fitting, assuming that the reaction obeys the first-order law. Since the reaction conditions for different type of photocatalysts are a little different, TiO_2 was used as the baseline to compare the photocatalytic activity of all materials. Therefore, the reaction constant k of one material was divided by that of TiO_2 (k_0) under identical reaction conditions, and the obtained relative reaction rate constant, i.e. k/k_0, was used in this chapter.

3.2 Photoisomerization of NBD over metal-doped TiO$_2$: Effect of metal dopants

TiO$_2$ is widely used in photocatalytic reactions due to its low cost and chemical stability, but suffers from the fast recombination of photoinduced electron-hole pairs. Doping with metal ions is regarded as an effective method to improve the efficiency of TiO$_2$ (Yang et al., 2007; Adán et al., 2007). So metal (Cu, Cr, Ce, V, Fe, Zn)-doped TiO$_2$ was studied firstly for the photoisomerization of NBD.

The structural parameters of prepared materials characterized using XRD, EDX, XPS and N$_2$-adsorption are shown in Table 1. According to the bulk composition from EDX data and surface composition from XPS data, V, Fe and Ce are dispersed in the inner part of prepared materials whereas Cu, Cr and Zn ions are enriched on the particle surface. Specifically, only a small amount of Cu is introduced into the material. Generally, there are three possible dispersion modes for dopants, namely substitutional, interstitial and surface positions. The local structure of dopants ions can be deduced based on their ionic radii, that is, Fe and V ions with radii close to Ti ions in substitutional sites, large Ce ions in interstitial positions, whereas Cu ions with largest radii on the surface. The surface enrichment of relatively small Cr and Zn ions that have comparable radii with Ti ions is a little surprising because they could enter the lattice, but consistent with results reported by other researchers (Zhu et al., 2010; Jing et al., 2006). The reason may be that these ions are originally inside the lattice but diffuse to the surface through oxygen vacancies during the calcination process, or the hydrolysis rate of these ions is much slower than that of Ti ions.

Materials	Grain size (nm)	S_{BET} (m$^2 \cdot$g^{-1})	Ti/M ratio	
			EDX	XPS
TiO$_2$	21.5	21.5	-	-
Cu-TiO$_2$(15)	19.9	13.1	90.4	3.8
Cr-TiO$_2$(15)	14.7	40.9	20.0	3.0
Ce-TiO$_2$(15)	11.4	64.3	16. 9	19.8
V-TiO$_2$(15)	9.9	102.7	19.0	15.6
Fe-TiO$_2$(15)	7.0	120.6	18.5	19.8
Zn-TiO$_2$(100)	8.1	84.9	-	7.1

Table 1. Structural characteristics of metal-doped TiO$_2$
(Pan et al., 2010; Zou et al., 2008a).

When metal dopants are dispersed in the substitutional site, some Ti-O-M structures are expected to form, which will cause a shift in the binding energy of Ti species because the difference in Pauling electronegativity can induce electron transfer from Ti to M ions. As shown in Fig. 1, the XPS signal (binding energy) of Ti is shifted to higher values after doping with V and Fe, while for other doping the shift is not so obvious because the metals are not located in the substitutional sites with no, or only a few, M-O-Ti structures formed.

Doping can restrain the growth of particle to some degree no mater what the doping mode is, but the mechanism may be different. Fe and Zn-doping produces considerably small particles, see Table 1 and Fig. 2. For the substitutional doping like Fe- and V-doping, dopants in the lattice can destroy the crystal structure and restrain its growth. For the surface deposition or interstitial mode, like Ce- and Zn-doping, dopants may prevent the direct contact of TiO$_2$ crystallites and retard them agglomerating into big particle.

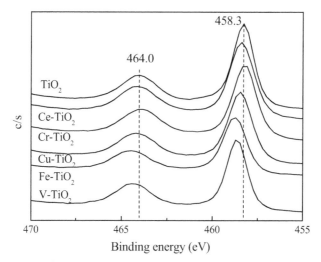

Fig. 1. Ti2p XPS spectra of metal-doped TiO_2. Reprinted with permission from Pan, L.; Zou, J.-J; Zhang, X. & Wang, L. (2010), *Industrial & Engineering Chemistry Research,* Vol.49, No.18, pp. 8526-8531. Copyright @ 2010 American Chemical Society.

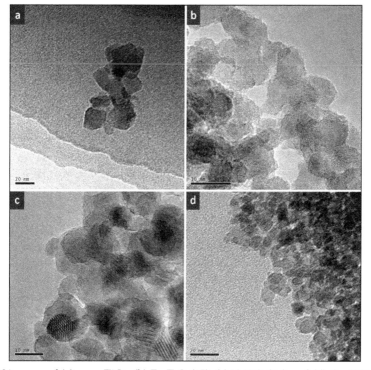

Fig. 2. TEM images of (a) pure TiO_2, (b) Fe-TiO_2(15), (c) V-TiO_2(15) and (d) Zn-TiO_2(100). (a) & (d) reprinted with permission from Zou, J.-J.; Zhu, B.; Wang, L.; Zhang, X. & Mi, Z.

The relative photocatalytic activity of doped TiO_2 (k/k_0) is also shown in Fig. 3. Except Cu, doping metal ions show positive effect on the photoisomerization of NBD, among which Zn-TiO_2 and Fe-TiO_2 are specifically active. The photoisomerization reaction is a complex process, and the physicochemical properties of photocatalyst such as grain size, type of dopant ions as well as their local structure are very important. Small particle is of course desired because it provides large active surface. It has been reported that the surface doping of Zn ions produces many surface OH groups that greatly enhance the intensity of surface photovoltage spectrum and photoluminescence and improve the photoactivity (Jing et al., 2006). As shown in Fig. 4, the activity of NBD photoisomerization is also closely relative to the concentration of surface OH.

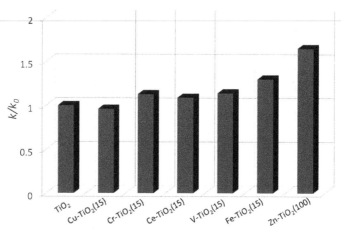

Fig. 3. Activity of metal-doped TiO_2 for the photoisomerization of norbornadiene (Pan et al., 2010; Zou et al., 2008a).

However, the role of surface OH seems invalid for the materials with substitutional doping. As shown in Fig. 5, the activity of Fe- and V-doped TiO_2 and their lattice oxygen concentration, not the surface OH, change in identical manner, strongly suggesting there is an inherent correlation between the photoisomerization and lattice oxygen. It is still not clear why two doping modes induce contrary result, probably because the reactant molecule is adsorbed on different site that will be discussed in section 4. As to the role of substitutional dopants, it has been reported that metal ions in substitutional sites can improve the photoinduced charge transfer and separation (Wang et al., 2009). It is believed that this process is very likely to occur through the M-O-Ti structure in which the metal dopants mainly serve as charge trapping and transferring center. Taking Fe-TiO_2 as example, the role of Fe is shown as follows: (1) Fe ions temporarily trap photoinduced charges in the neighboring Ti-O moiety:

$$Ti^{4+} - O^{2-} - Fe^{3+} - O^- - Ti^{3+} \rightarrow Ti^{4+} - O^{2-} - Fe^{2+} - O^- - Ti^{4+}$$

$$Ti^{4+} - O^{2-} - Fe^{3+} - O^- - Ti^{3+} \rightarrow Ti^{4+} - O^{2-} - Fe^{4+} - O^{2-} - Ti^{3+}$$

(2) The trapped charges are transferred to sideward Ti-O species, resulting in separated charges:

$$Ti^{4+} - O^{2-} - Fe^{2+} - O^- - Ti^{4+} \rightarrow Ti^{3+} - O^{2-} - Fe^{3+} - O^- - Ti^{4+}$$

$$Ti^{4+} - O^{2-} - Fe^{4+} - O^{2-} - Ti^{3+} \rightarrow Ti^{4+} - O^- - Fe^{3+} - O^{2-} - Ti^{3+}$$

In this way, the charge induced in one Ti-O moiety is quickly transferred to another Ti-O moiety through the Fe-O-Ti structure, thus effectively separating the charge and retarding the recombination.

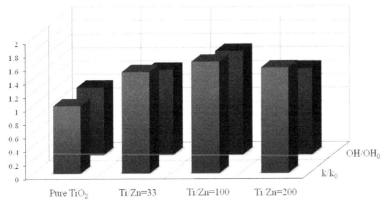

Fig. 4. Relationship of activity for the photoisomerization of norbornadiene and the relative surface OH concentration of Zn-TiO$_2$ (Zou et al., 2008a). OH, the content of surface OH; OH$_0$, the OH content of pure TiO$_2$.

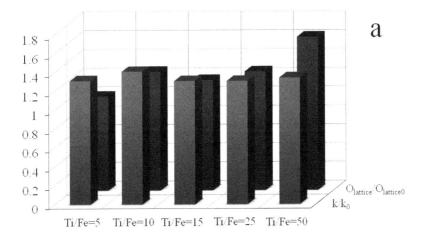

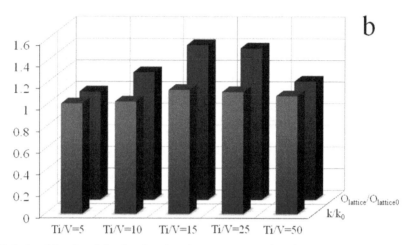

Fig. 5. Relationship of activity for the photoisomerization of norbornadiene and the relative lattice oxygen concentration of (a) $Fe-TiO_2$ and (b) $V-TiO_2$ (Pan et al., 2010).

3.3 Photoisomerization of NBD over Ti-containing MCM-41: Effect of Ti coordination

MCM-41 has uniform hexagonal mesopores with large internal surface area, exhibiting great potential as the supporting materials of TiO_2. It has been reported that incorporating Ti ions into framework or loading them on the wall of MCM-41 gives unique photocatalytic activity (Hu et al., 2003, 2006). So both Ti-incorporated and Ti-grafted MCM-41 materials were prepared for the photoisomerization of NBD.

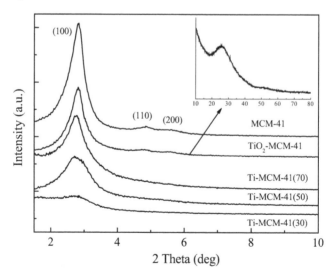

Fig. 6. XRD patterns of Ti-MCM-41 and TiO_2-MCM-41. Reprinted with permission from Zou, J.-J.; Zhang, M.-Y.; Zhu, B.; Wang, L.; Zhang, X. & Mi, Z. (2008), *Catalysis Letters*, Vol.124, No.12, pp. 139-145, Copyright @ 2008 Springer Netherlands.

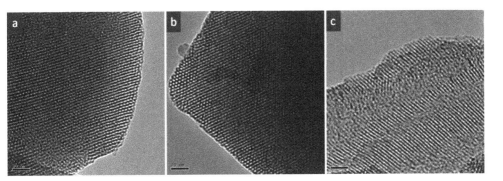

Fig. 7. TEM images of (a) MCM-41, (b) TiO_2-MCM-41 and (c) Ti-MCM-41(50). Reprinted with permission from Zou, J.-J.; Zhang, M.-Y.; Zhu, B.; Wang, L.; Zhang, X. & Mi, Z. (2008), *Catalysis Letters*, Vol.124, No.12, pp. 139-145, Copyright @ 2008 Springer Netherlands.

Grafting TiO_2 in the pore of MCM-41 does not influence the ordered hexagonal structure of support as its XRD patterns in the low-angle region are identical to MCM-41, see Fig. 6. An additional peak corresponding to the (101) reflex of anatase TiO_2 is observed at 25.5° but the intensity is extremely weak, so TiO_2 crystallites are highly dispersed in the pore of MCM-41. Incorporating Ti ions in the MCM-41 framework slightly impairs the structural integrity of MCM-41 but the ordered structure is well retained, shown by the weakened but obvious diffractive peaks. Also, the cell unit of Ti-MCM-41 is enlarged because the Ti-O bond distance is longer than the Si-O bond distance. TEM images in Fig. 7 further confirm the XRD result. No TiO_2 nanoparticles are observed for TiO_2-MCM-41 and its pore structure is identical to MCM-41, but some linear tubular pores of Ti-MCM-41 collapse into irregular pores.

The nature and coordination of Ti^{4+} ions was deduced according to the UV-vis diffuse reflectance spectra shown in Fig. 8. The absorption peak at 220 nm is ascribed to tetra-coordinated Ti whereas the peak at ~270 nm represents species in higher coordination environments (penta- or hexa-coordinated species). For Ti-MCM-41, most of the Ti species are dispersed in the framework (Ti-O-Si) when Ti content is low, but polymerized Ti species (Ti-O-Ti) present in case of higher Ti content. TiO_2-MCM-41 contains highly dispersed quantum-size TiO_2 nanodomains, see the blue-shifted absorption compared with bulk TiO_2.

The overall activity for the photoisomerization of NBD is Ti-MCM-41(30) > Ti-MCM-41(50) > TiO_2-MCM-41 > Ti-MCM-41(70) > TiO_2, see Fig. 9a. Since the amount of Ti species is different in these materials, the activity based on TiO_2 was also calculated to compare the inherent activity of different Ti species, with the order of Ti-MCM-41(50) ≈ Ti-MCM-41(70) > Ti-MCM-41(30) > TiO_2-MCM-41 > TiO_2, see Fig.9b. Considering the local structure of Ti, it can be seen that framework Ti species are most active in the photoisomerization of NBD, polymerized species follows and bulk TiO_2 has the lowest activity.

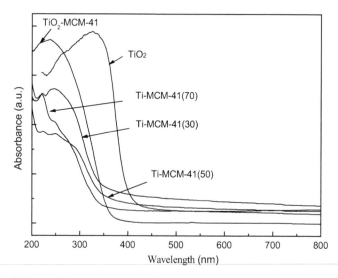

Fig. 8. UV-Vis diffuse reflectance spectra of Ti-MCM-41 and TiO₂-MCM-41. Reprinted with permission from Zou, J.-J.; Zhang, M.-Y.; Zhu, B.; Wang, L.; Zhang, X. & Mi, Z. (2008), *Catalysis Letters*, Vol.124, No.12, pp. 139-145, Copyright @ 2008 Springer Netherlands.

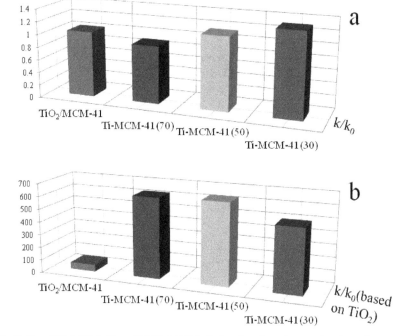

Fig. 9. Activity of Ti-MCM-41 and TiO₂-MCM-41 for the photoisomerization of norbornadiene (Zou et al., 2008b).

3.4 Photoisomerization of NBD over M-Ti-MCM-41: Combination of metal doping and framework Ti species

Transition-metal-incorporated MCM-41 generally shows high photocatalytic activity due to the high dispersion of photoactive sites and effective separation of electrons and holes (Hu et al., 2007; Yamashita et al., 2001; Matsuoka & Ampo, 2003; Davydov et al., 2001). Since Ti-MCM-41 produces highly active photocatalysts for the photoisomerization of NBD, it is expected that introducing second transition metal ion into Ti-MCM-41 may further enhance the activity. So series of transition-metal-incorporated (V, Fe and Cr) Ti-MCM-41 were synthesized for the photoisomerization of NBD, with Si/Ti ratio of 30.

According to the UV-vis spectra in Fig. 10, V and Fe ions are well dispersed in the materials whereas the dispersion of Cr ions is very poor. For V-Ti-MCM-41(150), V ions are highly dispersed in MCM-41 framework at atomic level with tetrahedral coordination, with some species in 6-fold (absorption around 370 nm) and higher coordination or even polymerized environments (absorption in >400 nm region) formed with the increase of V content. This tendency is also observed for Fe-Ti-MCM-41. However, for Cr-Ti-MCM-41, the absorption at

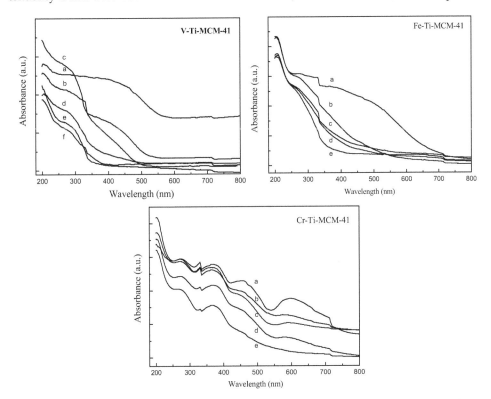

Fig. 10. UV-Vis diffuse reflectance spectra of M(V, Fe and Cr)-Ti-MCM-41 (a: Si/M=10, b: Si/M=33, c: Si/M=75, d: Si/M=100, e: Si/M=150, f: Ti-MCM-41). Reprinted with permission from Zou, J.-J.; Liu, Y.; Pan, L.; Wang, L. & Zhang, X. (2010), *Applied Catalysis B: Environmental*, Vol.95, No.3-4, pp. 439-445. Copyright @ 2010 Elsevier.

470 nm and 610 nm ascribed to poly- and bulk Cr_2O_3 is very intensive. The local structure of Cr ions are also testified by the IR spectra in Fig. 11. All Cr-Ti-MCM-41 samples show a shoulder band at 880-900 cm^{-1} assigned to Cr^{6+} species, according to the literature (Awate et al., 2005; Zhu et al., 1999). Specifically, Cr-Ti-MCM-41(10) has two bands at 630 and 570 cm^{-1} belonging to extra-framework Cr_2O_3 oxides.

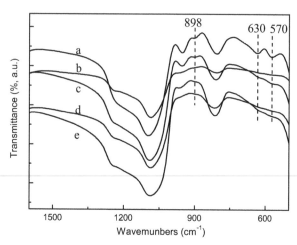

Fig. 11. IR spectra of Cr-Ti-MCM-41 (a: Si/M=10, b: Si/M=33, c: Si/M=75, d: Si/M=100, e: Si/M=150, f: Ti-MCM-41). Reprinted with permission from Zou, J.-J.; Liu, Y.; Pan, L.; Wang, L. & Zhang, X. (2010), *Applied Catalysis B: Environmental*, Vol.95, No.3-4, pp. 439-445. Copyright @ 2010 Elsevier.

The well dispersed V and Fe species show no obvious influence on the ordered structure of prepared materials, but the polymerized Cr species obviously impose negative effect on the structure, see Fig. 12. An extreme is observed for Cr-Ti-MCM-41(10), in which the characteristic diffractive peaks of ordered structure completely disappear, and a peak of bulk Cr_2O_3 appears. In TEM image, this material no longer possess hexagonal mesoporous structure, but agglomerate of many crystallites.

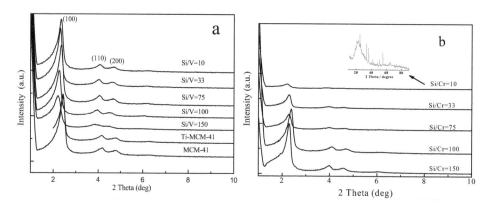

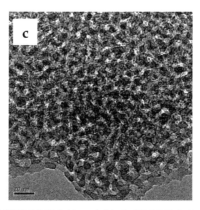

Fig. 12. XRD patterns of (a) V-Ti-MCM-41 and (b) Cr-Ti-MCM-41, and (c) TEM image of Cr-Ti-MCM-41(10). Reprinted with permission from Zou, J.-J.; Liu, Y.; Pan, L.; Wang, L. & Zhang, X. (2010), *Applied Catalysis B: Environmental*, Vol.95, No.3-4, pp. 439-445. Copyright @ 2010 Elsevier.

All the materials exhibit higher activity than Ti-MCM-41, see Fig. 13, indicating that introducing second metal is beneficial to the photoisomerization. Among the three metals, V-incorporation is most effective, Fe-incorporation follows, and Cr- incorporation is the least. The photocatalytic activity has nothing to do with the concentration of second transition metal ions, and the improvement in activity should be related to their state of dispersion and local structure. It has been reported that tetrahedrally coordinated M-oxide moieties dispersed in mesoporous materials can be easily excited under UV and/or visible-light irradiation to form corresponding charge-transfer excited states (Yamashita et al., 2001; Matsuoka & Anpo, 2003):

$$[M^{n+} - O^{2-}] \xrightarrow{\quad hv \quad} [M^{(n-1)+} - O^{-}]^{*} \quad \text{(M=V, Cr, Fe)}$$

Then M species can donate an electron to surrounding Ti-O moieties and O- can scavenge an electron from surrounding Ti-O moieties, inducing charge separation in Ti-O species (Davydov et al., 2001). Therefore, two different excitation mechanisms exist in M-Ti-MCM-41. One is direct excitation of Ti-O moieties by UV irradiation, and the other is indirect excitation via charge transition from $[M^{(n-1)+} - O^{-}]^{*}$ species. The second process should be responsible for the high photocatalytic activity of M-Ti-MCM-41 because of its high efficiency in charge formation and separation.

V-Ti-MCM-41(150) shows specifically high activity because majority of V ions are highly dispersed in 4-fold coordination, which brings up highly efficient excitation of Ti-O species. In addition, the well retained ordered structure and high surface area can enhance the adsorption of NBD molecules and provide more active sites. With the increase of V content, the activity is decreased because some 4-fold ions are transformed into undesirable highly-coordinated species and the damaged structure and small surface area may suppress the adsorption of reactants. The low activity of Cr-Ti-MCM-41 is due to poorly dispersed chromium ions and dramatically destroyed textural structure.

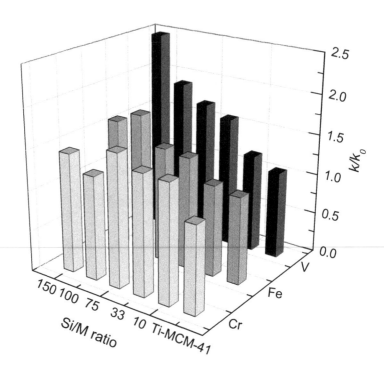

Fig. 13. Activity of M(V, Fe and Cr)-Ti-MCM-41 for the photoisomerization of norbornadiene (Zou et al., 2010).

Since some photocatalysts show absorption in visible-light region, one may wonder whether they can catalyze the isomerization under visible-light irradiation. However, there is no any observable conversion when the experiment was conducted using visible irradiation (>420 nm). This is different from the case of H_2 generation and organic degradation, where Cr-Ti-MCM-41 is reported to exhibit visible-light activity (Yamashita et al., 2001; Davydov et al., 2001; Chen & Mao, 2007). These results suggest that the reaction mechanism between the photoisomerization and other photocatalytic reactions may be very different.

4. Mechanism for NBD photoisomerization

Photoisomerization of NBD in the presence of sensitizers generally proceeds via triplet state mechanism (Bren' et al., 1991; Dubonosov et al., 2002), see Scheme 4. Under irradiation, the sensitizer is excited to triplet state (^3S) via single state (^1S), that subsequently transfers energy to NBD molecules and excites it to triplet state (^3NBD). Then ^3NBD undergoes adiabatic isomerization and forms triplet state of QC (^3QC) that rapidly decays to its ground state and produces QC.

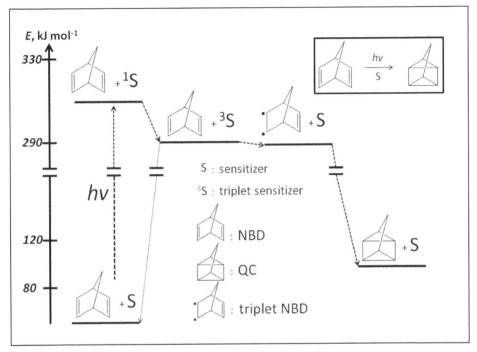

Scheme 4. Triplet sensitized photoisomerization of norbornadiene.

However, with the presence of Ti-containing photocatalyst, this mechanism is not suitable because the vertical triplet energy transfer from Ti-oxide species to NBD is very difficult. NBD molecules have to be firstly positively charged by photoinduced holes, but the free radical ion isomerization mechanism is ruled out because the energy of free NBD^{+} is significantly lower than free QC^{+}. In fact, the transformation of QC to NBD is through the $QC^{+} \rightarrow NBD^{+}$ free radical route (Ikezawa & Kutal, 1987). So the photoisomerization of NBD over semiconductors should be an adsorption-photoexcited process, which is very likely through the exciplex (charge-transfer intermediate), see Scheme 5. First, NBD molecule is adsorbed on the photoexcited Ti-oxides. Then surface-trapped hole is transferred to adsorbed molecule and a complex with NBD positively charged is formed. Subsequently the complex is transformed to structure with QC skeleton. Finally, QC is released into the liquid phase and the charge is recombined through reverse electron transfer. In this case the adsorption and charge transfer are two critical steps. The adsorptive site on different Ti-containing materials may be different. For Zn-TiO₂, surface OH very likely serves as the site because it plays an important role in the reaction, and the excited complex may be $TiO_{2}^{-} - OH \cdots NBD^{+}$. For Fe-TiO₂ and V-TiO₂, however, the lattice oxygen may work as the adsorbing site with the complex of $Ti^{4+} - [O^{2-}] - Ti^{4+} - O^{2-} \cdots NBD^{+}$. Any charge recombination process can deactivate the complex, so the function of dopants and framework Ti species is to retard the undesired recombination.

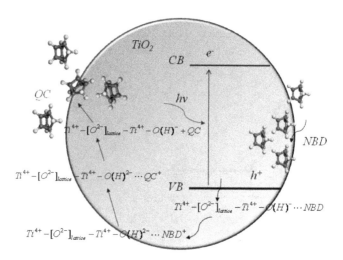

Scheme 5. Photoisomerization of norbornadiene via adsorption-photoexcitation over semiconductor.

5. Summary

The transform of norbornadiene is typical photoisomerization and of great importance for both solar energy harvesting and aerospace fuel synthesis. Our recent work shows that the heterogeneous Ti-containing materials show activity comparable to homogeneous sensitizers, along with many additional advantages in manipulation and scale-up. Ti-containing photocatalysts are extensively used in environmental and energy science and show many exciting and rapid progress, which will undoubtedly benefit the photoisomerization of alkenes like NBD. Specially, surface modulation may be very helpful because it can tune the adsorption and even charge transfer between reactant and catalyst. Even though, the photoisomerization shows some unique characteristics and further work is necessary to understand the mechanism and substantively improve the efficiency. It is expected that the heterogeneous photocatalysis may provide a new and promising pathway for photoisomerization of alkenes.

6. Acknowledgements

The authors greatly appreciate the supports from the Natural Science Foundation of China (20906069), the Foundation for the Author of National Excellent Doctoral Dissertation of China (200955), the Program for New Century Excellent Talents in University, and the Research Fund for the Doctoral Program of Higher Education of China (200800561011).

7. References

Adán, C.; Bahamonde, A.; Fernández-García, M. & Martínez-Arias, A. (2007). Structure and activity of nanosized iron-doped anatase TiO_2 catalysts for phenol photocatalytic degradation. *Applied Catalysis B: Environmental*, Vol.72, No. 1-2, (May 2007), pp. 11-17, ISSN 0926-3373

Al-Ekabi, H. & Mayo, P. (1985). Surface photochemistry: CdS photoinduced cis-trans isomerization of olefines. *Journal of Physical Chemistry*, Vol.89, No.26, (December 1985), pp. 5815-5812, ISSN 0022-3654

Awate, S.V.; Jacob, N.E.; Deshpande, S.S.; Gaydhankar, T.R. & Belhekar, A.A. (2005). Synthesis, characterization and photo catalytic degradation of aqueous eosin over Cr containing Ti/MCM-41 and SiO$_2$-TiO$_2$ catalysts using visible light. *Journal of Molecular Catalysis A Chemical*, Vol.226, No.2, (February 2005), pp.149-154, ISSN 1381-1169

Bren', V.A.; Dubonosov, A.D.; Minkin, V.I. & Chernoivanov, V.A. (1991). Norbornadiene-quadricyclane-an effective molecular system for the storage of solar energy. *Russian Chemical Reviews*, Vol.60, No. 5, (1991), pp. 451-469, ISSN 0042-1308

Cahill, P.A.; Steppel, R.N. (2004). Process of quadricyclane production. US Patent 20040054244A1

Chen, X. & Mao, S.S. (2007). Titanium dioxide nanomaterials: synthesis, properties, modifications, and applications, *Chemical Reviews*, Vol.107, No.7, (July 2007), pp. 2891-2959, ISSN 0009-2665

Chen, J.; Zhang, L.; Li, S.; Li, Y.-Y.; Chen, J.; Yang, G. & Li, Y. (2007). Valence isomerization in dendrimres by photo-induced electron transfer and energy transfer from the dendrimer backbone to the core. *Journal of Photochemistry and Photobiology A: Chemistry*, Vol.185, No.1, (January 2007), pp. 67-75, ISSN 1010-6030

Ciminelli, C.; Granucci, G. & Persico, M. The photoisomerization mechanism of azobenene: a semiclassical simulation of nonadiabatic dynamics, *Chemistry-A European Journal*, Vol.10, No.9, (May 2004), pp. 2327-2341, ISSN 0947-6539

Davydov, L.; Reddy, E. P.; France, P. & Smirniotis, P. G. (2001). Transition-metal-substituted titania-loaded MCM-41 as photocatalysts for the degradation of aqueous organics in visible light, *Journal of Catalysis*. Vol.203, No.1, (October 2001), pp. 157-167, ISSN 0021-9517

Dou, Y. & Allen, R.E. (2003). Detailed dynamics of a complex photochemical reaction: cis-trans photoisomerization of stilbene. *Journal of Chemical Physics*, Vol.119, No.20, (November 2003), pp. 10658-10666, ISSN 0021-9606

Dubonosov, A.D.; Bren, V.A. & Chernoivanov, V.A. (2001). Norbornadiene-quadricyclane as an abiotic system for the storage of solar energy. *Russian Chemical Reviews*, Vol.71, No.11, (2002), pp. 917-927, ISSN 0042-1308

Filley, J.; Miedaner, A.; Ibrahim, M.; Nimlos M.R. & Blake, D.M. (2001). Energetics of the 2+2 cyclization of limonene. *Journal of Photochemistry and Photobiology A: Chemistry*, Vol.139, No.1, (February, 2001), pp. 17-21, ISSN 1010-6030

Gao, G.; Deng, Y.& Kispert, L.D. (1998). Semiconductor photocatalysis: photodegradation and trans-cis photoisomerization of carotenoids. *Journal of Physical Chemistry B*, Vol.102, No.20, (May 2004), pp. 3897-3901, ISSN 1089-5647

Ghandi, M.; Rahimi, A. & Mashayekhi, G. (2006). Triplet photosensitization of myrcene and some dienes within zeolites Y through heavy atom effect. *Journal of Photochemistry and Photobiology A: Chemistry*, Vol.181, No.1, (July 2006), pp. 56-59, ISSN 1010-6030

Gu, L. & Liu, F. (2008). Photocatalytic isomerization of norbornadiene over zeolites. *Reaction Kinetics and Catalysis Letters*, Vol.95, No.1, (2008), pp. 143-151, ISSN 0133-1736

Hammond, G.S.; Wyatt, P.; DeBoer, C.D. & Turro, N.J. (1964). Photosensitized isomerization involving saturated centers. *Journal of the American Chemical Society*, Vol.86, No.12, (June 1964), pp. 2532-2533, ISSN 0002-7863

Hu, Y.; Higashimoto, S.; Martra, G.; Zhang, J.; Matsuoka, M.; Coluccia, S. & Anpo, M. (2003). Local structures of active sites on Ti-MCM-41 and their photocatalytic reactivity for the decomposition of NO, *Catalysis Letters*, Vol.90, No.3-4, (October 2003), pp. 161-163, ISSN 1011-372X

Hu, Y.; Martra, G.; Zhang, J.; Higashimoto, S.; Coluccia, S. & Anpo, M. (2006). Characterization of the local structures of Ti-MCM-41 and their photocatalytic reactivity for the decomposition of NO into N_2 and O_2. *Journal of Physical Chemistry B*, Vol.110, No.4, (February 2006), pp. 1680-1685, ISSN 1520-6106

Hu, Y.; Wada, N.; Tsujimaru, K. & Anpo, M. (2007). Photo-assisted synthesis of V and Ti-containing MCM-41 under UV light irradiation and their reactivity for the photooxidation of propane, *Catalysis Today*, Vol.120, No.2, (February 2007), pp. 139-144, ISSN 0920-5861

Ikezawa, H. & Kutal, C. (1987). Valence isomerization of quadricyclane mediated by illuminated semiconductor powders, *Journal of Organic Chemistry*, Vol.52, No.12, (July 1987), pp. 3299-3303, ISSN 0022-3263

Jiang, D.-L. & Aida, T. (1997). Photoisomerization in dendrimers by harvesting of low-energy photons. *Nature*, Vol.388, (July 1997), pp. 454-456. ISSN 0028-0836

Jing, L.; Xin, B.; Yuan, F.; Xue, L.; Wang, B. & Fu, H. (2006). Effects of surface oxygen vacancies on photophysical and photochemical processes of Zn-doped TiO_2 nanoparticles and their relationships. *Journal of Physical Chemistry B*, Vol.110, No.36, (September 2006), pp. 17860-17865, ISSN 1520-6106

Kokan, T.S.; Olds, J.R.; Seitzman, J.M. & Ludovice, P.J. (2009). Characterizing high-energy-density propellants for space propulsion applications. *Acta Astronautica*, Vol.65, No.7-8, (October-November 2009), pp. 967-986, ISSN 0094-5765

Lahiry, S. & Haldar, C. (1986). Use of semiconductor materials as sensitizers in a photo chemical energy storage reaction, norbornadiene to quadricyclane. *Solar Energy*, Vol.37, No.1, (1986), pp. 71-73, ISSN 0038-092X

Lu, Z.; Shen, M. & Yoon, T.P. (2011). [3+2] cycloaddition of aryl cyclopropyl ketones by visible light photocatalysis. *Journal of the American Chemical Society*, Vol.133, No.5, (February 2011), pp. 1162-1164, ISSN 0002-7863

Matsuoka, M. & Anpo, M. (2003). Local structures, excited states, and photocatalytic reactivities of highly dispersed catalysts constructed within zeolites, *Journal of Photochemistry and Photobiology C: Photochemistry Reviews*, Vol.3, No.3, (January 2003), pp. 225-252, ISSN 1389-5567

Minezawa, N. & Gordon, M.S. (2011). Photoisomerization of stilbene: a spine-flip density functional theory approach. *Journal of Physical Chemistry A*, Vol.115, No.27, (June 2011), pp. 7901-7911, ISSN 1089-5639

Mita, I.; Horie, K. & Hirao, K. (1989). Photochemistry in polymer solids. 9. photoisomerization of azobenen in a polycarbonate film. *Macromolecules*, Vol.22, No.2, (February 1989), pp. 558-563, ISSN 0024-9297

Pan, L.; Zou, J.-J.; Zhang, X. & Wang, L. (2010). Photoisomerization of norbornadiene to quadricyclane using transition metal doped TiO_2. *Industrial & Engineering Chemistry Research*, Vol.49, No.18, (September 2010), pp. 8256-8531, ISSN 0888-5885

Quenneville, J. & Martínez, T.J. (2003). Ab initio study of cis-trans photoisomerization in stilbene and ethylene. *Journal of Chemical Physics*, Vol.107, No.49, (February 2003), pp. 829-837, ISSN 1089-5639

Somekawa, K.; Odo, Y. & Shimo, T. (2009). Molecular simulations of photoaddition selectivity and chirality in challenging photochemical reactions. *Bulletin of the Chemical Society of Japan*, Vol.82, No.12, (December 2009), pp. 1447-1469, ISSN 0009-2673

Striebich, R. C.; & Lawrence, J. (2003). Thermal decomposition of high-energy density materials at high pressure and temperature. *Journal of Analytic and Applied Pyrolysis*, Vol.70, No.2, (December 2003), pp. 339-352, ISSN 0165-2370

Vlaar, M.J.M.; Ehlers, A.W.; Schakel, M.; Clendenning, S.B.; Nixon, J.F.; Lutz, M.; Spek, A.L. & Lammertsm, L. (2001), Norbornadiene-quadricyclane valence isomerization for a tetraphophorus derivative. *Angewandte Chemie International Edition*, Vol.40, No.23, (2001), pp. 4412-4415, ISSN 1433-7851

Waldeck, D. H. (1991). Photoisomerization dynamics of stilbenes. *Chemical Reviews*, Vol.91, No.91, (May 1991), pp. 415-436, ISSN 0009-2665

Wang, E.; Yang, W. & Cao, Y. (2009). Unique surface chemical species on indium doped TiO_2 and their effect on the visible light photocatalytic activity. *Journal of Physical Chemistry C*, Vol.113, No.49, (December 2009), pp. 20912-20917. ISSN 1932-7447

Xu, Y.; Smith, M.D.; Krause, J.A. & Shimizu, L.S. (2009). Control of the intramolecular [2+2] photocycloaddition in a bis-stilbene macrocycle. *Journal of Organic Chemistry*, Vol.74, No.13, (July 2009) , pp. 4874-4877, ISSN 0022-3263

Yamashita, H.; Yoshizawa, K.; Ariyuki, M.; Higashimoto, S.; Che, M. & Anpo, M. (2001). Photocatalytic reactions on chromium containing mesoporous silica molecular sieves (Cr-HMS) under visible light irradiation: decomposition of NO and partial oxidation of propane, *Chemical Communications*, Vol.5, (2001), pp. 435-436, ISSN 1359-7345

Yanagida, S.; Mizumoto, K. & Pac, C. (1986), Semiconductor photocatalysis. Cis-trans photoisomerization of simple alkenes induced by trapped holes at surface states. *Journal of the American Chemical Society*, Vol.108, No.4, (February 1986), pp. 647-654, ISSN 0002-7863

Yang, X.; Cao, C.; Hohn, K.; Erickson, L.; Maghirang, R.; Hamal, D.& Klabunde, K. (2007). Highly visible-light active C- and V-doped TiO2 for degradation of acetaldehyde. *Journal of Catalysis*, Vol.252, No.2, (December 2007), pp. 296-302, ISSN 0021-9517

Zhu, H.; Tao, J. & Dong, X. (2010). Preparation and photoelectrochemical activity of Cr-doped TiO_2 nanorods with nanocavities. *Journal of Physical Chemistry C*, Vol.114, No.7, (February 2010), pp. 2873-2879, ISSN 1932-7447

Zhu, Z.; Chang, Z. & Kevan, L. (1999). Synthesis and characterization of mesoporous chromium-containing silica tube molecular sieves CrMCM-41. *Journal of Physical Chemistry B*, Vol.103, No.14, (March 1999), pp.2680-2688, ISSN 1089-5647

Zou, J.-J.; Liu, Y.; Pan, L.; Wang, L. & Zhang, X. (2010). Photocatalytic isomerization of norbornadiene to quadricyclane over metal (V, Fe and Cr)- incorporated Ti-MCM-41. *Applied Catalysis B: Environmental*, Vol.95, No.3-4, (April 2010), pp. 439-445, ISSN 0926-3373

Zou, J.-J.; Zhang, M.-Y.; Zhu, B.; Wang, L.; Zhang, X. & Mi, Z. (2008). Isomerization of norbornadiene to quadricyclane using Ti-containing MCM-41 as photocatalysts. *Catalysis Letters*, Vol.124, No.12, (August 2008), pp. 139-145, ISSN 1011-372X

Zou, J.-J.; Zhu, B.; Wang, L.; Zhang, X. & Mi, Z. (2008), Zn- and La-modified TiO_2 photocatalysts for the isomerization of norbornadiene to quadricyclane. *Journal of Molecular Catalysis A: Chemical*, Vol.286, No.1-2 (May 2008), pp. 63-69, ISSN 1381-1169

Part 3

Photochemistry in Biology

5

UV Light Effects on Proteins:
From Photochemistry to Nanomedicine

Maria Teresa Neves-Petersen[1,2], Gnana Prakash Gajula[3]
and Steffen B. Petersen[1,2]
[1]International Iberian Nanotechnology Laboratory (INL), Braga
[2]Aalborg University, Aalborg,
[3]Materials and Metallurgy Group, Indira Gandhi Centre for Atomic Research,
Kalpakkam,
[1]Portugal
[2]Denmark
[3]India

1. Introduction

Throughout 4.5 billion year of molecular evolution, proteins have evolved in order to maintain the spatial proximity between aromatic residues (Trp, Tyr and Phe) and disulphide bridges (SS) (Petersen et al, 1999). Aromatic residues are the nanosized antennas in the protein world that can capture UV light (from ~250-298nm). Once excited by UV light they can enter photochemical pathways likely to have harmful effects on protein structures. However, disulphide bridges in proteins are excellent quenchers of the excited state of aromatic residues, contributing this way to protein stability and activity. UV light excitation of the aromatic residues is known to trigger electron ejection from their side chains (Bent & Hayon, 1975a; Bent & Hayon, 1975b; Bent & Hayon, 1975c; Creed, 1984a; Creed, 1984b; Kerwin & Rammele, 2007, Neves-Petersen et al., 2009a). These electrons can be captured by disulphide bridges, leading to the formation of a transient disulphide electron adduct radical, which will dissociate leading to the formation of free thiol groups in the protein. This observation lead to the development in our lab of a new photonic technology, Light Assisted Molecular Immobilization (LAMI), used to functionalize surfaces with biomolecules. This technology is being used in order to create a new generation of biosensors with unsurpassed density (number of spots per mm^2). This technology is also being used in order to create nanoparticles based drug delivery systems relevant to nanomedical applications (Parracino et al., 2011).

In this chapter we will describe the effects of UV excitation of proteins. Furthermore, we will also describe the specific, conserved structural motif in protein molecules that can be activated by UV light, leading to the formation of reactive free thiol groups. The dynamics of formation and the lifetimes of transient species will be described. Afterwards, the different applications of LAMI will be shown. LAMI has been successfully used for the creation of protein microarrays and in order to immobilize proteins on a surface according to any desired pattern, with submicrometer and nanometer resolution. LAMI has also been used for the creation of nanoparticle based drug delivery systems. An overview on different protein immobilization technologies will be given and the advantages of the new photonic technology will be highlighted. An overview of the use of nanoparticles in nanomedicine will also be given. Furthermore, a new light based cancer therapy which makes use of the knowledge derived from the effects of UV light on proteins will be described.

Light can change the properties of biomolecules and the number of drugs found to be photochemically unstable is steadily increasing. The effects of light on drugs include not only degradation reactions but also other processes, such as the formation of radicals, energy transfer, and luminescence. Adequate protection for most drug products during storage and distribution is needed. Indeed, proper storage conditions that secure protection from UV and visible radiation are essential for the efficacy of many common dermatologic drugs. If a drug is exposed to fluorescent tubes and/or filtered daylight for several weeks or months before it is finally administered to the patient, the drug may be altered. The most common consequence of drug photodecomposition is loss of potency with concomitant loss of therapeutic activity. It is therefore of interest to be aware of light induced reaction in biomolecules.

2. UV light induced photochemical reactions

Cells, their proteins and genes are sensitive to light. The vision process itself is initiated when photoreceptor cells are activated by light (photo-isomerization). Several papers report effects of UV light in cells and their proteins/genes. For example, UV-light is known to inhibit photosystem II activity in cyanobacterium and to enhance the transcription of particular genes (310nm light) (Vass et al., 2000). It is also known that near UV (290nm) exposed prion protein fails to form amyloid fibrils (Thakur & Rao, 2008). Nucleic acids in living cells are associated with a large variety of proteins. Therefore, it is logical to assume that the ultraviolet (UV) irradiation of cells could lead to reactive interactions between DNA and the proteins that are in contact with it. One reaction that does occur is the cross-linking between the amino acids in these associated proteins and the bases in DNA. Such reaction appears to be an important process that photoexcited DNA and proteins undergo *in vivo*, as well as in DNA-protein complexes *in vitro*. Since the crosslinking of DNA and protein by UV radiation is many times more sensitive than is thymine dimer formation, it was suggested that DNA-protein crosslinks may play a significant role in the inactivation of bacteria by UV radiation (Smith, 1962). The first amino acid shown to photochemically add to uracil was cysteine, to form 5-S-cysteinyl-6-hydrouracil (Smith and Aplin, 1966). The structure of the mixed photoproduct of thymine and cysteine was also determined (Smith, 1970). The first survey performed determined the ability of the 22 common amino acids to bind photochemically (upon 254nm excitation) to uracil. The 11 reactive amino acids were glycine, serine, phenylalanine, tyrosine, tryptophan, cystine, cysteine, methionine, histidine, arginine and lysine. The most reactive amino acids were phenylalanine, tyrosine and cysteine.

The three amino acid residues which side chains absorb in the UV range are the aromatic residues tryptophan (Trp), tyrosine (Tyr) and phenylalanine (Phe). Several reviews have been published on the photochemistry and photophysics of Trp (Bent & Hayon, 1975a; Creed, 1984b), Tyr (Bent & Hayon, 1975b; Creed, 1984a), Phe (Bent & Hayon, 1975c), and Cystine (name given to each bridged cysteine in a disulphide bridge) (Creed, 1984a).

Excitation to higher energy states is followed by relaxation to ground state (*e.g.* fluorescence, phosphorescence) or to excited state photochemical or photophysical processes, such as photoionization (Creed, 1984b).

Tryptophan

Flash photolysis studies have revealed two non-radiative relaxation channels from the singlet excited state of Tryptophan (Bent & Hayon, 1975a):

1. Electron ejection to the solvent, yielding solvated electrons, e^-_{aq}, which have a broad absorption peak centred at ~720 nm and the tryptophan radical cation Trp$^{•+}$ which has its maximum absorption at ~560 nm. Trp$^{•+}$ deprotonates rapidly, yielding the neutral radical Trp$^•$ that has its maximum absorption at ~510 nm.

$$Trp + hv \rightarrow Trp^{•+} + e^-_{aq} \tag{1}$$

$$Trp^{•+} \rightarrow Trp^• + H^+ \tag{2}$$

2. Intersystem crossing, yielding the triplet-state ^{3}Trp which has its maximum absorption at ~450 nm. The triplet state tryptophan can transfer an electron to a nearby disulphide bridge to give Trp$^{•+}$ and the disulphide bridge electron adduct RSSR$^{•-}$, where the latter has its maximum absorption at ~420 nm (Bent & Hayon, 1975a).

$$^1Trp + hv \rightarrow {}^1Trp^* \tag{3}$$

$$^1Trp^* \rightarrow {}^3Trp \tag{4}$$

$$^3Trp + RSSR \rightarrow Trp^{•+} + RSSR^{•-} \tag{5}$$

Tyrosine

Another aromatic residue with non-negligible absorption in the near-UV region is tyrosine (Tyr-OH). At neutral pH tyrosine has absorption maxima at 220nm (ϵ~9000 M^{-1}cm^{-1}) and 275nm (ϵ~1400 M^{-1}cm^{-1}) (Creed, 1984a). At alkaline pH the OH group of tyrosine side chain deprotonates. The resulting tyrosinate (Tyr-O$^{•-}$) has a slightly red-shifted absorption compared to tyrosine, with maxima at 240nm (ϵ~11000 M^{-1}cm^{-1}) and 290nm (ϵ~2300 M^{-1}cm^{-1}) (Creed, 1984a). Photoexcited tyrosine can fluoresce, decay non-radiatively, or undergo intersystem crossing to the triplet state, from which most of the photochemistry proceeds. Alternatively, at neutral pH, tyrosine can be photoionized through a biphotonic process that involves absorption of a second photon from the triplet state. This results in a solvated electron (e$^-_{aq}$) and a radical cation (Tyr-OH$^{•+}$) that will rapidly deprotonate to create the neutral radical (Tyr-OH$^•$). Photoionization of tyrosinate at high pH is monophotonic and results in a neutral radical (Tyr-O$^•$) and a solvated electron (e$^-_{aq}$).

$$^3Tyr - OH + hv \rightarrow Tyr - OH^{•+} + e^-_{aq} \tag{6}$$

$$Tyr - OH^{•+} \rightarrow Tyr - OH^• + H^+ \tag{7}$$

$$Tyr - O^{\bullet -} + hv \rightarrow Tyr - O^{\bullet} + e^{-}_{aq} \tag{8}$$

The triplet state tyrosine is rapidly quenched by molecular oxygen or nearby residues like tryptophan or disulphide bridges (Bent & Hayon, 1975b):

$$^{3}Tyr - OH + RSSR \rightarrow Tyr - O^{\bullet} + H^{+} + RSSR^{\bullet -} \tag{9}$$

2.1 Important photochemical mechanism in disulphide bridge containing proteins

An important photochemical mechanism in proteins involves reduction of disulphide bridges (SS) upon UV excitation of Trp and Tyr side chains (Kerwin & Rammele, 2007, Neves-Petersen et al., 2002 & 2009a). As shown above, UV-excitation of tryptophan or tyrosine can result in their photoionization and to the generation of solvated electrons (Bent & Hayon, 1975a & 1975b; Creed, 1984b, Kerwin & Rammele, 2007, Neves-Petersen et al., 2009a). The generated solvated electrons can subsequently undergo fast geminate recombination with their parent molecule, or they can be captured by electrophillic species like molecular oxygen, H_3O^+ (at low pH), and cystines as summarized below:

$$e^{-}_{aq} + O_2 \rightarrow O_2^{\bullet -} \tag{10}$$

$$e^{-}_{aq} + H_3O^+ \rightarrow H^{\bullet} + H_2O \tag{11}$$

$$e^{-}_{aq} + RSSR \rightarrow RSSR^{\bullet -} \tag{12}$$

In the case where the electron is captured by the cystine, the result can also be the breakage of the disulphide bridge (Hoffman & Hayon, 1972):

$$e^{-}_{aq} + RSSR \rightarrow RSSR^{\bullet -} \tag{10}$$

$$RSSR^{\bullet -} \Leftrightarrow RS^{\bullet} + RS^{-} \tag{11}$$

$$RSSR^{\bullet -} + H^{+} \Leftrightarrow RS^{\bullet} + RSH \tag{12}$$

The resultant free thiol radicals/groups can then subsequently react with other free thiol groups to create a new disulphide bridge. Reduction of SS upon UV excitation of aromatic residues has been shown for proteins such as cutinase and lysozyme (Neves-Petersen et al., 2009a, 2006 & 2002), bovine serum albumin (Skovsen et al., 2009a; Parracino et al., 2011) prostate specific antigen (Parracino et al., 2010), and antibody Fab fragments (Duroux et al., 2007). As mentioned in the introduction, this phenomenon has led to a new technology for protein immobilization (LAMI, light assisted molecular immobilization) since the created thiol groups can bind thiol reactive surfaces leading to oriented covalent protein immobilization (Neves-Petersen et al., 2006;Snabe et al., 2006; Duroux et al., 2007a, 2007b & 2007c; Skovsen et al., 2007, 2009a & 2009b; Neves-Petersen et al., 2009b; Parracino et al., 2010 & 2011).

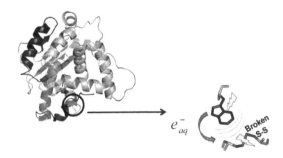

Fig. 1. The spatial proximity between aromatic residues and disulphide bridges (SS) has been conserved throughout molecular evolution. Trp is the preferred spatial neighbor of SS (Petersen et al., 1999). UV excitation of the side chain of aromatic residues leads to electron ejection. The electron can then be captured by disulphide bridges, leading to their dissociation.

2.2 Observing the solvated electron and other transient species formed upon UV excitation of proteins

In the flash photolysis experiments on lysozyme by Grossweiner and Usui (Grossweiner & Usui, 1971) it was shown that the initial photoproducts upon UV excitation of lysozyme are the photo-oxidized tryptophan residue, solvated electrons, and the cystine residue (disulphide bridge) electron adduct. In a more recent paper by Zhi Li *et al.* (Z. Li et al., 1989), experiments on a model system demonstrated that the fast electron transfer is consistent with direct electron transfer between the tryptophan triplet state and a nearby disulphide bridge (this is a very short range interaction that decays exponentially as the distance between the donor and acceptor increases). This process will result in a RSSR•⁻ radical, which again can result in breakage of the disulphide bridge, as shown above (scheme 12).

It is clear from the above that there are many possible pathways for the breakage of intra-molecular disulphide bridges in proteins upon UV excitation of aromatic residues, even in the absence of molecular oxygen. Breakage of the disulphide bridge can lead to conformational changes in the protein, not necessarily resulting in inactivation of the protein. Transient absorption data of *Fusarium solani pisi* cutinase has also been acquired, with supplemental experimental data on tryptophan and lysozyme as a reference (Neves-Petersen et al., 2009a). Cutinase is a good model protein for studying the UV induced breakage of disulphide bridges is since it contains only one tryptophan that is within van der Waals contact of a disulphide bridge (closest distance ~3.8 Å).

Data showed that UV excitation of cutinase lead to the formation of the solvated electron (transient species with absorption maximum around 710-720nm, see Fig. 2) and of the disulphide bridge electron adduct radical, RSSR•⁻ (transient species with absorption maximum around 420nm, see Fig. 2) (Neves-Petersen et al., 2009a). Figs. 2 and 3 show the kinetics of formation of the solvated electron. The increase in absorption of light at 710nm can clearly be seen following excitation at 266nm, which coincides with time zero in Figure 2. The data displayed in Fig. 3 is the intensity of absorbed light at 710nm (the intense peak displayed in Fig. 2) during the initial 43ns after excitation of cutinase with

266nm laser light. Light is absorbed at 710nm due to the presence of a new transient species created upon 266nm excitation of cutinase: the solvated electron. Solvated electrons are transient, i.e., short lived. One group that will capture them are disulphide bridges (RSSR). The capture of the solvated electrons by disulphide bridges leads to a decrease of the concentration of the solvated electron, and therefore, to a decrease in the intensity of absorbed light at 710nm. Such decay is displayed in Fig. 4. Furthermore, the combination of the solvated electron with the disulphide bridge leads to the formation of the disulphide-electron adduct radical, RSSR•-, a group which has its maximum absorption around 420nm. The presence of such group can be seen in Fig. 2, since absorption around 420nm can be observed.

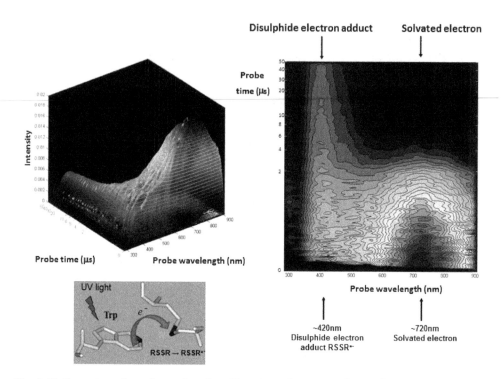

Fig. 2. Cutinase transient absorption data. Transient absorption data collected at probe times from 0 to 50µs. The transient solvated electron absorbs maximally light at ~710-720nm (intense peak) and the disulphide bridge electron adduct radical has its maximum absorption at ~420nm. The intensity of the peaks displayed in the 2D image to the right can be seen in the 3D image to the left.

Determination of the lifetimes of the different transient species formed upon UV excitation of proteins can be carried out by fitting the kinetic data displayed in Figs. 3 and 4. Upon fitting the initial increase of absorption at a 710nm and 420nm one recovers the rate of formation of solvated electrons and of the disulphide bridge electron adduct radical, respectively.

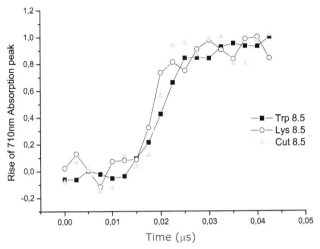

Fig. 3. Normalized transient absorption data at 710nm for 0-43ns probe times displaying the kinetics of formation of the solvated electron for tryptophan (Trp), lysozyme (Lys) and cutinase (Cut) samples at pH 8.5

Likewise, fitting the decay in the absorption peaks at 710nm and 420nm will allow us to recover the lifetime of the solvated electrons and of the disulphide bridge electron adduct radical, respectively. The lifetimes of the solvated electron in lysozyme and cutinase samples at different pH values can be found in Table I. Below is shown the decay kinetics of the solvated electron within 10 µs after 266nm excitation of cutinase (Neves-Petersen et al., 2009a).

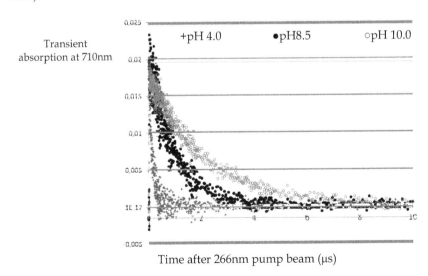

Fig. 4. Decay kinetics of the solvated electron absorption peak at 710nm within 10µs after excitation.

The governing equations for the time-resolved intensity decay data were assumed to be a sum of exponentials as in

$$Abs(t) = \sum \alpha_i \cdot exp\left(\frac{-t}{\tau_i}\right) \tag{13}$$

where $Abs(t)$ is the intensity decay, α_i is the amplitude (pre-exponential factor), τ_i the lifetime of the i-th component and $\sum \alpha_i = 1.0$. Data was analysed using a global analysis approach.

The fractional intensity f_i of each decay time is given by

$$f_i = \frac{\alpha_i \tau_i}{\sum_j \alpha_j \tau_j} \tag{14}$$

and the mean lifetime is

$$\langle \tau \rangle = \sum_i f_i \tau_i \tag{15}$$

It was observed that the solvated electron average lifetime is shorter at acidic pH values, which is correlated with the fact that H_3O^+ captures the solvated electron. Furthermore, the solvated electron lifetime is significantly shorter in protein systems as compared to from Trp alone in solution, thus indicating that a protein offers other pathways involving capture of the solvated electron.

pH	Tryptophan <t> (µs)	Lysozyme <t> (µs)	Cutinase <t> (µs)
4.0	1.0	0.3	0.3
7.5	1.1	0.7	
8.5	2.3	0.8	1.0
10.0	2.8	1.6	2.1

Table 1. Mean lifetime of the solvated electron in all samples at different pH values.

Data analysis shows that the solvated electron has ns and sub µs decay lifetimes (Neves-Petersen et al., 2009a). These different lifetimes can be explained due to different recombination pathways of the solvated electron: recombination with the parent molecule (geminate recombination), with the hydronium ion present in the solvent or with other electron acceptor, such as disulphide bridges and positively charged groups. The intensity of the solvated electron peak is clearly pH dependent (Neves-Petersen et al., 2009a). This is correlated with the fact that the hydronium ion H_3O^+ is an electron scavenger. Recombination happens according to the reaction (Spanel & Smith, 1995):

$$e_{aq}^- + H_3O^+ \rightarrow H^{\bullet} + H_2O \tag{16}$$

Therefore, the lower the pH the faster the rate of decay of solvated electrons formed upon UV excitation of Trp molecules. In a protein, besides H_3O^+, different groups can act as electron scavengers, e.g. positively charged residues, the carbonyl group of the peptide chain (Faraggi & Bettelheim, 1977) as well as disulphide bridges, according to the reaction below:

$$e_{aq}^- + RSSR \rightarrow RSSR^{\bullet -} \tag{17}$$

Data shows that the higher the pH the longer time it takes for the solvated electron to recombine with the parent molecule (geminate recombination) or another electron scavenger molecule, such as H_3O^+. The observed lifetime increase with pH can be explained since the lower the pH, the higher the concentration of H_3O^+ and therefore the larger the probability of recombination of the solvated electron with the hydronium ion. Furthermore, for proteins, the higher the pH of the solution, the larger the number of basic titratable residues that have lost their positive charge and became neutral (His, Lys, Arg) and the larger the number of acidic titratable residues that have acquired a negative charge (Asp, Gly, Tyr, Cys not bridged). This means that an increase of pH leads to a loss of positive charge in the protein and a gain of neutral and negative charged residues in the protein. This will lead to an increase of the areas in the protein that carry a negative electrostatic potential. Therefore, an increase in pH will decrease the efficiency of electron recombination with the molecule due to electrostatic repulsion. This will lead to an increase of the solvated electron lifetime, as observed in Fig. 4.

3. Protein Immobilization onto surfaces: An overview

The importance of immobilisation technology is demonstrated by the recent development of DNA microarrays, where multiple oligonucleotide or cDNA samples are immobilised on a solid surface in a spatially addressable manner. These arrays have revolutionised genetic studies by facilitating the global analysis of gene expression in living organisms. Similar approaches have been developed for protein analysis, where as little as one picogram of protein need be bound to each point on a microarray for subsequent analysis. The proteins bound to the microarrays, can then be assayed for functional or structural properties, facilitating screening on a scale and with a speed previously unknown. The biomolecules bound to the solid surface may additionally be used to capture other unbound molecules present in the mixture. Development of this technology, with the goal of immobilising a biomolecule on a solid surface in a controlled manner, with minimal surface migration of the bound moiety and with full retention of its native structure and function, has been the subject of intensive investigation in recent years (Veilleux & Duran, 1996). The simplest type of protein immobilisation exploits the high inherent binding affinity of surfaces to proteins in general. For example, proteins will physically adsorb to hydrophobic substrates via numerous weak contacts, comprising van der Waals and hydrogen bonding interactions. The advantage of this method is that it avoids modification of the protein to be bound. On the other hand, adsorbed proteins may be distributed unevenly over the solid support and/or inactivated since, e.g., their clustering may lead to steric hindrance of the active site/binding region in any subsequent functional assay.

Molecules can be immobilised on a carrier or solid surface either passively through hydrophobic or ionic interactions, or covalently by attachment to surface groups. In response to the enormous importance of immobilisation for solid phase chemistry and biological screening, the analytical uses of the technology have been widely explored. The technology has found broad application in different areas of biotechnology, e.g. diagnostics, biosensors, affinity chromatography and immobilisation of molecules in ELISA assays. Alternative methods of immobilisation rely on the use of a few strong covalent bonds to bind the protein to the solid surface (Wilson & Nock , 2001). Examples include immobilisation of biotinylated proteins onto streptavidin-coated supports, and immobilisation of His-tagged proteins, containing a poly-histidine sequence, to Ni^{2+}-chelating supports. Other functional groups on the surface of proteins which can be used for attachment to an appropriate surface include reacting an amine with an aldehyde via a Schiff-base, cross-linking amine groups to an amine surface with gluteraldehyde to form peptide bonds, cross-linking carboxylic acid groups present on the protein and support surface with carbodiimide, cross-linking based on disulphide bridge formation between two thiol groups and the formation of a thiol-Au bond between a thiol group and a gold surface. Amine groups in proteins are widely used for protein covalent immobilization via NHS (N-hydroxysuccinimide)-EDC (N-ethyl-N'-(dimethylaminopropyl) carbodiimide hydrochloride) chemistry (Johnsson et. al., 1991). Following immobilisation, un-reacted N-hydroxysuccinimide esters on the support are deactivated with ethanolamine hydrochloride to block areas devoid of bound proteins. The method is laborious since the reagents, used at each step of a chemical immobilisation method, usually need to be removed prior to initiating the next step.

Methods for the immobilization of biomolecules via disulphide bridges are described by Veilleux J (1996). Protein samples are treated with a mild reducing agent, such as dithiothreitol, 2-mercaptoethanol or tris(2-carboxyethyl)phosphine hydrochloride to reduce disulphide bonds between cysteine residues, which are then bound to a support surface coated with maleimide. Alternatively primary amine groups on the protein can be modified with 2-iminothiolane hydrochloride (Traut's reagent) to introduce novel sulfhydryl groups, which are thereafter immobilized to the maleimide surface. Immobilization of proteins on a gold substrate via SH groups formed upon intracellular reduction of surface engineered disulphide bridges is shown for cupredoxin protein plastocyanin (Andolfi et al., 2002). The disulphide bridge has been engineered upon mutating the solvent accessible residues Ile21 and Glu25. An alternative approach to engineering thiol-groups into a protein has been described for ribonuclease (RNaseA), which has four essential cystines (Sweeney et al., 2000). In this case a single cysteine residue was substituted for Ala19, located in a surface loop near the N-terminus of RNase A. The cysteine in the expressed RNase was protected as a mixed disulphide with 2-nitro-5-thiobenzoic acid. Following subsequent de-protection with an excess of dithiothreitol, the RNase was coupled to the iodoacetyl groups attached to a cross-linked agarose resin, without loss of enzymatic activity. Again, preparation of the protein for immobilization requires its exposure to both protecting and de-protecting agents, which may negatively impact its native structure and/or function.

Light-induced immobilization techniques have also been explored, leading to the use of quinone compounds for photochemical linking to a carbon-containing support (European patent EP0820483). Activation occurs following irradiation with non-ionizing UV and visible light. Masks can be used to activate certain areas of the support for subsequent attachment of biomolecules. Following illumination, the photochemically active compound anthraquinone will react as a free radical and form a stable ether bond with a polymer surface. Since anthraquinone is not found in native biomolecules, appropriate ligands have to be introduced into the biomolecule. In the case of proteins, this additional sample preparation step may require thermochemical coupling to the quinone and may not be site specific. A further development of light-induced immobilization technology is disclosed in US patents US 5,412,087 and US 6406844, which describe a method for preparing a linker bound to a substrate. The terminal end of the linker molecule is provided with a reactive functional group protected with a photo-removable protective group, e.g. a nitro-aromatic compound. Following exposure to light, the protective group is lost and the linker can react with a monomer such as an amino acid at its amino or carboxy-terminus. The monomer, furthermore, may itself carry a similar photo-removable protective group which can also be displaced by light during a subsequent reaction cycle. The method has particular application to solid phase synthesis, but does not facilitate orientated binding of proteins to a support. Bifunctional agents possessing thermochemical and photochemical functional substituents for immobilizing an enzyme are disclosed in US patent US 3,959,078. Derivatives of arylazides are described which allow light mediated activation and covalent coupling of the azide group to an enzyme, and substituents which react thermochemically with a solid support. The orientation of the enzyme molecules is not controlled. A method for oriented, light-dependent, covalent immobilization of proteins on a solid support, using the heterobifunctional wetting-agent N-(m-(3-(trifluoromethyl)diazirin-3-yl)phenyl)-4-maleimidobutyramine, is described by Collioud et al. (Collioud et al., 1993). The aryldiazirine function of this cross-linking reagent facilitates light-dependent, carbene-mediated, covalent binding to either inert supports or to biomolecules such as proteins, carbohydrates and nucleic acids. The maleimide function of the cross-linker allows binding to a thiolated surface by thermochemical modification of cysteine thiols. However this treatment may modify the structure and activity of the target protein. Light-induced covalent coupling of the cross-linking reagent to a protein via the carbene function, however, has the disadvantage that it does not provide controlled orientation of the target protein.

Common for most of the described immobilization methods is their use of one or more thermochemical/chemical steps, sometimes with hazardous chemicals, some of which are likely to have a deleterious effect on the structure and/or function of the bound protein. The available methods are often invasive, whereby foreign groups are introduced into a protein to act as functional groups, which cause protein denaturation, as well as lower its biological activity and substrate specificity. There is a need in the art of protein coupling and immobilization to improve the method of coupling, where the structural and functional properties of the coupled or immobilized component are preserved and the orientation of coupling can be controlled. We believe that LAMI represents a significant step in this direction.

3.1 Light Assisted Molecular Immobilization technology (LAMI)

Photochemistry, biosensor microarrays and drug delivery systems

Light assisted molecular immobilization technology provides a photonic method for coupling a protein or a peptide on a carrier via stable bonds (covalent bond or thiol-Au bond) while preserving the native structural and functional properties of the coupled protein or peptide. This technology avoids the use of one or more chemical steps, in contrast with traditional coupling methods for protein immobilisation, which typically involve several chemical reactions. That can be costly, time-consuming as well as deleterious to the structure/function of the bound protein. Furthermore the orientation of the protein or peptide, coupled according to the method of the present invention, can be controlled, such that their functional properties, e.g. enzymatic, may be preserved. In comparison, the majority of known protein coupling methods lead to a random orientation of the proteins immobilised on a carrier, with the significant risk of lower biological activity and raised detection limits.

LAMI technology exploits an inherent natural property of proteins and peptides, whereby a disulphide bridge in a protein or peptide, located in close proximity to an aromatic amino acid residue, is disrupted following excitation of aromatic amino acids. The aromatic residues are actually the preferred spatial neighbours of disulphide bridges (Petersen et al., 1999). The thiol groups created by light induced disulphide bridge disruption in a protein or peptide are then used to immobilise the protein or peptide to a carrier. The formed free thiol groups in the protein can afterwards attach the protein onto a thiol reactive surface, such as gold, thiol derivatized glass and quartz, or even plastics (see Fig. 5). The new protein immobilization technology has led to the development of (Neves-Petersen et al., 2006;Snabe et al., 2006; Duroux et al., 2007a, 2007b & 2007c; Skovsen et al., 2007, 2009a & 2009b; Neves-Petersen et al., 2009b; Parracino et al., 2010 & 2011):

- microarrays of active biosensors and
- biofunctionalization of thiol reactive nanoparticles, aiming at engineering drug delivery systems

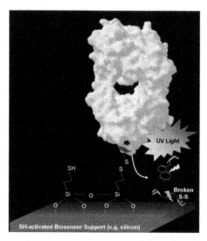

Fig. 5. The principle of light assisted molecular immobilization (LAMI) sketched with tryptophan near a disulphide bridge on a protein molecule. UV illumination of aromatic residues leads to disulphide bridge (SS) opening and to the formation of free SH groups, which will react to thiol reactive surfaces.

3.2 Bio-functionalization of surfaces with micrometer-resolution

With a beam of UV laser light we are able to open disulphide bonds in most SS-containing proteins. If this happens at or close to a thiol reactive surface, such as thiol derivatized glass, quartz or a gold surface, the protein is immobilized onto the surface. Since this happens where the UV photons are present, the size of the focal spot e.g. in a simple focusing setup determines where immobilization takes place. We are able to control this process such that spot size is ~3-5 micron. The process is relatively fast, being determined by physical chemical parameters as well as the light fluency (power per unit area). Currently we are operating with 100 ms illumination per spot with ~1mW 280nm 8MHz femtosecond pulses (Duroux et al, 2007). With a pitch of 10 micron and spot size of 5 micron, this allows for about 40.000 spots per mm². We have verified that Fab anti prostate specific antigen can be immobilized with our technology and still remains biologically active (Parracino et al., 2010).

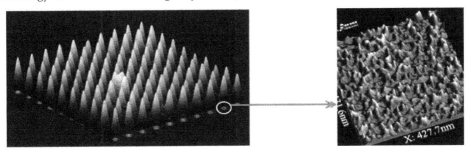

Fig. 6. (Left) Protein microarray engineered with LAMI using 280nm laser light focused to a spot size of 5 micron. The displayed array has 5 micron spots, 10 micron pitch leading to an array density equivalent to 10⁴ spots per 1mm² of sensor surface. (Right) Atomic Force Microscopy visualization of protein immobilized using LAMI, of the central area of 1 spot of the displayed array.

Although light assisted immobilization technology has obvious applications in the area of biosensor microarraying, any pattern of immobilized proteins can be generated. As such, we have successfully transferred several different bitmaps to selected surfaces. As seen in the figure a bitmap of a fullerene was printed into a protein pattern that retained almost all graphical details of the original bitmap (Neves-Petersen et al., 2009b). The size of the protein printed surface is 1mm².

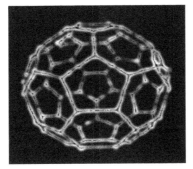

Fig. 7. Fluorescence is emitted from proteins that have been immobilized with LAMI technology. A film of protein has been illuminated according to a fullerene pattern. The illuminated proteins (280nm excitation) were immobilized onto the thiol reactive surface. The image has 10μm resolution and dimensions ~1260μm x 1220 μm

In Fig. 8 is displayed the optical setup used in order to immobilize the proteins according to a particular bitmap:

- The slide surface is covered with a protein film
- A bit map is loaded into the computer
- The surface is illuminated according to the bitmap, i.e., light will hit the surface reproducing the image in the bitmap
- Molecules will only be immobilized on the surface if they have been illuminated
- The slide is washed, removing the non-illuminated and therefore non-immobilized proteins
- The fluorescence of the immobilized molecules can be observed

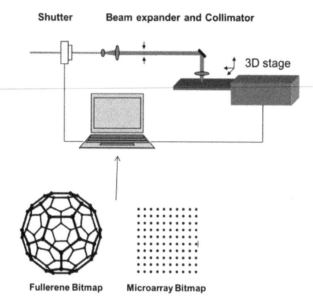

Fig. 8. Optical setup used in order to immobilize the proteins according to a particular bitmap. The slide were the protein film is placed on is optically flat and derivatised with thiol (SH) groups using silane chemistry.

3.3 Biofunctionalization of surfaces with nm/submicrometer-resolution

The technique of UV-light assisted immobilization of disulphide containing proteins has been combined with the Fourier transforming properties of lenses as well as with a simple mm scale feature size spatial mask. Theory predicts that when light passes through a spatial mask placed in the back focal plane of a focusing lens, we should obtain an intensity pattern in the front focal plane corresponding to the Fourier transform of the spatial mask (see Fig. 9).

A spatial mask consisting of eight holes arranged in a square with a hole in each corner and in the middle of each side is displayed below (Fig. 10). This pattern was then Fourier transformed in order to evaluate the diffraction pattern that the simplified eight-hole mask would generate after being Fourier transformed by a lens. The diffraction patter is displayed in Fig. 10

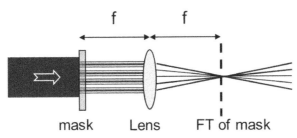

f   f

mask   Lens   FT of mask

Fig. 9. Consider a thin lens illuminated by a monochromatic plan wave. According to the theory of Fourier optics, an aperture (transmission mask) placed in the back focal plane of a lens will generate a diffraction pattern in the front focal plane that will be identical to the FT of the aperture.

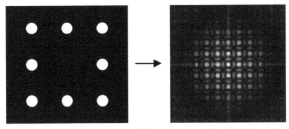

Fig. 10. A spatial mask of eight holes arranged in a square (total dimension of mask is 1cmx1cm) placed in the back focal plane of a lens, will give rise to a diffraction pattern displayed to the right (central part of the diffraction pattern).

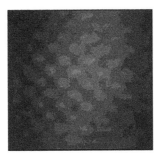

Fig. 11. Fluorescence image of the fluorescently labeled proteins immobilized using the simple eight-hole mask, depicting the fluorescence emission from FITC. The peaks in the array are interspaced by ~1.5µm and have a FWHM of 750nm. The pitch of 1.5µm corresponds to a spot density of ~4.5x10^5 spots per mm².

Using the light pattern produced by this simple mask with 8 x 1mm sized holes, in a single shot we produce multiple spots with immobilized protein with a spot size of ~700 nm (Fig. 11 and Skovsen et al., 2009b). With this spot density, we can populate 1 mm² with close to 1 million sensor spots, which represents an improvement of 10 fold over existing commercially available high density protein arraying methods. Our approach bypasses the use of micro dispenser techniques – and the technical difficulties associated with the use of such. It is simple, and fast. With our current ability to generate immobilized patterns with a spot size down to 700 nanometer, we believe that we can design spatial patterns of binding proteins which could identify and bind, e.g., specific cells such as stem cells.

Our previous works report that we do see patterns similar to what theory predicts (Skovsen et al., 2009b; Petersen et al., 2010) but not always identical (Petersen et al., 2010). We have also shown that the presence of biomolecules on the slide surface can break the diffraction pattern of light (Petersen et al., 2010).

4. Biofunctionalization of thiol reactive nanoparticles, aiming at the development of drug delivery systems

For the last few decades, research efforts have been focused on the development of new materials at the atomic, molecular and macromolecular levels, on the length scale of approximately 1-100nm in order to build materials at the nanoscale, and to explore structures, devices and systems that have novel properties. It has been shown that it is possible to tune precisely the physico-chemical properties of nano-materials by modifying the crystal size, shape and composition. Among various nano materials, magnetic core-shell particles have been broadly used in many technological applications, especially for biological applications such as drug targeting and delivery, cell labeling and separation, cancer therapy, magnetic resonance image (MRI) contrast agents, bio-sensors and bio-imaging. Core-shell particles result from the combination of different metals that together display new properties compared to their monometallic counterparts. Therefore, the combination of both metal's properties such as optical, electrical, magnetic and catalytic can be used for technological applications. Among the core-shell nanoparticles, Fe_3O_4@Au and Fe_3O_4@SiO_2 particles are widely used not only for its magnetic, optical and chemical properties but also for chemical stability, good biocompatibility, low toxicity, easy dispersibility, affinity towards biomolecules with amine/thiol/carboxylic terminal groups and convenient preparation techniques. Magnetite nanoparticles of size below 26.1 nm are super-paramagnetic in nature which means that these particles can be controlled by external magnetic field but retain no coercivity value (no residual magnetism) once the field is removed (Gnanaprakash et al., 2007). This magnetic property is being used extensively for biosciences in various applications, including bioseparation and imaging. Since biomolecules are highly sensitive to pH, temperature and chemical environment, immobilization protocols should be developed in order to secure high molecular activity and stability.

4.1 New photonic methodology used to create functional nanoparticles

Recently, many protocols have been proposed for the immobilization of various biomolecules on to the particle surfaces for novel properties and various applications. The review paper on "Chemical Strategies for Generating Protein Biochips" by Jonkheijm et al. (Jonkheijm et al., 2008) describes different approaches using covalent and non-covalent immobilization chemistry are reviewed. Recent studies demonstrated that the incorporation of chiral molecules onto nanoparticles provides new opportunities for achieving specificity in the recognition of protein surfaces (You et al. 2008). Immobilization of trypsin on super-paramagnetic nanoparticles allows using higher enzyme concentrations, leading to shorter digestion time than free enzyme molecules and easy separation from the solution (Y.Li et al. 2007). Earlier studies have shown that immobilization of biomolecules not only makes separation easier but also increases the stability of enzyme towards pH, temperature, chemical denaturants and organic solvents (Z. Yang et al. 2008, H.Yang et al., 2004). Water soluble carbodiimide was used to activate the direct adsorption of glucose oxidase, streptokinase, chymotrypsin, dispase, BSA and alkaline phosphatase on magnetic particles (Koneracka et al. 2002). Recently, decanthiol capped gold nanoparticles were modified with dithiobis (succinimidyl propionate) for BSA coupling by ligand exchange (H.-Y.Park et al. 2007). Ma and colleagues used condensation product of 3-glycidoxypropyltrimethoxysilane and iminodiacetic acid charged with Cu^{2+} to immobilize BSA onto the silica coated magnetic nanoparticles through metal ion affinity towards protein (Ma et al., 2006).

We are tagging biomolecules directly onto magnetic nanoparticles using our new photonic technology, light assisted molecular immobilization. The surface of these nanoparticles is thiol reactive, being gold or thiol derivatised silica. These particles provide very high surface area to tag protein efficiently and also there is no need to use other reagents to enhance the binding. The surface affinity towards the thiol groups present in the protein will be used to immobilize the protein molecule onto the nanoparticle. Bovine serum albumin (BSA, a carrier protein) and insulin have been successfully immobilized with our new photonic technology. Recently, LAMI technology has been used to create free and active thiol functional groups in BSA to be linked to Fe_3O_4@Au core-shell nanoparticle (Parracino et al., 2011).

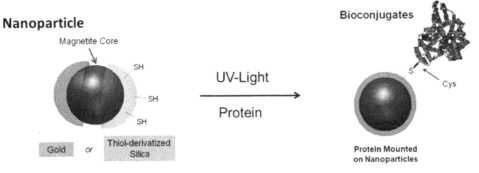

Fig. 12. Covalent immobilization of proteins onto the thiol reactive surface of nanoparticles using light assisted molecular immobilization (LAMI) technology, in order to engineer drug delivery systems.

We have developed the necessary technology that allows us to produce a variety of nanoparticles, from gold and silica nanoparticles to core-shell superparamagnetic nanoparticles. Furthermore, we can further derivatise the silica outer layer of those nanoparticles with chemical functional groups, such as thiol, amino and carboxylic groups. The combination of such knowledge with our new photonic immobilization technology allows us to build protein bioconjugates in a new way. Our new photonic immobilization technology is ideal to couple drugs, proteins, peptides, DNA and other molecules to nanoparticles such as gold or biopolymer nanospheres, which can subsequently be used as molecular carriers into cells for therapeutic purposes.

4.2 Medical applications of bio-functionalized nanoparticles

Nanomedicine is the medical application of nanotechnology and related disciplines, which mostly include biocompatible nanoparticle platforms that contain both therapeutic and/or imaging components. Since biomolecules such as individual cells, mRNA, DNA and proteins are nanoscale sized, probes of equivalent dimensions can provide very effective detection of individual chemical interactions of biomolecules, understanding of the chemical reactions and manipulation of the same. Therefore, the integration of nanotechnology and medical sciences has led to fundamental understanding in molecular biology, as well as new advanced technological applications such as drug targeting and delivery, cell labeling and separation, cancer therapy, magnetic resonance image (MRI) contrast agents, bio-sensors and bio-imaging. Nanoparticles may carry chemo-, radio-, and gene therapeutics or combination of these. They can be inorganic nanoparticles (noble metal, metal oxide, silica, mesoporous silica and combination of these components), lipid aggregates, and synthetic surfactant-polymer systems (such as vesicles, micelles). Inorganic nanoparticles that have unique physicochemical properties allow applications in nanomedicine after proper synthesis, coating, surface functionalization and bioconjugation. These nanosized materials provide a robust framework in which two or more components can be incorporated to give multifunctional capabilities (Wang et al., 2008; Salgueiriño-Maceira & Correa-Duarte, 2007). The combination of metals or polymer molecules should provide suitable and tunable magnetic, optical and chemical properties, chemical stability, low toxicity, easy dispersibility, affinity towards biomolecules with amine/thiol/carboxylic terminal groups and convenient preparation techniques.

Among the various compositions, gold composites are used for the development of various clinical diagnosis methods (Baptista et al., 2008; Raj et al., 2011) because of its size dependent optical properties. Time-resolved single-photon counting fluorescence studies on porphyrin monolayer-modified gold clusters revealed resonant energy transfer between the porphyrin and the gold surface, which is a phenomenon of considerable interest in biophotonics (Imahori & Fukuzumi, 2001). The ability to control the size and shape of gold nanoparticles and their surface conjugation with antibodies allow for both selective imaging and photo-thermal killing of cancer cells by using light with longer wavelengths for tissue penetration (Gobin et al., 2007). Similar success was also demonstrated with polymer-coated superparamagnetic iron oxide nanoparticles conjugated with, e.g., fluorescent molecules, tumor-targeting moieties and anticancer drugs which aim targeting human cancers. Imaging inside the body can either be done using magnetic resonance or fluorescence imaging (Kohler et al., 2006). Quantum dots optical properties including bright emission, photostability, size dependant luminescence and long fluorescence lifetimes make them also suitable for bioimaging applications. In combination with superparamagnetic nanoparticles and surface modification with peptides or other functional groups, these multifunctional particles are being used in bioimaging, bioseparation and in order to understand the behaviour of nanoparticles in cells such as tracking the particles, cell uptake of particles, drug dose evolution at targeted site (Janczewski et al., 2011; Summers H.D et al., 2011). Many biosensors and bioseparation protocols were also demonstrated using such multifunctional nanoparticles (Rossi et al., 2006; Liu and Xu, 1995; Fan et al., 2003). Figure 13 shows the example of functional nanoparticles. Silica nanoparticles can be linked to functional groups like carboxylic, thiol, amine or hydroxide, in order to attach dye molecules and therapeutic molecules to track particles inside the human body and targeted drug delivery. However, intense research studies are needed in order to overcome the barriers in the human body that challenge the efficiency of nanoparticle delivery such as walls of blood vessels, physical entrapment of particles in organs and removal of particles by phagocytic cells. Therefore, the ideal nanoparticle system to be used in nanomedicine should not only overcome such barriers but also allow for real time visualisation of particles, detection of the damaged tissues, selective and rapid accumulation at diseased tissue, effective drug delivery or effective therapy (in the case of hyperthermia). Hence, pioneering works are being carried out at related interdisciplinary fields such as chemistry, biology, pharmacy, nanotechnology, medicine and imaging.

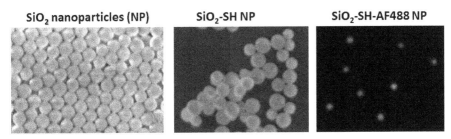

Fig. 13. The surface of silica nanoparticles can be functionalized with different chemical groups which can be used to bind fluorophores (e.g. in this example the fluorophore AF 488 has been coupled to the SH groups introduced on the surface of bare silica nanoparticles), proteins or other molecules.

Nanoparticles based drug delivery is particularly advantageous when planning cancer therapy because the leaky blood cells in tumours mean that particles of certain size tend to accumulate more in cancer tissues than in normal tissues (Peer D et al, 2007). By coating nanoparticles with biomolecules that recognize receptors on cell membranes, a wide range of drugs and imaging agents can enter cells (endocytosis). Drug loaded magnetic particles can be localized to the specific site by an external magnetic field (Hu et al., 2008). This allows more concentrated doses of the anticancer drugs to be delivered to the cancer cells and keep them on site for longer periods of time. In order to prevent dangerous agglomeration of the particles in the blood stream, the particles must be of a small size relative to the dimensions of the capillaries, monodisperse and spherical in shape. In addition, the particles must have a high magnetic moment and switch their magnetisation quickly at low fields (Ankamwar et al., 2010; Shah et al., 2011). Nanoparticles surface modified with amphiphilic polymeric surfactants such as poloxamers, poloxamines or polyethylene glycol derivatives (S.-M. Lee et al., 2010), folic acid derivatives (Das et al, 2008; Landmark et al., 2008), silica derivatized with different functional groups or with porous structure (Wang et al., 2008; Slowing et al., 2007) are used for drug targeting and delivery. Moreover, triggerable drug delivery systems enable on-demand controlled release of drugs that may enhance therapeutic effectiveness and reduce systemic toxicity. Recently, a number of new materials have been developed that exhibit, e.g., sensitivity to UV and visible light, such that irradiation can release covalently bound drugs from dendrimers or dendrons with photocleavable cores or photoactivated surfaces (C. Park et al., 2008), temperature (Bikram et al., 2007; Peng et al., 2011), near-infrared light (S.-M.Lee et al., 2010), pH (Benarjee & Chen, 2008), ultrasound (Bawa et al.,

2009), or external magnetic fields (Hu et al., 2008). Long-wavelength light such as radio frequency radiation and microwave radiation have also been used to trigger drug release. This responsiveness can be triggered remotely to provide flexible control of dose magnitude and timing. Mann et al. (Mann et al., 2011) demonstrated enhanced delivery of therapeutic liposomes carried in E-selectin thioaptamer conjugated porous silica particles to the bone marrow tissue. These particles can also be utilised to deliver imaging agents and growth factors (eg. colony stimulating factor) for the protection of bone marrow against chemotherapy and radiation.

The use of biocompatible iron oxide particles for hyperthermia is increasing in cancer therapy. Iron oxide magnetic nanoparticles exposed to an alternating magnetic field act as localized heat sources at certain target regions inside the human body. The heating of magnetic oxide particles with low electrical conductivity in an external alternating magnetic field is mainly due to either loss processes during the reversal of coupled spins within the particles or due to frictional losses if the particles rotate in an environment of appropriate viscosity. Magnetic nanoparticles coated with amphipathic polymer pullulan acetate (food additive) were examined for their cytotoxicity and cellular uptake. Moreover, in vitro hyperthermia treatment of KB cells produced therapeutic efficacies of 56% and 78% at 45°C and 47°C, respectively, indicating the great potential of surface modified magnetic nanoparticles as magnetic hyperthermia mediators (Gao et al., 2010). Gonzalez-Fernandez and co-workers have presented a study on the magnetic properties of bare and silica-coated ferrite nanoparticles with sizes between 5 and 110 nm (Gonzalez-Fernandez et al., 2009). Their results show a strong dependence of the power absorption with particle size, with a maximum around 30 nm, as expected for a Neel relaxation mechanism in single-domain particles. Recently, in order to enhance the heat conversion capacity of nanoparticles from electromagnetic energy into heat, core-shell nanoparticles were designed to exploit the advantage of exchange coupling between a magnetically hard core and magnetically soft shell to maximize the specific loss power (J.H. Lee et al., 2011). However, in order to avoid the risk of overheating during the hyperthermia effect, curie temperature tuning is done by designing the composition of the core magnetic nanoparticles (Kaman et al., 2009).

MRI is one of the most useful diagnostic tools for medical sciences. MRI contrast agents are chemical substances introduced to the region being imaged to increase the contrast between different tissues or between normal and abnormal tissue, by altering the relaxation times. Generally, gadolinium or manganese salts as well as superparamagnetic iron-oxide based particles are by far the most commonly used materials as MRI contrast agents. Superparamagnetic iron oxide based contrast agents have the advantage of producing an enhanced proton relaxation in MRI better than those produced by paramagnetic ions. Consequently, lower doses are needed which reduce to a great extent the secondary effects in the human body. Particle's negligible remanence after removing the magnetic field (minimizes the particles aggregation) and low toxicity makes them beneficial for in vivo applications (Kinsella et al., 2011; Y.Park et al., 2009).

Various biosensors were developed exploiting novel properties of nanomaterials. Rossi and co-workers have shown the utility of fluorescent nanospheres to detect the breast cancer marker HER2/neu in a glass slide based assay (Rossi et al., 2006). For the detection of cholesterol, plasmon resonance properties of gold nanoparticles were used after conjugating digitonin onto the surface of gold nanoparticles (Raj et al., 2011).

4.3 Self-organization of magnetic nanoparticles

The magnetic properties and self-organization of magnetic nanoparticles has to be understood in order to use these particles for biological applications. Extensive studies have been carried out in this direction. Superparamagnetic nanoparticles consisting of single-domain magnetite nanoparticles are randomly dispersed in liquid in the absence of external magnetic field. Under an external magnetic field, these particles are assembled to form chain-like structures along the magnetic field lines. An attractive magnetic force due to the magnetic dipole is balanced by repulsive electrostatic and solvation forces. Upon removal of the external magnetic field, thermal energy dominates and the particles disintegrate (Fig. 14). A wide range of studies were carried out for quantitative investigation of the temporal self-organization of superparamagnetic composite particles in the presence of an external magnetic field. The kinetics of field-induced self-organization into linear chains, time-dependent chain-size distribution, resolved growth steps condensation, polarization, colinearity, and concatenation, the average chain growth rate, and inter-particle interaction length were calculated in the presence of external magnetic fields (Gajula et al., 2010). These studies give us valuable information relevant to hyperthermia treatment, MRI contrast agents, bioseparation, drug targeting studies and is relevant for making 2D and 3D biocompatible structures for tissue engineering. Moreover, the LAMI technique will enhance the nanomedical applications by creating strong, covalent bond interactions between the nanoparticles and therapeutical biomolecules.

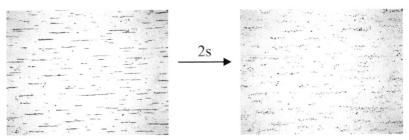

2s

Fig. 14. Bright field microscope images of superparamagnetic nanoparticles in the (left) presence and (right) absence of an external magnetic field (2s after field removal).

Most tissues in organisms are composed of repeating basic cellular structures that are embedded in an extracellular 3D matrix. Tissue functionality arises from these components and the relative spatial locations of these components (Nichol & Khademhosseini, 2009). Tissue engineering enables to recreate the native 3D architecture *in vitro*. Artificial biocompatible 3D structures enable researchers to create organs for transplantation and to understand the structure/function relationship, to characterize cell-membrane mechanical properties, enables the theoretical analyses and to model cellular events and diseases. Xu and co-workers fabricated magnetic nanoparticles loaded cell-encapsulating microscale hydrogels and assembled these gels into 3D multilayer constructs using magnetic fields (Xu et al., 2011). A three-dimensional tissue culture based on magnetic levitation of cells in the presence of a hydrogel consisting of gold, magnetic iron oxide nanoparticles and filamentous bacteriophage has been exploited for direct cell manipulation and cell sorting. By spatially controlling the magnetic field, the geometry of the cell mass is manipulated (Souza et al., 2010). Ito and co-workers have demonstrated the successful construction and delivery of human retinal pigment epithelial cell sheets using magnetic nanoparticles (Ito et al., 2005). Biomimetic nanopatterns can also be explored for analysis and control of live cells growth (Kim et al., 2010). Richert and co-workers observed the control over the growth of the osteogenic cells, fibroblastic cells and smooth muscle cells using nano-porous titanium alloy explaining that nanotopography may modulate cytoskeletal organization and membrane receptor organization (Richert et al., 2008). This work indicates the impact of nanoscale engineering in controlling cell-material interfaces, which can have profound implications for the development of tissue engineering and regeneration medicine.

Structural colour originates from physical configurations of materials e.g. upon light interaction the lattice spacing of the melanin rods generate various colours in the feathers of a peacock. Structural colour is free from photobleaching, unlike traditional pigments or dyes. Owing to its unique characteristics, there have been many attempts to make artificial structural colour through various technological approaches such as colloidal crystallization, dielectric layer stacking and direct lithographic patterning. However, these techniques are time consuming, cost, and great effort is needed to produce multicoloured patterns on a substrate. Using the self-organization of nanoparticles, external magnetic field tunable bandgaps can be created on various substrates within a few seconds (Philip et al., 2003; H. Kim et al., 2009). During the chaining process under external magnetic field, the combination of attractive and repulsive forces determines the interparticle distance, and the interparticle distance in a chain determines the colour of the light diffracted from the chain, which can be explained by Bragg diffraction theory. Thus, the colour can be tuned by simply varying the interparticle distance using external magnetic fields. The spacing between the particles is sensitive towards charges over the particle's surface. This property can be used to develop biosensors. For example, the specific molecular recognition of cholesterol by digitonin on nanoparticles may result in a reduction of the inter-particle spacing due to the enhanced hydrophobicity of the surface after cholesterol binding. Therefore, the light diffracted from the chain varies indicating the adsorption of cholesterol onto the nanoparticles surface. Cassagneau et al., have shown that the reversible colour tuning of a colloidal crystal could be potentially adopted for biosensing. Biospecific binding of avidin with biotin is demonstrated by monitoring changes in the bandgap spectral peak position caused by Bragg-diffraction of electromagnetic waves within the structure (Cassagneau & Caruso, 2002). The same chaining technique is explored for the DNA separation (Doyle et al., 2002).

5. Photonic cancer therapy

The epidermal growth factor receptor (EGFR), also known as HER1/Erb-B1, belongs to the ErbB family of receptor tyrosine kinases (RTKs) (Riese & Stern, 1998; Yarden & Sliwkowski, 2001; Olayioye et al., 2000). Binding of ligands such as EGF and TGF, leads to homo- and heterodimerization of the receptors (Olayioye et al., 20003). Dimerization in the case of EGFR leads to autophosphorylation of specific tyrosine residues in the intracellular tyrosine kinase domain. EGFR activation results in cell signaling cascades that promote tumor cell proliferation, survival and inhibits apoptosis (Fig. 14). EGFR is expressed or highly expressed in non-small-cell lung cancer (NSCLC) and in a variety of common solid tumors, and has also been associated with poor prognosis. High EGFR expression is generally associated with invasion, metastasis, late-stage disease, chemotherapy resistance, hormonal therapy resistance and poor outcome. Therefore, inhibition of EGFR function is a rational treatment approach.

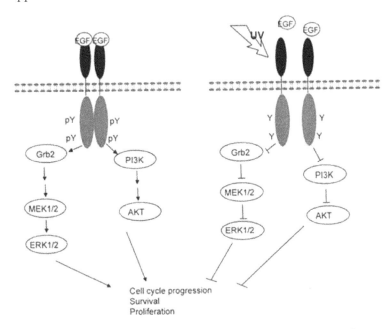

Fig. 15. Overview of the cellular pathways affected by the laser-pulsed UV illumination of the EGF (epidermal growth factor) receptor leading to attenuation of the EGFR signalling cascade. Photoactivation of aromatic residues within the extracellular domain of the EGF receptor, leads to disruption of nearby disulphide bridges. This prevents the ligand, e.g. EGF, from binding to the receptor and activating the EGFR pathways. In addition, it is possible that laser-pulsed UV illumination targets the intracellular domain of the EGF receptor causing photodegradation of phosphorylation-targeted tyrosine residues again preventing the proteins from binding to the phosphorylated tyrosine residues.

Bioinformatics studies show that the extracellular domain of human HER3 (ErbB3), a member of the epidermal growth factor receptor (EGFR) family, is exceedingly rich in SS bridges and aromatic residues (Petersen et al, 2008, Fig. 15). The structure of the extracellular domain consists of four domains. Tethered domains II and IV are displayed. A total of 22 disulphide bridges can be seen in one domain and 25 in the other domain. On the other hand, biophysical studies show that UV illumination of aromatic residues nearby disulphide bridges leads to the SS disruption (Neves-Petersen et al, 2002, Fig. 1). Therefore, it is likely that we induce structural changes in the 3D structure of EGFR upon prolonged UV excitation. If the UV light fluency (power per unit area) is above a certain threshold it is likely that 3D structural changes occur that will prevent the correct binding to, e.g., EGF. We have shown that relatively low intensity UV light (0.273 mW 200fs femtosecond pulses at 280nm diffused onto a petri dish area) can be used to induce cancer cell death (apoptosis) in skin cancer cells in culture (Olsen B.B. et al., 2007, Petersen S.B. et al., 2008). We have also shown that most likely the epidermal growth factor receptor in cancer cell membranes upon absorption of the UV photons changes its molecular structure as a consequence. The net result is that phosphorylation and cell signaling is abolished, which in turn leads to apoptosis. The technology may lead to an important new modality in the treatment of various cancers. Pulsed UV illumination can halt activation of cancer cell membrane receptors and thereby all downstream reactions that would lead to cancer, shutting down the cells' biological functions. Moreover, this new treatment activated the cell's own cell death program. This has been documented on two human epidermal cancer cell lines (Olsen B.B. et al., 2007). The photonic dosage necessary for therapeutical results has additionally been determined.

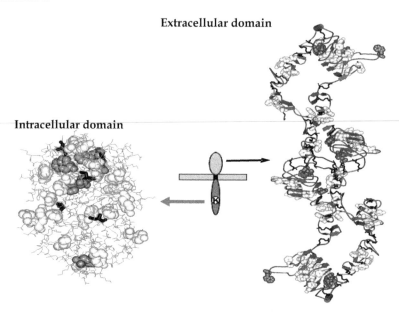

Extracellular domain

Intracellular domain

Fig. 16. Molecular structure of the intracellular and extracellular domains of human HER3 (ErbB3), a member of the epidermal growth factor receptor (EGFR) family. Tethered extracellular domains II and IV are displayed. The pdb code of the extracellular domain is 1m6b.pdb. The extracellular domain is extremely rich in disulphide bridges. Tethered extracellular domains II and IV are displayed. A total of 22 disulphide bridges can be seen in each of the displayed domains. The intracellular domain is a protein tyrosine kinase. In the extracellular domain, disulphide bridges are depicted as black sticks, the tryptophan amino acids as rendered grey CPK models, and the tyrosine and phenylalanine residues are depicted as white CPK models. As expected, no disulphide bridges are observed in the intracellular domain due to the reduction environment of cells. However, this domain is also rich in aromatic residues.

This technology is applicable to the treatment of various forms of cancer. Using optical fibers it is possible to illuminate localized areas in any region of the human body. Therefore, both external and internal tumors can be treated. This method may also offer a better treatment of a surgical wound cavity prior to its closure in order to prevent cancer reappearance. This new photonic method differs from the classical photodynamic therapy

(PDT) which requires the use of a photosensitizer molecule that upon excitation and interaction to molecular oxygen leads to the formation of singlet oxygen, which kills the cells. The new photonic cancer therapy does not require the use of photosensitizer molecules. UV light induced 3D structural changes in the EGFR protein prevents binding/activation by EGF, halting this way the phosphorylation of the intracellular domain of EGFR and stopping metabolic pathways that lead to cancer proliferation. The new photonic cancer therapy can be used in combination to PDT and other cancer therapies.

Several reports describe how UV light can activate the EGF receptor hence activating the AKT and MAPK pathway (Warmuth et al., 1994; Coffer PJ at al., 1995; Huang et al., 1996; Katiyar, 2001; Wan et al., 2001; Iordanov et al., 2002; Matsumura & Ananthaswamy, 2004; El-Abaseri at al., 2006). Our observations seem to point at another effect of UV light, in apparent contrast with those results. The reason for this discrepancy could be found in the illumination power per unit of illuminated area (fluency). In our experiments the total integrated power over a second is significantly less than the average solar UV output but comparing the actual output during a pulse event we have 1000-fold higher intensity during the 200 femtosecond long pulse event (Olsen et al., 2007, Petersen et al., 2008).

6. Conclusion

Our work on the interplay between the protein molecule and UV light has resulted in new basic science insights. It has also led to the development of a new protein covalent immobilization technique. The new photonic technology has been used successfully to design and engineer drug delivery systems and biosensors at the micro and nanoscale relevant to nanomedicine. The new engineering principle is made possible due to the presence of a conserved structural motif in proteins conserved by nature throughout evolution.

As a surprising spin-off, our work has resulted in new knowledge concerning how UV light can stop skin cancer. Pulsed UV illumination can halt activation of cancer cell membrane receptors and thereby stop all downstream reactions that would lead to cancer, shutting down the cells' biological functions. Moreover, this new treatment activated the cell's own cell death program (apoptosis). In particular we have realized that UV light chemically modifies the same receptor protein that many cancer therapeutic treatments are trying to target chemically. We believe that this holds promise for a totally new approach to treat some types of localized cancer. We will strive to develop an in-depth understanding of how and why nanometer-sized protein structures respond to light exposure.

7. Acknowledgment

Maria Teresa Neves-Petersen acknowledges the leave of absence granted by AAU.

8. References

Andolfi, L.; Cannistraro, S.; Canters, G.W.; Facci, P.; Ficca, A.G.; Amsterdam, I.M.C.V. & Verbeet, M.Ph. (2002). A poplar plastocyanin mutant suitable for adsorption onto gold surface via disulphide bridge, *Arch. Biochem. Biophys.* Vol. 399, No.1, (March 2002), pp. 81-88, ISSN 0003-9861.

Ankamwar, B.; Lai, T.C.; Huang, J.H.; Liu, R.S.; Hsiao, M.; Chen, C. H. & Hwu Y. K. (2010). Biocompatibility of Fe3O4 nanoparticles evaluated by in vitro cytotoxicity assays using normal, glia and breast cancer cells, *Nanotechnology*, Vol. 21, No.7, (February 2010), pp.075102, ISSN(print) 0957-4484.

Banerjee, S.S. & Chen D.-H. (2008). Multifunctional pH-sensitive magnetic nanoparticles for simultaneous imaging, sensing and targeted intracellular anticancer drug delivery, *Nanotechnology*, Vol.19, No.50, (December 2008), pp.505104, ISSN(print) 0957-4484.

Baptista, P.; Pereira, E.; Eaton, P.; Doria, G.; Miranda, A.; Gomes, I.; Quaresma, P. & Franco, R. (2008). Gold nanoparticles for the development of clinical diagnosis methods. *Ana. Bioanal.Chem.*, Vol.391, No.3, (June 2008), pp.943-950, ISSN(print) 1618-2642.

Bawa, P.; Pillay, V.; Choonara, Y. E. & du Toit, L.C. (2009). Stimuli-responsive polymers and their applications in drug delivery, *Biomed. Mater.*, Vol.4, No.2, (April 2009), pp.022001, ISSN(Print) 1748-6041

Bent D.V. & Hayon, E. (1975a). Excited-state chemistry of aromatic amino-acids and related peptides .III. tryptophan, *J. Am. Chem. Soc.*, Vol.97, No.10, (May 1975), pp.2612–2619, ISSN(Print): 0002-7863.

Bent D.V. & Hayon, E. (1975b). Excited-state chemistry of aromatic amino-acids and related peptides .I. tyrosine, *J. Am. Chem. Soc.*, Vol.97, No.10, (May 1975), pp.2599–2606, ISSN(Print): 0002-7863.

Bent D.V. & Hayon, E. (1975c). Excited-state chemistry of aromatic amino-acids and related peptides .II. phenylalanine, *J. Am. Chem. Soc.*, Vol.97, No.10, (May 1975), pp.2606–2612, ISSN(Print): 0002-7863.

Bikram, B.; Gobin, A.M.; Whitmire, R.E. & West, J.L. (2007). Temperature-sensitive hydrogels with SiO_2–Au nanoshells for controlled drug delivery, *J. Controlled Release*, Vol.123, No.3 (November 2007), pp.219-227, ISSN: 0168-3659.

Cassagneau, T. & Caruso, F. (2002). Inverse Opals for Optical Affinity Biosensing. *Adv. Mater.*, Vol.14, No.22 (November 2002), pp.1629–1633, ISSN 1521-4095 (online).

Coffer, P.J.; Burgering, B.M.; Peppelenbosch, M.P.; Bos, J.L. & Kruijer, W. (1995). UV activation of receptor tyrosine kinase activity. *Oncogene*, Vol.11: pp. 561-569.

Collioud, A.; Clemence, J.F.; Saenger, M. & Sigrist, H. (1993). Oriented and covalent immobilization of target molecules to solid supports—Synthesis and application of a light-activatable and thiol-reactive cross-linking reagent. *Bioconjugate Chem.* Vol. 4, No.6, (November 1993), pp. 528-536, ISSN(print) 1043-1802.

Creed, D. (1984a). The photophysics and photochemistry of the near-uv absorbing amino-acids .2. Tyrosine and its simple derivatives, *Photochem. Photobiol.*, Vol. 39, No.4, (April 1984), pp.563–575, ISSN (Online)1751-1007.

Creed, D.(1984b). The photophysics and photochemistry of the near-uv absorbing amino-acids .1. Tryptophan and its simple derivatives, *Photochem. Photobiol.*, Vol. 39, No.4, (April 1984), pp.537–562, ISSN (Online)1751-1007.

Das, M.; Mishra, D.; Maiti, T.K.; Basak, A. & Pramanik, P. (2008). Bio-functionalization of magnetite nanoparticles using an aminophosphonic acid coupling agent: new, ultradispersed, iron-oxide folate nanoconjugates for cancer-specific targeting, *Nanotechnology*, Vol.19, No.41, (October 2008), pp.415101, ISSN(print) 0957-4484.

Doyle, P. S.; Bibette, J.; Bancaud, A. & Viovy, J-L. (2002). Self-Assembled Magnetic Matrices for DNA Separation Chips, *Science*, Vol. 295, No. (2002)2237, ISSN 1095-9203 (online).

Duroux, M.; Duroux, L.; Neves-Petersen, M.T.; Skovsen, E. & Petersen S.B. (2007a). Novel photonic technique creates micrometer resolution protein arrays and provides a new approach to coupling of genes, peptide hormones and drugs to nanoparticle carriers. *Applied Surface Science*, Vol. 253, No. 19, (July 2007), pp. 8125-8129, ISSN: 0169-4332.

Duroux, M.; Skovsen, E.; Neves-Petersen, M.T.; Duroux, L.; Gurevich, L. & Petersen, S.B. (2007b), Light-induced immobilisation of biomolecules as a replacement for present nano/micro droplet dispensing based arraying technologies, *Proteomics*, Vol 7, No.19, (October 2007), pp.3491-3499, ISSN 1615-9861.

Duroux, M.; Gurevich, L.; Neves-Petersen, M.T.; Skovsen, E.; Duroux, L. & Petersen, S.B. (2007c). Using light to bioactivate surfaces: A new way of creating oriented, active immunobiosensors, *Applied Surface Science*, Vol. 254, No. 4, (December 2007), pp. 8125-8129, ISSN: 0169-4332.

El-Abaseri, T.B.; Putta, S. & Hansen, L.A. (2006). Ultraviolet irradiation induces keratinocyte proliferation and epidermal hyperplasia through the activation of the epidermal growth factor receptor. *Carcinogenesis*, Vol.27, pp. 225-231.

Fan, J.; Lu, J.; Xu, R.; Jiang, R. & Gao, Y. (2003). Use of water-dispersible Fe2O3 nanoparticles with narrow size distributions in isolating avidin, *J. Colloid Interface Sci.* Vol. 266, No.1 (October 2003), pp.215-218, ISSN: 0021-9797.

Faraggi, M. & Bettelheim, A. (1977). The reaction of the hydrated electron with amino acids , peptides, and proteins in aqueous solutions: tryptophyl peptides, *Radiation Research*, Vol. 72, No. 1, (October 1977), pp.81-88, ISSN(print) 0033-7587.

Gajula, G.P.; Neves-Petersen, M.T. & Petersen; S.B. (2010). Visualization and quantification of four steps in magnetic field induced two-dimensional ordering of superparamagnetic submicron particles, *Appl. Phys. Lett.*, Vol 97, No.10, (September 2010), pp.103103, ISSN(Print) 0003-6951.

Gao, F.; Cai, Y.; Zhou, J.; Xie, X.; Ouyang, W.; Zhang, Y.; Wang, X.; Zhang, X.; Wang, X.; Zhao, L. & Tang, J. (2010). Pullulan acetate coated magnetite nanoparticles for hyper-thermia: Preparation, characterization and in vitro experiments. *Nano Res.* Vol.3, No.1, (January 2010), pp.23-31, ISSN(Print)1998-0124.

Gnanaprakash, G.; Philip, J. & Raj, B. (2007). Effect of divalent metal hydroxide solubility product on the size of ferrite nanoparticles, *Mater. Lett.*, Vol. 61, No. 23-24, (September 2007), pp. 4545-4548, ISSN: 0167-577X.

Gobin, A. M.; Lee, M. H.; Halas, N. J.; James, W. D.; Drezek, R. A. & West, J. L. (2007). Near-Infrared Resonant Nanoshells for Combined Optical Imaging and Photothermal

Cancer Therapy. *Nano Lett.* Vol. 7, No.7, (July 2007), pp.1929–1934, ISSN(print) 1530-6984.

Gonzalez-Fernandez, M. A.; Torres, T.; Andres-Verges, M.; Costo, R.; de la Presa, P.; Serna, C. J.; Morales, M. P.; Marquina, C.; Ibarra, M. R. & Goya, G. F. (2009). Magnetic nanoparticles for power absorption: Optimizing size, shape and magnetic properties. *J. Solid Stat. Chem.*, Vol.182, No.10, (October 2009), pp.2779-2784, ISSN 0022-4596.

Grossweiner, L.I. & Usui, Y. (1971). Flash Photolysis and inactivation of aqueous lysozyme, *Photochem. Photobiol.* Vol. 13, No.3, (March 1971), pp. 195-214, ISSN (Online) 1751-1097.

Hoffman, M. Z. & Hayon, E. (1972). One-electron reduction of the disulphide linkage in aqueous solution. Formation, protonation, and decay kinetics of the RSSR- radical, *J. Am. Chem. Soc.* Vol. 94, No.23, (November 1972), pp 7950–7957, ISSN(Print): 0002-7863.

Huang, R.P.; Wu, J.X.; Fan, Y. & Adamson, E.D. (1996). UV activates growth factor receptors via reactive oxygen intermediates. *J Cell Biol* Vol.133: pp. 211-220.

Hu, S.-H.; Liu, D.-M.; Tung, W.-L.; Liao, C.-F. and Chen, S.-Y. (2008). Surfactant-Free, Self-Assembled PVA-Iron Oxide/Silica Core–Shell Nanocarriers for Highly Sensitive, Magnetically Controlled Drug Release and Ultrahigh Cancer Cell Uptake Efficiency. *Adv. Funct. Mater.*, Vol.18, No.19 (October 2008), pp. 2946–2955, ISSN (online) 1616-3028.

Imahori, H. & Fukuzumi, S. (2001). Porphyrin Monolayer-Modified Gold Clusters as Photoactive Materials. *Adv. Mater.*, Vol. 13, No.15, (Auguest 2001), pp.1197-1199, ISSN 1521-4095 (online).

Iordanov, M.S., Choi, R.J., Ryabinina, O.P., Dinh, T.H., Bright, R.K. & Magun, B.E. (2002). The UV (Ribotoxic) stress response of human keratinocytes involves the unexpected uncoupling of the Rasextracellular signal-regulated kinase signaling cascade from theactivated epidermal growth factor receptor. *Mol Cell Biol*, Vol.22, pp.5380-5394.

Ito, A; Hibino, E.; Kobayashi, C.; Terasaki, H.; Kagami, H.; Ueda, M.; Kobayashi, T. & Honda H. (2005). Construction and Delivery of Tissue-Engineered Human Retinal Pigment Epithelial Cell Sheets, Using Magnetite Nanoparticles and Magnetic Force, *Tissue Engineering*, Vol.11, No.3-4, (March/April 2005), pp. 489-496, ISSN(print) 2152-4947.

Jańczewski, D.; Zhang, Y.; Das, G. K.; Yi, D. K.; Padmanabhan, P.; Bhakoo, K. K.; Tan, T. T. Y. & Selvan, S. T. (2011). Bimodal magnetic–fluorescent probes for bioimaging. *Microscopy Research and Technique*, Vol.74, No.7, (July 2011), pp563-576, ISSN (online) 1097-0029.

Johnsson, B.; Löfås, S. & Lindquist, G. (1991). Immobilization of proteins to a carboxymethyldextran-modified gold surface for biospecific interaction analysis in surface plasmon resonance sensors, *Anal. Biochem.*, Vol 198, No.2, (November 1991), pp.268-77, ISSN: 0003-2697.

Jonkheijm, P.; Weinrich, D.; Schröder, H.; Niemeyer, C. and Waldmann, H. (2008). Chemical Strategies for Generating Protein Biochips. *Angewandte Chemie International Edition*, Vol. 47, No.50, (December 2008), pp. 9618–9647, ISSN (online) 1521-3773.

Kaman, O.; Pollert, E.; Veverka, P.; Veverka, M.; Hadová, E.; Knížek, K.; Maryško, M.; Kašpar, P.; Klementová, M.; Grünwaldová, V; Vasseur, S.; Epherre, R.; Mornet, S.; Goglio, G. & Duguet, E. (2009), Silica encapsulated manganese perovskite nanoparticles for magnetically induced hyperthermia without the risk of overheating, *Nanotech*. Vol. 20, No.27, (July 2009), pp.275610, ISSN(print) 0957-4484.

Katiyar, S.K. (2001). A single physiologic dose of ultraviolet light exposure to human skin in vivo induces phosphorylation of epidermal growth factor receptor. *Int J Oncol*, Vol. 19: 459-464.

Kerwin, B. A. & Jr. Remmele, R. L. (2007), Protect from Light: Photodegradation and Protein Biologics, *J. Pharmac. Sci.*, Vol. 96, No.6, (June 2007), pp.1468-1479, ISSN(Online) 1520-6017.

Kim, D.-H.; Lee, H.; Lee, Y. K.; Nam, J.-M. & Levchenko, A. (2010). Biomimetic Nanopatterns as Enabling Tools for Analysis and Control of Live Cells. *Adv. Mater.*, Vol.22, No.41, (November 2010), pp. 4551–4566, ISSN 1521-4095 (online).

Kim, H.; Ge, J.; Kim, J.; Choi, S-e.; Lee, H.; Lee, H.; Park, W.; Yin, Y. & Kwon, S. (2009). Structural colour printing using a magnetically tunable and lithographically fixable photonic crystal, *Nature Photonics*, Vol. 3, No9, (September 2009), pp.534-540, ISSN: 1749-4885.

Kinsella, J. M.; Ananda, S.; Andrew, J. S.; Grondek, J. F.; Chien, M.-P.; Scadeng, M.; Gianneschi, N. C.; Ruoslahti, E. & Sailor, M. J. (2011). Enhanced Magnetic Resonance Contrast of Fe3O4 Nanoparticles Trapped in a Porous Silicon Nanoparticle Host. *Adv. Mater.*, Vol.23, No.36, (September 2011), pp.H248–H253, ISSN 1521-4095 (online).

Kohler, N.; Sun, C.; Fichtenholtz, A.; Gunn, J.; Fang, C. & Zhang, M. (2006). Methotrexate-Immobilized Poly(ethylene glycol) Magnetic Nanoparticles for MR Imaging and Drug Delivery. *Small*, Vol. 2, No.6, (June 2006), pp.785–792, ISSN(online) 1613-6829.

Koneracka, M.; Kopcansky, P.; Timko, M.; Ramchand, C.N.; de Sequeira, A. & Trevan, M. (2002). Direct binding procedure of proteins and enzymes to fine magnetic particles. *J. Mol. Catal. B: Enzym.* Vol.18, No.1-3, (September 2002), pp.13-18, ISSN 1381-1177.

Landmark, K.J.; DiMaggio, S.; Ward, J.; Kelly, C.; Vogt, S.; Hong, S.; Kotlyar, A.; Myc, A.; Thomas, T.P.; Penner-Hahn, J.E.; Baker, J.R. Jr.; Holl, M.M.B. & Orr, B.G. (2008). Synthesis, Characterization, and in Vitro Testing of Superparamagnetic Iron Oxide Nanoparticles Targeted Using Folic Acid-Conjugated Dendrimers, *ACS Nano*, Vol.2, No.4, (April 2008), pp.773-783, ISSN(print) 1936-0851.

Lee, J.-H.; Jang, J-T.; Choi, J.-S.; Moon, S. H.; Noh, S-h.; Kim, J-w.; Kim, J-G; Kim, I-S; Park, K.I. & Cheon, J. (2011). Exchange-coupled magnetic nanoparticles for efficient heat induction, *Nature Nanotech*. Vol.6, No.7, (July 2011), pp.418-422, ISSN 1748-3387.

Lee, S.-M.; Park, H. and Yoo, K.-H. (2010). Synergistic Cancer Therapeutic Effects of Locally Delivered Drug and Heat Using Multifunctional Nanoparticles. *Adv. Mater.*, Vol.22, No.36 (September 2010), pp.4049–4053, ISSN 1521-4095 (online).

Li, Y.; Xu, X.; Deng, C.; Yang, P. & Zhang, X. (2007). Immobilization of Trypsin on Superparamagnetic Nanoparticles for Rapid and Effective Proteolysis. *J. Proteome Res.*, Vol.6, No.9, (September 2007), pp.3849-3855, ISSN 1535-3893.

Li, Z.; Lee, W.E. & Galley, W.C. (1989). Distance dependent of the tryphtophan-disulphide interaction at the triplet level from pulsed phosphorescence studies on a model system, *Biophys. J.*, Vol. 56, No.2, (Auguest 1989), pp.361-367, ISSN (printed): 0006-3495.

Liu, Q. X. & Xu, Z. H. (1995). Self-Assembled Monolayer Coatings on Nanosized Magnetic Particles Using 16-Mercaptohexadecanoic Acid, *Langmuir*, Vol. 11, No.12, (December 1995), pp.4617-4622, ISSN(print) 0743-7463.

Ma, Z.; Guan, Y. & Liu, H. (2006). Superparamagnetic silica nanoparticles with immobilized metal affinity ligands for protein adsorption. *J. Magn. Magn. Mater.* Vol. 301, No.2, (June 2006), pp.469-477, ISSN 0304-8853.

Mann, A. P.; Tanaka, T.; Somasunderam, A.; Liu, X.; Gorenstein, D. G. and Ferrari, M. (2011). E-Selectin-Targeted Porous Silicon Particle for Nanoparticle Delivery to the Bone Marrow. *Adv. Mater.*, Vol.23, No.36, (September 2011), pp.H278–H282, ISSN 1521-4095 (online).

Matsumura, Y & Ananthaswamy, H.N. (2004). Toxic effects of ultraviolet radiation on the skin. *Toxicol Appl Pharmacol* Vol.195, pp. 298-308.

Neves-Petersen, M. T.; Gryczynski, Z; Lakowicz, J.; Fojan, P.; Pedersen, S.; Petersen, E. & Petersen, S. B. (2002). High probability of disrupting a disulphide bridge mediated by an endogenous excited tryptophan residue, *Protein Sci.*, Vol.11, No. 3, (March 2002), pp. 588-600, ISSN 1469-896X.

Neves-Petersen, M. T.; Snabe, T.; Klitgaard, S.; Duroux, M. & Petersen, S. B. (2006). Photonic activation of disulphide bridges achieves oriented protein immobilization on biosensor surfaces, *Protein Sci.*, Vol 15, No.2, pp. 343–351, ISSN 1469-896X.

Neves-Petersen, M. T.; Klitgaard, S.; Pascher, T.; Skovsen, E.; Polivka, T.; Yartsev, A.; Sundström, V. & Petersen, S.B. (2009a). Flash photolysis of cutinase: identification and decay kinetics of transient intermediates formed upon UV excitation of aromatic residues, *Biophys. J.*, Vol. 97, No.1 (January 2009), pp.211-226, ISSN (printed): 0006-3495.

Neves-Petersen, M.T.; Duroux, M.; Skovsen, E.; Duroux, L. and Petersen, S.B. (2009b). Printing novel architectures of nanosized molecules with micrometer resolution using light, *J. Nanosci. Nanotechn.* Vol. 9, No. 6, (June 2009), pp.3372–3381, ISSN: 1533-4880 (Print).

Nichol, J. W & Khademhosseini, A. (2009). Modular tissue engineering: engineering biological tissues from the bottom up, *Soft Matter*, Vol. 5, No. 7, (February 2009), pp.1312-1319, ISSN (print) 1744-683X.

Olayioye, M.A.; Neve, R.M.; Lane H.A. & Hynes, N.E. (2000). The ErbB signaling network: receptor heterodimerization in development and cancer, *EMBO J*, Vol. 19, No. 13, (July 2000), pp. 3159 - 3167, ISSN (printed): 0261-4189.

Olsen, B.B.; Neves-Petersen, M.T.; Klitgaard, S.; Issinger O.-G. & Petersen, S.B. (2007). UV light blocks EGFR signalling in human cancer cell lines, *International Journal of Oncology*, Vol. 30. No.1, (January 2007), pp 181-185, (Print) ISSN: 1019-6439.

Park, C.; Lim, J.; Yun, M. and Kim, C. (2008). Photoinduced Release of Guest Molecules by Supramolecular Transformation of Self-Assembled Aggregates Derived from Dendrons. *Angewandte Chemie International Edition*, Vol.47, No.16, (April 2008), pp.2959-2963, ISSN(online) 1521-3773.

Park, H. -Y.; Schadt, M. J.; Wang, L.; Lim, I. -I. S.; Njoki, P. N.; Kim, S. H.; Jang, M. -Y.; Luo, J. & Zhong, C.-J. (2007). Fabrication of Magnetic Core@Shell Fe Oxide@Au Nanoparticles for Interfacial Bioactivity and Bio-separation. *Langmuir*, Vol.23, No.17, (Auguest 2007), pp.9050-9056, ISSN(print) 0743-7463.

Park, Y. I.; Kim, J. H.; Lee, K. T.; Jeon, K.-S.; Na, H. B.; Yu, J. H.; Kim, H. M.; Lee, N.; Choi, S. H.; Baik, S.-I.; Kim, H.; Park, S. P.; Park, B.-J.; Kim, Y. W.; Lee, S. H.; Yoon, S.-Y.; Song, I. C.; Moon, W. K.; Suh, Y. D. & Hyeon, T. (2009). Nonblinking and Nonbleaching Upconverting Nanoparticles as an Optical Imaging Nanoprobe and T1 Magnetic Resonance Imaging Contrast Agent. *Adv. Mater.*, Vol. 21, No.44, (November 2009), pp.4467–4471, ISSN 1521-4095 (online).

Parracino, A.; Neves-Petersen, M.T.; di Gennaro, A.K.; Pettersson, K.; Lövgren, T. & Petersen, S.B. (2010). Arraying prostate specific antigen PSA and Fab anti-PSA using light-assisted molecular immobilization technology, *Protein Sci.* Vol. 19, No.9, (september, 2010), pp.1751-1759, ISSN 1469-896X.

Parracino, A.; Gajula, G.P.; Kold, A.; Correia, M.; Neves-Petersen, M.T.; Rafaelsen, J. & Petersen; S.B. (2011). Photonic immobilization of BSA for nanobiomedical applications: Creation of high density microarrays and superparamagnetic bioconjugates. *Biotech. Bioengineering*, Vol. 108, No.5, (May 2011), pp.999-1010, ISSN(online) 1097-0290.

Peer, D.; Karp, J. M.; Hong, S.; Farokhzad, O.C.; Margalit R & Langer R. (2007), Nanocarriers as an emerging platform for cancer therapy. *Nature Nanotech*. Vol.2, No.12, (December 2007), pp.751-760, ISSN 1748-3387.

Peng, C.-L.; Tsai, .H-M.; Yang, S.-J.; Luo, T.-Y.; Lin, C.-F.; Lin, W.-J. & Shieh, M.-J. (2011). Development of thermosensitive poly(n-isopropylacrylamide-co-((2-dimethylamino) ethyl methacrylate))-based nanoparticles for controlled drug release, *Nanotechnology*, Vol.22, No.26, (July 2011), pp.265608, ISSN(print) 0957-4484.

Petersen, M.T.N.; Jonson, P.H. & Petersen, S. B. (1999). Aminoacid neighbours and detailed conformational analysis of cysteins in proteins. *Protein engineering design and selection*, Vol.12, No.7, (July 1999), pp. 535-548, ISSN 1741-0126

Petersen S.B.; Neves-Petersen, & Olsen W. (2008). The EGFR family of receptors sensitizes cancer cells towards UV light. Optical Interactions with Tissue and Cells XIX. *Proceedings of the SPIE*, Vol. 6854, (February 2008), pp. 68540L-68540L-10, ISBN: 9780819470294.

Petersen, S. B.; di Gennaro, A.K.; Neves-Petersen, M.T.; Skovsen, E. & Parracino, A. (2010). Immobilization of biomolecules onto surfaces according to ultraviolet light diffraction patterns, *Applied Optics*, Vol. 49, No. 28, (October 2010), pp. 5344-5350, ISSN: 1559-128X.

Philip, J.; Jaykumar, T.; Kalyanasundaram, P. & Raj, B. (2003). A tunable optical filter, *Meas. Sci. Technol.* Vol.14, No.8, (August 2003), pp.1289-1294, ISSN(Print) 0957-0233.

Raj, V.; Jaime, R.; Astruc, D. & Sreenivasan, K. (2011), Detection of cholesterol by digitonin conjugated gold nanoparticles, *Biosensors and Bioelectronics*, Vol.27, No.1, (September 2011), pp.197–200, ISSN: 0956-5663.

Richert, L.; Vetrone, F.; Yi, J.-H.; Zalzal, S. F.; Wuest, J. D.; Rosei, F. & Nanci, A. (2008). Surface Nanopatterning to Control Cell Growth. *Adv. Mater.*, Vol. 20, No.8, (April, 2008), pp. 1488–1492, ISSN 1521-4095 (online).

Riese, D. J. & Stern, D. F. (1998). Specificity within the EGF family/ErbB receptor family signaling network. *BioEssays*, Vol. 20, No.1, (January 1998), pp. 41–48, ISSN (online) 1521-1878.

Rossi, L.M.; Shi, L.; Rosenzweig, N. & Rosenzweig, Z, (2006). Fluorescent silica nanospheres for digital counting bioassay of the breast cancer marker HER2/nue, *Biosensors and Bioelectronics*, Vol.21, No.10, (April 2006) pp.1900–1906, ISSN: 0956-5663.

Salgueiriño-Maceira, V. & Correa-Duarte, M. (2007). Increasing the Complexity of Magnetic Core/Shell Structured Nanocomposites for Biological Applications. *Adv. Mater.*, Vol.19, No.23, (December 2007), pp. 4131–4144, ISSN 1521-4095 (online).

Shah, N.B.; Dong, J. & Bischof J.C. (2011). Cellular Uptake and Nanoscale Localization of Gold Nanoparticles in Cancer Using Label-Free Confocal Raman Microscopy, *Molecular Pharm.* Vol.8, No.1, (Febrrary 2011), pp.176–184, ISSN(print) 1543-8384.

Skovsen, E.; Duroux, M.; Neves-Petersen, M.T.; Duroux, L. & Petersen, S.B. (2007). Molecular Printing Using UV-Assisted Immobilization of Biomolecules, International *Journal of Optomechatronics*, Vol. 1, No. 4, (November 2007), pp 383-391, Print ISSN: 1559-9612.

Skovsen, E.; Neves-Petersen, M.T.; Kold, A.; Duroux, L. & Petersen, S.B. (2009a). Immobilizing Biomolecules Near the Diffraction Limit. *J. Nanosci. Nanotechn.* Vol. 9, No. 7, (July 2009), pp.4333–4337, ISSN: 1533-4880.

Skovsen, E; Kold,A.; Neves-Petersen, M. T. and Petersen, S. B. (2009b). Photonic immobilization of high-density protein arrays using Fourier optics, *Proteomics*, Vol 9, No.15, (August 2009), pp.3945-3958, ISSN 1615-9861.

Slowing, I.; Trewyn, B.; Giri, S. & Lin, V.-Y. (2007). Mesoporous Silica Nanoparticles for Drug Delivery and Biosensing Applications. *Adv. Funct. Mater.*, Vol.17, No.8, (May 2007), pp.1225–1236, ISSN (online) 1616-3028.

Smith, K.C. & Aplin, R.T. (1966). A mixed photoproduct of uracil and cysteine (5-S-cysteine-6-hydrouracil). A possible model for the in vivo crosslinking of deoxyribonucleic acid and protein by ultraviolet light. *Biochemistry*, Vol. 5, No. 6, (June 1966), pp.:2125-2130, Print ISSN: 0006-2960.

Smith, K.C. (1962). Dose dependent decrease in extractability of DNA from bacteria following irradiation with ultraviolet light or with visible light plus dye. *Biochem. Biophys. Res. Commun.*, Vol. 8, No.3, (July), pp.157-163, ISSN: 0006-291X.

Smith, K.C., (1970). A mixed photoproduct of thymine and cysteine: 5-S-cysteine, 6-hydrothymine. *Biochem. Biophys. Res. Commun.*, Vol. 39, No.6, (June1970), pp.:1011-1016, ISSN: 0006-291X.

Snabe, T.; Røder, G.A.; Neves-Petersen, M.T.; Petersen, S.B. & Buus, S. (2006), Oriented Coupling of Major Histocompatibility Complex (MHC) to Sensor Surfaces using Light Assisted Immobilisation Technology, *Biosens. Bioelectron.* Vol.21, No. 8, (February 2006), pp. 1553-1559, ISSN: 0956-5663.

Souza, G. R.; Molina, J. R.; Raphael, R. M.; Ozawa, M. G.; Stark, D. J.; Levin, C. S.; Bronk, L. F.; Ananta, J. S.; Mandelin, J.; Georgescu, M-M.; Bankson, J. A.; Gelovani, J. G.; Killian, T. C.; Arap, W & Pasqualini, R. (2010). Three-dimensional tissue culture based on magnetic cell levitation, *Nature Nanotech.* Vol.5, No.4, (April 2010), pp.291-296, ISSN 1748-3387.

Spanel, P.; & Smith, D. (1995). Recent studies of electron attachment and electron-ion recombination at therma energies, *Plasma Sources Sci. Techno.* Vol. 4, No.2, (May 1995), pp. 302-306. ISSN 0963-0252 (Print).

Summers, H. D.; Rees, P.; Holton, M. D.; Brown, M. R.; Chappell, S. C.; Smith, P.J. & Errington R.J. (2011). Statistical analysis of nanoparticle dosing in a dynamic cellular system, *Nature Nanotech.*, Vol.6, No.3, (March 2011), pp.170-174, ISSN 1748-3387.

Sweeney, R.Y.; Kelemen, B.R.; Woycechowsky, K.J. & Raines R.T. (2000). A highly active immobilized ribonuclease. *Anal Biochem.*, Vol. 286, No.2, (November 2000), pp.312-314, ISSN 0003-2697.

Thakur, A. K. & Mohan Rao Ch. (2008). UV-Light Exposed Prion Protein Fails to Form Amyloid Fibrils, *Plos one*, Vol 3, No. 7, (July 2008), pp. E2688, eISSN 1932-6203.

Vass, I.; Kirilovsky, D.; Perewoska, I.; Máté, Z.; Nagy, F. & Etienne, A.-L. (2000). UV-B radiation induced exchange of the D1 reaction centre subunits produced from the psbA2 and psbA3 genes in the cyanobacterium Synechocystis sp. PCC 6803. *Eur J Biochem.*, Vol. 267, No.9, (May 2000), pp. 2640–2648, ISSN (online) 1432-1033.

Veilleux, J.K. & Duran L.W. (1996). Covalent Immobilization of Biomolecules to Preactivated Surfaces. *IVD Technology magazine*, No.2, (March 1996), pp.26-31, ISSN 1093-5207.

Wan, Y.S.; Wang, Z.Q.; Shao, Y.; Voorhees, J.J. & Fisher, G.J. (2001). Ultraviolet irradiation activates PI 3-kinase/AKT survival pathway via EGF receptors in human skin in vivo. Int J Oncol, Vol.18, pp. 461-466.

Wang, L.; Zao, W. & Tan, W. (2008). Bioconjugated silica nanoparticles: Development and applications. *Nano Res.*, Vol.1, No.2, (May 2008), pp.99-115, ISSN 1998-0124 (Print).

Warmuth, I.; Harth, Y.; Matsui, M.S.; Wang, N. & De Leo, V.A. (1994). Ultraviolet radiation induces phosphorylation of the epidermal growth factor receptor. *Cancer Res,* Vol.54, pp. 374-376.

Wilson D.S. & Nock S. (2001). Functional protein microarrays, *Current Opinion in Chemical Biology,* Vol 6, No.1, (February 2001), pp.81-85, ISSN: 1367-5931.

Xu, F.; Wu, C.-a. M.; Rengarajan, V.; Finley, T. D.; Keles, H. O.; Sung, Y.; Li, B.; Gurkan, U. A. & Demirci, U. (2011). Three-Dimensional Magnetic Assembly of Microscale Hydrogels. *Adv. Mater.*, Vol.23, No.37, (October 2011), pp.4254–4260, ISSN 1521-4095 (online).

Yang, H.; Zhang, S.; Chen, X.; Zhuang, Z.; Xu, J. & Wang, X. (2004). Magnetite-Containing Spherical Silica Nanoparticles for Biocatalysis and Bioseparations. *Anal. Chem.*, Vol. 76, No.5, (March 2004), pp.1316-1321, ISSN (print) 0003-2700.

Yang, Z.; Si, S. & Zhang, C. (2008). Magnetic single-enzyme nanoparticles with activity and stability, *Biochem. Biophys. Res. Commun.* Vol.367, No.1, (February 2008), pp.169-175, ISSN 0006-291X.

Yarden, Y & Sliwkowski M.X. (2001). Untangling the ErbB signalling network, *Nat. Rev. Mol. Cell. Biol.*, Vol. 2, No.2, (February 2001), pp.127-137, ISSN. 1471-0072.

You, C. -C.; Agasti, S. S. & Rotello, V. M. (2008). Isomeric control of protein recognition with amino- and dipeptide- functionalized gold nanoparticles. *Chem. Eur. J.*, Vol.14, No.1, (December 2008), pp.143-150, ISSN 1521-3765.

6

Photo-Induced Proton Transfers of Microbial Rhodopsins

Takashi Kikukawa[1], Jun Tamogami[2], Kazumi Shimono[2],
Makoto Demura[1], Toshifumi Nara[2] and Naoki Kamo[2]
[1]Faculty of Advanced Life Science, Hokkaido University,
[2]College of Pharmaceutical Sciences, Matsuyama University,
Japan

1. Introduction

Microbial rhodopsins are photoactive membrane proteins that are widely distributed over the microbial world. They commonly consist of seven transmembrane helices forming an internal pocket for a chromophore retinal, whose photo-induced isomerization triggers the respective photochemical reactions of these proteins. They are generally classified into photosensors or ion-pumps, but the members of both classes have individualities. In the case of photosensors, there exist a variety of signal-transduction modes, including interaction with other membrane proteins, interaction with cytoplasmic proteins, and light-gated ion channel activity. For ion-pumps, there exist outwardly directed H^+ pumps and inwardly directed Cl^- pumps. In spite of these functional diversities, most microbial rhodopsins show photo-induced proton-transfer reactions among amino acid residues and the external medium. These reactions reflect the pKa changes of some residues induced by the protein conformational changes during the respective photochemical reactions. To analyze these reactions, it is indispensable to detect the small pH changes of the external medium due to the proton release/uptake during the photoreaction cycles. Electrochemical cells using indium-tin oxide (ITO) or tin oxide (SnO_2) transparent electrodes are a powerful and convenient tool that enables such measurement in external media under a variety of pH conditions (Robertson & Lukashev, 1995; Wang et al., 1997; Koyama et al., 1998a; Tamogami et al., 2009; Wu et al., 2009). Here, we will describe the rapidly expanding family of the microbial rhodopsins and the application of the ITO method to their photoresponses.

2. Visual rhodopsins and archaeal rhodopsins

Rhodopsin is a protein in the retina of animals that works as a light sensor. Rhodopsin contains retinal (vitamin A aldehyde) as a chromophore and its absorption maximum is ~500 nm. Retinal binds with a specific lysine residue *via* a protonated Schiff base. Absorption of photons induces the isomerization of 11-*cis* to all-*trans* retinal and subsequent conformational changes including the deprotonation of the Schiff base (J.L. Spudich et al., 2000). Rhodopsin had been considered to be confined to animals until the findings from haloarchaea.

In the early 1970s, however, a retinal protein was discovered in the membrane of the highly halophilic archaeon *Halobacterium salinarum* (formally *halobium*). The natural habitats of *H. salinarum* are the Dead Sea, the Great Salt Lake and salt ponds. The newly discovered rhodopsin was named bacteriorhodopsin (BR), and it was shown to act as a light-driven proton pump (Oesterhelt & Stoeckenius, 1971). By using light energy, BR transports a proton from a cytoplasmic to an extracellular space, which produces the proton-electrochemical potential difference across the membranes, which in turn drives the synthesis of ATP *via* H$^+$-ATPase. Light irradiation to BR induces the retinal isomerization from all-*trans* to 13-*cis*, while the isomerization of visual rhodopsins is from 11-*cis* to all-*trans*. In addition, the most significant difference from the visual rhodopsins is the existence of a so-called photocycle: light absorption leads to the excitation of the pigment, which decays to the original pigment *via* a variety of photo-intermediates. This linear cyclic photochemistry is called a photocycle. On the other hand, for most visual rhodopsins, the Schiff base linkage with the retinal is disrupted as a consequence of the photochemical reaction. The proton-transfer mechanism of BR has been intensively investigated so far. During the photocycle, the protonated Schiff base affords its proton to the extracellular space and receives another proton from the other side, *i.e.*, the cytoplasmic space. Thus, this cycle involves the alternation between the protonated and deprotonated states of the Schiff base.

Later, three additional retinal proteins were discovered in *H. salinarum*. These were halorhodopsin (HR) (Matsuno-Yagi & Mukohata, 1977; Schobert & Lanyi, 1982; Mukohata et al., 1999; Váró et al., 2000; Essen, 2002), sensory rhodopsin I (SRI) (Bogomolni & J.L Spudich, 1982; Hazemoto et al., 1983; J.L Spudich & Bogomolni, 1984) and sensory rhodopsin II (SRII, also called phoborhodopsin) (Takahashi et al., 1985; Tomioka et al., 1986; Wolff et al., 1986; E.N. Spudich et al., 1986; Marwan & Oesterhelt, 1987). BR and HR are light-driven ion pumps: HR is an inwardly directed Cl$^-$-pump and BR is an outwardly directed H$^+$-pump as described above. On the other hand, SRI and SRII act as receptors of phototaxis. In the cell membranes, these receptors form firm complexes with their cognate transducers. By utilization of these sensing systems, the cell moves toward light of the preferred wavelength ($\lambda > 520$ nm) where BR and HR can work, and escapes from the shorter wavelength light ($\lambda < 520$ nm) which may contain dangerous UV light. These retinal proteins also exhibit their own functions during the respective photocycles. Under the physiological states, HR and two SRs do not exhibit the proton-pumping activities. However, the photocycles of two SRs involve alterations between the protonated and deprotonated states of the Schiff bases, resulting in the proton releases and uptakes at the extracellular sides (Bogomolni et al., 1994; Sasaki & J.L. Spudich, 1999, 2000).

3. Microbial rhodopsins

About 30 years after the discovery of BR, archaeal rhodopsin homologues began to be identified in various microorganisms, including proteobacteria, cyanobacteria, fungi, dinoflagellates, and alga (J.L. Spudich & Jung, 2005). Thus, the microbial species containing the retinal protein genes inhabit a broad range of environments. At present, these rhodopsin homologues are called type 1 rhodopsins or microbial rhodopsins, and they define a large phylogenetic class spreading to all three domains of life, *i.e.*, archaea, bacteria and eukarya.

To distinguish them from rhodopsins in the microbial world, the rhodopsins in animals are called type 2 rhodopsins.

Studies on BR, HR, SRI and SRII have shown that two amino acid residues corresponding to Asp85 and Asp96 in BR are key residues for the functional difference among these archaeal rhodopsins. As mentioned above, the Schiff bases of BR and two SRs become deprotonated during the early halves of the photocycles. Asp85[BR] and the corresponding aspartates of SRs function as the proton acceptors from the Schiff bases. For HR, on the other hand, Asp85[BR] is replaced by Thr, and thus HR can bind Cl-, a transportable ion, to the vicinity of the Thr and the protonated Schiff base. The other residue, Asp96[BR], functions as a proton donor for the reprotonation process of the Schiff base and contributes to acceleration of the turnover rate of the photocycle. This residue is not conserved in the two SRs and consequently their photocycles are much slower than that of BR. HR also undergoes a fast photocycle in the manner of BR but does not conserve this aspartate, because HR undergoes its photocycle without the deprotonation of the Schiff base.

On the basis of these findings, the physiological functions of newly found microbial rhodopsins have been deduced from the conservation states of the residues corresponding to Asp85[BR] and Asp96[BR]. In addition to these clues, the functions of some proteins have been confirmed experimentally by using the purified proteins and/or by observation of their photo-induced behaviors. From these analyses, the newly found microbial rhodopsins are now categorized as either ion-pumps (H+ or Cl-) or photoreceptors. In paticular, many newly found pigments are categorized as H+ pumps. This fact suggests that the utilization of light energy *via* H+ pumps is widely adopted in the microbial world. For many H+ pumps, the pumping activities have been experimentally confirmed. Representative examples include proteorhodopsin (PR) from proteobacteria living throughout the world's oceans (Béjà et al., 2000); xanthorhodopsin (XR) from *Salinibacter ruber*, a highly halophilic eubacterium (Balashov et al., 2005); *Leptosphaeria* rhodopsin (LR) from *Leptosphaeria maculans*, a fungal pathogen to a plant (Waschuk et al., 2005); *Gloeobacter* rhodopsin (GR) from *Gloeobacter violaceus*, a cyanobacterium living in fresh water (Miranda et al., 2009); and *Acetabularia* rhodopsin (AR) from a gigantic unicellular marine algae, *Acetabularia acetabulum*, which reaches to 10 cm in height (Tsunoda et al., 2006). Recently, two clones of AR, named ARI and ARII, were isolated from the same organism (Lee et al., 2010). These were somewhat different from the original AR. For ARII, the H+-pumping activity was confirmed using *Xenopus* oocytes and the detailed photochemistry was examined with the help of a cell-free expression system (Wada et al., 2011; Kikukawa et al., 2011). On the other hand, photosensing rhodopsins have also been found in bacteria and eukarya. Although these represent only a minority of the newly found microbial rhodopsins, they exhibit a variety of signal-transduction mechanisms. Two SRs in the archaeal membrane relay the photosignals to the cognate transducer proteins embedded in the membrane. From the eubacterium *S. ruber*, SRI itself was found and has been extensively characterized (Kitajima-Ihara et al., 2008). Unlike these SRs, the photosensor called *Anabaena* sensory rhodopsin (ASR) from *Anabaena sp.* PCC7120, a cyanobacterium living in fresh water, is considered to relay the signal to the soluble protein (Jung et al., 2003). *Anabaena* does not have a flagellum and so does not show the phototaxis. Instead, *Anabaena* shows a photoresponse called chromatic adaptation. Thus, it is considered that ASR controls the biosynthesis of chromoproteins forming the light-harvesting complex. In addition to these, a new type of photosensing rhodopsin called channelrhodopsin was found in *Chlamydomonas reinhardtii*, a green flagellate alga (Sineshchekov et al., 2002; Nagel et al., 2002; Suzuki et al., 2003). This is a light-gated ion channel and induces the photomotile behavior of the cell. Thus, the world of microbial rhodopsins is now rapidly expanding.

4. The photo-induced proton transfer associated with the photocycle of microbial rhodopsins

Microbial rhodopsins have linear cyclic photochemical reactions called photocycles. For all microbial rhodopsins examined so far, the retinal Schiff bases are protonated in their dark states under physiological conditions. Except in the case of HR, a Cl⁻ ion pump, the Schiff bases become deprotonated during the photocycles independent of H⁺-pumping or photosensing rhodopsins. As described below, these primary and the subsequent proton-transfer reactions are closely related to the functions of both microbial rhodopsins.

4.1 H⁺-pumping rhodopsins

The illumination of the pigment protein leads to the excited state, which is relaxed thermally to the original pigment *via* various photochemical intermediates (Fig. 1A). The best-studied rhodopsin is BR (Haupts, 1999; Balashov, 2000; Heberle, 2000; Lanyi, 2004, 2006), and the H⁺-pumping mechanism of BR is described below. BR at the ground state and the intermediates K, L, M, N and O have been investigated with various biophysical methods. The photocycle comprises stepwise reactions of the thermal reisomerization of the photoisomerized 13-*cis* retinal to the initial all-*trans*, and the proton is transferred toward the higher pKa residue accompanied with pKa changes during the photocycle. Reflecting the differences in protein conformation and protonation states of some residues, the intermediates assume the respective absorption spectra. The photoisomerization from initial all-*trans* to 13-*cis* retinal is completed until the formation of K-intermediate. The subsequent proton transfer observed at around neutral pH occurs as the following sequence (see Fig. 1A and B). First, the deprotonation of the protonated Schiff base occurs in the formation of M-intermediate. The proton from the Schiff base is transferred to its counterion Asp85[BR] and the subsequent proton release to the extracellular (EC) space occurs from the proton-releasing complex (PRC) consisting of Glu194[BR], Glu204[BR], Arg82[BR] and water molecules. Next, the proton of protonated Asp96[BR] locating at the cytoplasmic (CP) channel transfers to the Schiff base in the M-N transition, which lead to the reprotonation of the Schiff base. Then the deprotonated Asp96[BR] uptakes a proton from the CP space in the N-O transition, which is accompanied by the reisomerization of retinal to the initial all-*trans* state. Finally, the proton of protonated Asp85[BR] is transferred to PRC during the decay of O-intermediate, and then the protein returns to the original state. This series of proton-transfer reactions accomplishes the net proton transport from the CP to EC side.

As mentioned above, a light-driven proton pump has been found in many microorganisms belonging to bacteria and eukarya. These rhodopsins also undergo the photocycle including the intermediates similar to those of BR. However, there are several differences in the photocycles and proton transfers between BR and other H⁺-pumping rhodopsins. Examples are as follows. (1) PR from marine bacteria also goes through K, L, M, N and O (or PR') intermediates at around neutral pH (Dioumaev et al., 2002; Friedrich et al., 2002; Váró et al.,

2003). However, the proton movement from the proton donor residue (Glu108[PR] corresponding to Asp96[BR]) to the Schiff base and its subsequent proton uptake from the CP space occur simultaneously in the M-N transition (Dioumaev et al., 2002). For BR, these two proton movements occur separately in the M-N and N-O transitions. (2) For ARII from marine algae, the photocycle includes much larger reverse reactions between L, M, N and O than are seen in BR (Kikukawa et al., 2011). Although the reverse reactions are also present in BR, their rates in ARII appear to be much larger than in BR. A rapid reverse reaction might be disadvantageous for the unidirectional ion pumping. (3) For all microbial rhodopsins, the proton conduction channel is bisected by retinal. For BR, these two channels are severely isolated and, in the dark state, the CP half channel is kept under a highly hydrophobic condition. This asymmetric structure had been believed to be important for the ion pumping function. For several proton pumps such as XR and GR, however, this characteristic structure of BR is not conserved (Luecke et al., 2008; Miranda et al., 2009). Therefore, the newly found proton pumps seem to adopt a mechanism that is at least partly different from that in BR. Thus, it would be an interesting subject to clarify the essential mechanisms for the respective proton pumps.

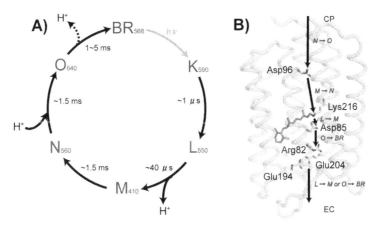

Fig. 1. The photocycle scheme (A) and the structure of BR (B) with the important residues involved in the proton-transfer reactions.

In (A), BR, K, L, M, N and O represent the unphotolyzed state and intermediates, respectively. Their λ_{max}'s are given in the subscripts, and the lifetimes of the intermediates are also shown. The photoisomerization of the retinal by illumination triggers the stepwise photoreactions accompanied with the proton movements. In (B), these proton transfers at respective steps are indicated with arrows (see section 4.1 for the details). The timing of proton transfer depends on the pH of the medium. Above pH 5, the proton release occurs in L-decay while below at about pH 5, the proton release occurs in O-decay (arrow with a broken line) instead of L-decay. The structure in (B) was drawn from the PDB coordinate file 1C3W.

4.2 Photosensing rhodopsins

During the photocycles of almost all photosensing rhodopsins examined so far, the deprotonations of the Schiff bases occur much as for the H⁺-pumping rhodopsins. In many cases, these deprotonations result in the proton release/uptake reactions with the external medium. Thus, the proton-transfer reactions also control the decay rates of some intermediates of photosensing rhodopsins.

The best characterized photosensing rhodopsins are SRI and SRII from archaea and bacteria. Some homologues of these SRs show the outwardly directed proton pumping activities when they exist alone in the membrane (Bogomolni et al., 1994; Sasaki & J.L. Spudich, 2000; Sudo et al., 2001; Schmies et al., 2001). Upon the complex formations with the cognate transducers, both the proton release and uptake occur at only the extracellular side (this is so-called proton circulation) instead of the vectorial proton transport. These facts mean that SRs possess proton-transfer machinery like BR and this machinery is probably sensitive to the protein conformational change relating with the signal transduction mechanism. Like the H^+-pumping rhodopsins, the states having deprotonated Schiff bases are also called M-intermediates, and are the putative signaling states for SRs. The longer lifetimes of M-intermediates are considered to increase the signaling efficiencies. Thus, the reprotonations of the Schiff bases influence the signaling efficiencies.

Studies on two SRIIs from *Natronomonas pharaonis* (NpSRII) (Kamo et al., 2001; J.L. Spudich & Luecke, 2002; Pebay-Peyroula et al., 2002; Klare et al., 2004) and *H. salinarum* (HsSRII) (Sasaki & J.L. Spudich, 1998, 1999) have revealed the differences in the proton-transfer reactions associated with their M-intermediate decays. For NpSRII, the reprotonation of the Schiff base occurs by uptaking a proton directly from the bulk due to the lack of a proton donor to the Schiff base (corresponding to Asp96[BR]). Therefore the M-decay in NpSRII is very slow as compared with BR and depends on the pH of the medium (Miyazaki et al., 1992). For HsSRII, on the other hand, there are two proton-transfer pathways in the decay of the M-intermediate. One is the pathway in which the proton comes directly from the bulk to the Schiff base, and the other is the pathway in which the proton comes from an unidentified X-H residue. Which proton pathway becomes the major component is dependent on the pH of the medium (Sasaki & J.L. Spudich, 1999; Tamogami et al., 2010). In addition, it has been reported that NpSRII possesses the H^+-pumping activity (Sudo et al., 2001; Schmies et al., 2001) but HsSRII does not (Sasaki & J.L. Spudich, 1999, 2000). Thus, despite their identical physiological functions, NpSRII and HsSRII have several differences with respect to their photocycles and proton transfers. For these photosensing rhodopsins, therefore, it would be of interest to investigate the photo-induced proton-transfer mechanisms as well as their relations with the photosignaling transductions to the cognate transducers.

5. Importance of measurements of the photo-induced proton transfer of microbial rhodopsins

As described above, the proton movements between the residues inside the protein as well as between the residue and the external space occur during the photocycles of most microbial rhodopsins. The protein conformational changes during the photocycles alter the pKa of the residues and thereby cause the proton-transfer reactions. For H^+-pumping rhodopsins, these proton movements directly couple with their functional mechanisms, and for the photosensing rhodopsins, these movements affect the signal transduction efficiencies by controlling the decay rates of some intermediates and reflect the conformational alterations by the complex formation with the cognate transducers. For respective proton-transfer reaction, a proton moves from an amino acid residue having smaller pKa to one having higher pKa. Then, if the pKa of the residue from which the proton is released to the external medium is larger than the pH in the medium, the proton cannot be released. For such a case, a subsequent proton movement (*e.g.*, a proton uptake from the medium) occurs prior to the release. This "traffic jam" of the proton movement actually occurs in the photocycle of BR (Balashov, 2000). The pKa of its PRC at the proton-releasing state is about 6 (Zimányi et al., 1992; Balashov, 2000). Under an acidic pH sufficiently lower than pH 6, proton uptake is observed prior to the release (see Figs. 1A and 2B). Therefore, the extent of the pH-dependence of the proton release/uptake could be utilized to estimate the pKa value of a residue that is important for the proton-transfer reaction. Similarly, the proton-transfer rate may afford information about some important amino acid residues. To analyze these reactions, it is necessary to detect a small pH change in the external medium due to the proton movements caused during a single photocycle.

6. Necessity of a device to measure rapid pH changes or proton-transfer rates at any pH

The photocycle of the ion-pumping rhodopsins completes in ~50-100 ms, and many photosensing rhodopsins have slower photocycle of ~sec. A pH glass electrode is too slow to respond to such pH changes. For such a rapid reaction induced by a flash, the pH change is usually obtained using pH-sensitive dyes whose absorbance depends on pH (Heberle, 2000). This is a convenient method and the rapid change is measurable. However, there is a weak point in that the medium pH should be restricted at pH near the pKa values of the dyes. In other words, the measurements under various medium pH values cannot be performed with a single dye. In addition, the subtraction of the signal in the co-presence of a dye and a rhodopsin from that of the rhodopsin alone should be performed. If one wants to estimate the pKa of an important amino acid residue, the pH profile of the magnitude and/or the rate of the proton transfer are indispensable. Hence, a device is needed for the detection of pH changes or proton-transfer rates at any pH. In the following sections, we will describe an electrochemical cell using an indium-tin oxide (ITO) or tin oxide (SnO$_2$) electrode and its application to the microbial rhodopsins. This electrochemical cell is a useful device for these measurements due to its high sensitivity and rapid time-resolution.

7. An ITO (or SnO$_2$) electrode works as a pH-sensitive electrode

Koyama and his coworkers first developed this method using BR (Miyasaka & Koyama, 1991; Miyasaka et al., 1992). They constructed a photo-electrochemical cell in which BRs were absorbed on a SnO$_2$-coated transparent glass electrode, and detected the electric current evoked by constant illumination by using another SnO$_2$ electrode as a reference electrode. The origin of this electric current was assumed to be the charge displacement by BR (Koyama et al., 1994). On the other hand, Robertson and Lukashev suggested that the origin of this signal is the medium pH change caused by the photoinduced proton release and uptake in BR (Robertson & Lukashev, 1995). Actually, our group confirmed this suggestion by the following results of three experiments (Tamogami et al., 2009). 1) The equilibrium potential of an ITO electrode showed a linear relationship to pH (see Fig. 2A). 2) The amplitudes of photoelectrical signals decreased with increasing buffer concentration (see Fig. 2B). 3) HR from *N. pharaonis* (NpHR), an inwardly directed Cl$^-$ pump, did not cause a photoelectrical signal. HR is known to be converted into an H+ pump in the presence of azide (Váró et al., 1996). In accordance with this, a photoelectrical signal was evoked from the HR-adsorbed electrode by the addition of azide (see Fig. 2C and Koyama et al., 1998b). Thus it was established that an ITO (or SnO₂) electrode works as a pH-sensitive electrode.

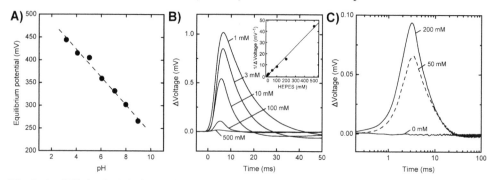

Fig. 2. An ITO (or SnO₂) electrode works as a pH-sensitive electrode.

A) The relationship between the equilibrium potential of ITO and pH. B) The effect of buffer on the photo-induced signals in BR by the ITO electrode. The inset shows the plot of the reciprocal of the amplitudes of ITO signals against buffer concentrations. The experimental medium contains 400 mM NaCl and HEPES of each concentration (1, 3, 10, 100 and 500 mM) at pH 7.5. The excitation light (2 ms duration) was > 520 nm. C) The photo-induced ITO signals in NpHR in the presence and absence of azide. Experiments were performed in medium containing 133 mM Na₂SO₄, azide at each concentration (0, 50 and 200 mM) and 1 mM HEPES at pH 7.0. The excitation light (2 ms duration) was > 440 nm. NpHR was expressed in the *E. coli* expression system, and then was reconstituted with Egg L-α-phosphatidylcholine. Other experimental setups are described in Fig. 3 and previous report (Tamogami et al., 2009). Panels A and B were adapted with permission from Tamogami et al., 2009, *Photochem. Photobiol.* Copyright 2009 The authors, Journal Compilation, The American Society of Photobiology.

8. Measurements of proton release/uptake by BR with a rapid time resolution

BR has been investigated in great detail. Thus, BR and its mutants are good references to show the relevance of our measurements using this electrochemical cell. Figure 3A shows a schemata of our electrochemical cell. This cell is essentially the same as that previously constructed by Koyama and coworkers with some modifications (Miyasaka et al., 1992). Illumination on the electrode-attached proteins induces the electrochemical potential change between the working and the counter ITO electrodes. Figure 3B indicates the flash-induced signals in BR. In this figure, the upward shift signifies the acidification of the medium near the working electrode. Thus, this shift signified the proton release from proteins to the bulk, while the downward shift signifies the proton uptake from the bulk to the protein. For BR, as described above, the sequence of the proton release and uptake can be altered depending on the external medium pH. For example, at pH 6.0 and 9.0, the proton release is followed by uptake, since the proton can be released from PRC at the early step of the photocycle (in the formation of M). The subsequent uptake occurs in the decay of N. On the other hand, at pH 3.0, the proton uptake is followed by release. This may be interpreted as meaning that the proton cannot be released from PRC under this pH condition because the medium pH is lower than the pKa of PRC in the proton-releasing state (M-intermediate). Thus, the proton uptake in the decay of N is observed first and the release occurs in the decay of O concomitantly with the deprotonation of Asp85[BR] (see the arrow with a broken line in Fig. 1A). The same proton-transfer sequence was also observed in the PRC-lacking mutants, as shown in Fig. 4, where the data for a mutant of E194Q/E204Q[BR] are summarized. For these PRC-lacking mutants, the proton uptake occurred first even at the neutral pH (see the trace labeled with "ITO signal" in Fig. 4). For wild-type BR at pH 4.5 (see Fig. 3B), the proton uptake occurred first and then overshoot of the signal was observed at around 35 ms. This may be interpreted as the mixture of the two populations of BR. The molecules exhibiting proton uptake first constitute the major population. Other molecules constituting a minor population exhibit proton release first. The final proton transfer corresponding to proton uptake (decay of the positive signal) caused by the minor population is slower than the proton release of the major population. Thus, the signal amplitude of this system can reflect the ratio of two kinds of molecules having different proton-transfer sequence.

In this measurement system, the time course of the photo-induced proton transfer can also be determined. Figure 4 shows the comparison of three types of signals measured for the E194Q/E204Q[BR] mutant. These are (1) flash-induced absorbance changes of the mutant itself measured at three typical wavelengths; (2) a photo-induced pH change measured by the ITO system and; (3) the corresponding signal measured by a pH-sensitive dye, pyranine. The sign of the pyranine signal is opposite that of ITO. For pyranine, the upward shift signifies the proton uptake, while the downward shift signifies the proton release. As shown in this figure, the proton uptake and release agree well with the formation and decay of O, which are represented by the absorbance change at 660 nm. This is the typical proton-transfer sequence of the PRC-disabled mutant (Brown et al., 1995; Balashov et al., 1997; Dioumaev et al., 1998, Koyama et al., 1998a). The response of the dye is very fast and completely follows the pH change within this time range. The time course of the dye signal almost coincides with that of ITO signal. Thus, our current ITO measurement system can monitor the pH change in the time range of *ca.* 10 ms to several hundred milliseconds (Tamogami et al., 2009). This could be utilized to identify the intermediate accompanying the proton-transfer reaction.

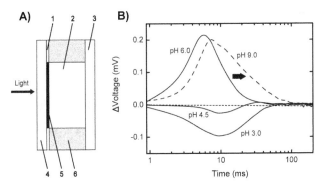

Fig. 3. The measurement of the photo-induced proton transfer in BR by the ITO transparent electrode.

(A) The structure of the photoelectrochemical cell constructed by using ITO electrodes. (1) A silicon sheet; (2) electrolyte plus buffer; (3) a counter ITO electrode; (4) a working ITO electrode; (5) a thin layer of the sample dried on the electrode surface; (6) a Lucite chamber. The emf change between the two ITO electrodes was picked up by an AC amplifier with a low-cut filter of 0.08 Hz which eliminated the baseline fluctuation. This limited the duration of observation time. (B) Flash light (2 ms)-induced signals in BR under varying pH values. Measurements were carried out in a solution containing 400 mM NaCl and 1 mM 6-mixed buffer (citrate/MES/MOPS/HEPES/CHES/CAPS) adjusted to the desired pH with HCl or NaOH. BR indicates the purple membrane prepared by a standard method (Becher & Cassim, 1975). The 6-mixed buffer was used because of its almost constant buffer capacities in a wide pH range. Panel B was adapted with permission from Tamogami et al., 2009, *Photochem. Photobiol.*. Copyright 2009 The authors, Journal Compilation, The American Society of Photobiology.

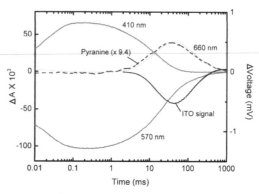

Fig. 4. Comparison between the flash-induced absorbance changes and proton-transfer signals in E194Q/E204Q[BR].

The red, blue and green lines represent absorbance changes of BR at 410, 570 and 660 nm, where the M-intermediate, unphotolyzed state and O-intermediate are mainly monitored, respectively. These traces were obtained by the flash photolysis spectroscopy performed by the procedure as described previously (Sato et al., 2003). The black broken and solid lines are the pyranine's signal (monitored at 450 nm) and ITO signal, respectively. Measurements of the absorbance changes of BR and pyranine were performed in the solution containing 400 mM NaCl plus 0.5 mM HEPES at pH 7.1 as described elsewhere (Tamogami et al, 2009). On the other hand, the ITO experiment was performed as shown in Fig. 3 in a solution containing 400 mM NaCl plus 1 mM 6-mixed buffer at pH 7.1. The protein sample was the purple membrane isolated from *H. salinarum* expressing this mutant.

9. Estimation of the p*K*a values of important residues involved in the photo-induced proton transfer of BR

It is a pronounced advantage of the ITO electrode method that the measurements can be performed over a wide pH region. Since the buffer capacity is kept constant for the measuring pH range by mixing the buffering agents and the detected pH changes are quite small due to the very faint amount of the adsorbed protein, the measured voltage changes are proportional to the numbers of protons moved by the photo-induced transfer. By analyzing these voltage changes, the pKa values concerned with the proton transfer can be estimated as described below. Figure 5A shows the peak magnitude of the voltage changes by BR (peak values in Fig. 3B) measured under different medium pH values. As shown in this figure, below pH ~ 5, the proton uptake was followed by the proton release. On the other hand, above pH ~ 5, the proton release was followed by the proton uptake. Corresponding to these two pH ranges, the pH profile consists of two bell-shaped functions having opposite signs. This suggests the contribution of the four pKa's of residues involved in the proton-transfer reactions. Then, this pH profile can be expressed by the following equation:

$$\Delta Voltage = -A\left(\frac{1}{1+10^{pKa_1-pH}}\right)\left(\frac{1}{1+10^{pH-pKa_2}}\right) + B\left(\frac{1}{1+10^{pKa_3-pH}}\right)\left(\frac{1}{1+10^{pH-pKa_4}}\right) \tag{1}$$

where A and B are constants used to adjust the magnitude of the response, and the four pKa values from pKa_1 to pKa_4 are assumed to increase in this order. A fitting analysis using Eq. 1 gave the following pKa values: $pKa_1 = 2.6$, $pKa_2 = 4.1$, $pKa_3 = 6.1$ and $pKa_4 = 9.0$. For BR, the pKa values involved in the photo-induced proton transfer had been estimated mainly by various spectroscopic measurements (Balashov, 2000). The pKa's determined by the ITO method agreed well with those previously reported. Therefore, the origins of the estimated pKa's were verified on the basis of previous reports. As a result, the pKa's from pKa_1 to pKa_4 were identified as the pKa's of Asp85[BR] in the dark, PRC in the O-intermediate, PRC in the M-intermediate and PRC in the dark, respectively. In addition to this analysis using the peak voltages, the rates of the voltage changes were also informative. The proton uptake by Asp96[BR] from the CP space, which coincides with N decay, becomes slow as the medium pH increases (see the traces at pH 6 and pH 9 in Fig. 3B). The rate constants of the proton uptake obeyed the Henderson-Hasselbalch equation with a single pKa, and the value of Asp96[BR] during N decay was estimated at 7.8 (Tamogami et al., 2009). This value also agreed with the previously reported value (Balashov, 2000). These pKa values determined by the ITO method are considered those of key amino acid residues for the proton pumping function of BR. Therefore this method is useful for detecting the proton transfer directly and deducing the important pKa values.

10. Application to other H$^+$-pumping rhodopsins

This ITO method has been successfully applied to the newly found H$^+$-pumping rhodopsins (Tamogami et al., 2009; Kikukawa et al, 2011). The panels B and C in Fig. 5 are the results for ARII from marine algae and PR from marine bacteria, respectively. Interestingly, the pH profiles of these three proton pumps are quite different. The prominent differences are as follows: (1) The pH profiles of the three rhodopsins commonly have bell-shaped negative peak areas, indicating that the proton uptake occurs prior to the release. However, the negative peak of PR is located at a quite higher pH (pH~8) than the other two rhodopsins (pH 3~4). This reflects the difference of the most acidic pKa's, governing the pH where the proton uptake starts to occur. These pKa's correspond to those of aspartates (Asp85[BR], Asp81[ARII] and Asp97[PR]), the counterions of the respective protonated Schiff bases. This pKa for PR is about 7, which is much higher than the pKa's of about 2.6 of the other two rhodopsins. This high pKa for PR reported by using spectroscopic methods (Dioumaev et al., 2002; Friedrich et al., 2002; Lakatos et al., 2003; Imasheva et al., 2004; Partha et al, 2005). Thus, we obtained the same result *via* direct measurements of the proton-transfer reactions. (2) As the pH increases, all three rhodopsins begin their proton releases prior to their proton uptakes.

The positive areas of the pH profiles correspond to this proton-transfer sequence. However, the starting pH's, which reflect the pKa's of the proton-releasing residues, are different. These pKa values are about 6.1 for BR, 8 for ARII and 10 for PR, respectively. The higher pKa's of ARII and PR might reflect the absence of residues constituting the PRC of BR. ARII lacks a residue corresponding to Glu194[BR], one of two glutamates constituting the PRC. By using the mutant of ARII, we confirmed that another glutamate, Glu199[ARII], which corresponds to Glu204[BR], functions as the proton-releasing residue (unpublished data). On the other hand, PR lacks both glutamates. For PR, therefore, an unknown residue works as the proton-releasing residue. (3) The pH profile of ARII has a surprising feature: the magnitude of the proton release again increases with a further increase in pH above 10. This indicates that a certain residue, other than Glu199[ARII], starts to work as the proton-releasing residue at this pH range. These observations suggest that, despite the identical function, these proton pumps possess partially different mechanisms. Thus various interesting phenomena have been discovered by the experiments using the ITO method.

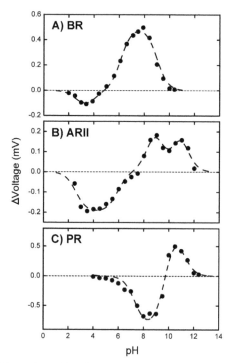

Fig. 5. Comparison of the pH profile of the photo-induced proton transfer among various microbial rhodopsins: (A) BR; (B) ARII; (C) PR.

The peak values of the photo-induced signals were plotted against the medium pH. The experimental conditions were identical to those described in Fig. 3. For BR, the purple membrane prepared by a standard method (Becher & Cassim, 1975) was used. ARII and PR were expressed in the cell-free system and *E. coli* expression system, respectively, and then they were reconstituted with Egg L-α-phosphatidylcholine. Panels A and C was adapted with permission from Tamogami et al., 2009, *Photochem. Photobiol.* Copyright 2009 The authors, Journal Compilation, The American Society of Photobiology. Panel B was adapted with permission from Kikukawa et al., 2011, *Biochemistry.* Copyright 2011 American Chemical Society.

11. Application to photosensing rhodopsins

This ITO method is also applicable to the proton transfers of photosensing rhodopsins. We have adopted this method for the archaeal sensory rhodopsins, NpSRII (Iwamoto et al., 1999), HsSRII (Tamogami et al., 2010) and a putative new class of photosensing rhodopsin called sensory rhodopsin III from *Haloarcular marismortui* (HmSRIII) (Nakao et al., 2011).

Consequently, we successfully determined the timings of the proton uptake/release during their respective photocycles. These results are attributed to the high sensitivity of this method as compared with an alternative method using a pH-sensitive dye. Most photosensing rhodopsins have slow photocycles (~sec). This slow turnover rate of the photocycle makes it difficult to adopt the pH-sensitive dye method for this measurement. The absorbance change due to the pH-sensitive dye is very small, and so a slight baseline fluctuation of the absorbance change results in a significant artifact. Especially for a long-term measurement corresponding to the slow photocycle, this baseline fluctuation becomes prominent.

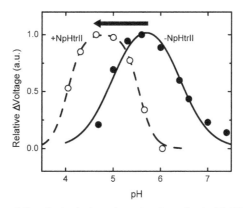

Fig. 6. The pH profiles of the photo-induced proton transfer in NpSRII in the presence and absence of NpHtrII$_{1-159}$.

The values of the data points at 10 ms after flash light (the duration of 4 ms) excitation were plotted against pH. The closed and open symbols are the plots in the absence and presence of NpHtrII$_{1-159}$, respectively. The NpSRIIs reconstituted with L-α-phosphatidylcholine were employed for the measurements. Added NpHtrIIs were truncated from the 1st to the 159th amino acid. The other experimental conditions were identical to those described in Fig. 3.

Adapted with permission from Iwamoto et al., 2004, *Biochemistry*. Copyright 2004 American Chemical Society.

For the three sensory rhodopsins examined so far, we confirmed that the rates of proton uptake reactions determined the decay of their respective M-intermediates, the putative signaling states. In addition to these, we also detected the alterations in the proton-transfer reactions of NpSRII by the complex formation with its cognate transducer, called NpHtrII (Iwamoto et al., 2004). Figure 6 shows the pH profile of the photo-induced proton transfer in NpSRII in the presence and absence of NpHtrII. As shown here, NpSRII releases a proton prior to the uptake at this pH range. From the mutation analyses, the proton-releasing residue was identified as Asp193[NpSRII] (corresponding to Glu204[BR]), which is located at the end of the EC channel. Then, the lower and higher pKa's governing the bell-shaped pH profile were attributed to the pKa's of Asp193[NpSRII] at the M-intermediate for release and at the dark state, respectively. As shown in this figure, the pH profile in the presence of NpHtrII shifts to a lower pH compared to that in the absence of NpHtrII, implying that the complex formation induces the conformational change of NpSRII and leads to the pKa changes of Asp193[NpSRII]. In the crystal structures, the significant conformational change around Asp193[NpSRII] was not observed by the complex formation (Gordeliy et al., 2002). Thus, the pKa's of the proton-transfer reactions could be responsive to a small perturbation of the conformation.

The bell-shaped pH profile of NpSRII, showing the first proton release, is shifted to acidic pH regions as compared with the positive peaks of the proton pumps shown in Fig. 5. This reflects the lower pKa's of Asp193[NpSRII], the proton-releasing residue. For NpSRII, the binding of Cl- to the vicinity of Asp193[NpSRII] was suggested by the ITO method (Iwamoto et al., 2004), ATR-FTIR measurement (Kitade et al., 2009) and the crystal structure (Royant et al., 2001), while this Cl- binding is not known for the proton pumps examined so far. The physiological meaning of the much lower pKa of Asp193[NpSRII] should be examined in a future study.

12. Conclusion and future perspectives

The electrochemical cell using ITO (or SnO$_2$) electrodes is a powerful and convenient device to detect the proton movements associated with photo-induced reactions of microbial rhodopsins and probably other photoactive pigments. Due to the high sensitivity and rapid response, this system enables us to follow the proton movements during a single photocycle under various buffer conditions. As described above, our current system cannot follow a reaction faster than 10 ms. The first proton movements of microbial rhodopsins appear to occur within 0.1-1 ms. Thus, the system response should be improved. On this point, we have already confirmed that the combination of nsec laser pulse with a homemade amplifier can accelerate the response to about 20 µs.

Early studies on the microbial rhodopsins concerned exclusively four rhodopsins in *H. salinarum*. BR in particular attracted much interest and was investigated in great detail. Thus, BR has been considered a kind of prototype of microbial rhodopsins. However, newly found microbial rhodopsins would seem to challenge the prototype status of BR, since they possess features not seen in BR, as described above. Microbial rhodopsins have been found in a wide variety of microorganisms living in various environments. Thus, it is reasonable to consider that the newly found rhodopsins acquired their original mechanisms to adapt to their living environments. In this report, we showed the pH-dependent proton movements of BR, ARII, PR and NpSRII (Figs. 5 and 6). Even though only four rhodopsins were considered, their pH dependences were quite different. This might reflect the mechanical divergence of microbial rhodopsins. In the future, a detailed analysis of each rhodopsin will certainly be important. This should be achieved by using various amino acid mutants. Moreover, the proton movements of many more microbial rhodopsins should be examined using this electrochemical cell. From these studies, we could obtain deeper insights into the mechanistic principles of individual rhodopsins.

13. References

Balashov, S.P.; Imasheva, E.S.; Ebrey, T.G.; Chen, N.; Menick, D.R. & Crouch, R.K. (1997). Glutamate-194 to cysteine mutation inhibits fast light-induced proton release in bacteriorhodopsin. *Biochemistry*, Vol.36, No.29, pp. 8671-8676, ISSN 0006-2960

Balashov, S.P. (2000). Protonation reactions and their coupling in bacteriorhodopsin. *Biochim. Biophys. Acta.*, Vol.1460, No.1, pp. 75-94, ISSN 0005-2728

Balashov, S.P.; Imasheva, E.S.; Boichenko, V.A.; Antón, J.; Wang, J.M. & Lanyi, J.K. (2005). Xanthorhodopsin: a proton pump with a light-harvesting carotenoid antenna. *Science*, Vol.309, No.5743, pp. 2061-2064, ISSN 0036-8075

Becher, B. & Cassim, J.T. (1975). Improved isolation procedures for the purple membrane of *Halobacterium halobium*. *Prep. Biochem.*, Vol.5, No.2, pp. 161-178, ISSN 0032-7484

Béjà, O.; Aravind, L.; Koonin, E.V.; Suzuki, M.T.; Hadd, A.; Nguyen, L.P.; Jovanovich, S.B.; Gates, C.M.; Feldman, R.A.; Spudich, J.L.; Spudich, E.N. & DeLong, E.F. (2000). Bacterial rhodopsin: Evidence for a new type of phototrophy in the sea. *Science*, Vol.289, No.5486, pp. 1902-1906, ISSN 0036-8075

Bogomolni, R.A. & Spudich J.L. (1982). Identification of a third rhodopsin-like pigment in phototactic *Halobacterium halobium*. *Proc. Natl. Acad. Sci. USA*, Vol.79, No.20, pp. 6250-6254, ISSN 0027-8424

Bogomolni, R.A.; Stoeckenius, W.; Szundi, I.; Perozo, E.; Olson, K.D. & Spudich, J.L. (1994). Removal of transducer HtrI allows electrogenic proton translocation by sensory rhodopsin I. *Proc. Natl. Acad. Sci. USA*, Vol.91, No.21, pp. 10188-10192, ISSN 0027-8424

Brown, L.S.; Sasaki, J.; Kandori, H.; Maeda, A.; Needleman, R. & Lanyi, J.K. (1995). Glutamic acid 204 is the terminal proton release group at the extracellular surface of bacteriorhodopsin. *J. Biol. Chem.*, Vol.270, No.45, pp. 27122-27126, ISSN 0021-9258

Dioumaev, A.K.; Richter, H-T.; Brown, L.S.; Tanio, M.; Tuzi, S.; Saito, H.; Kimura, Y.; Needleman, R. & Lanyi, J.K. (1998). Existence of a proton transfer chain in bacteriorhodopsin: Participation of Glu-194 in the release of protons to the extracellular surface. *Biochemistry*, Vol.37, No.8, pp. 2496-2506, ISSN 0006-2960

Dioumaev, A.K.; Brown, L.S.; Shih, J.; Spudich, E.N.; Spudich, J.L. & Lanyi, J.K. (2002). Proton transfers in the photochemical reaction cycle of proteorhodopsin. *Biochemistry*, Vol.41, No.17, pp. 5348-5358, ISSN 0006-2960

Essen L.O. (2002). Halorhodopsin: Light-driven ion pumping made simple? *Curr. Opinion Struct. Biol.*, Vol.12, No.4, pp. 516-522, ISSN 0959-440X

Friedrich, T.; Geibel, S.; Kalmbach, R.; Chizhov, I.; Ataka, K.; Heberle, J.; Engelhard, M. & Bamberg, E. (2002). Proteorhodopsin is a light-driven proton pump with variable vectoriality. *J.Mol.Biol.*, Vol.321, No. 5, pp. 821-838, ISSN 0022-2836

Gordeliy, V.I.; Labahn, J.; Moukhametzianov, R.; Efremov, R.; Granzin, J.; Schlesinger, R.; Büldt, G.; Savopol, T.; Scheidig, A.J.; Klare, J.P. & Engelhard, M. (2002). Molecular basis of transmembrane signalling by sensory rhodopsin II-transducer complex. *Nature*, Vol.419, No.6906, pp. 484-487, ISSN 0028-0836

Haupts, U.; Tittor, J. & Oesterhelt, D. (1999) Closing in on bacteriorhodopsin: Progress in understanding the molecule. *Annu. Rev. Biophys. Biomol. Struct.*, Vol.28, pp. 367-399, ISSN 1056-8700

Hazemoto, N.; Kamo, N.; Terayama, Y.; Kobatake, Y. & Tsuda, M. (1983). Photochemistry of two rhodopsinlike pigments in bacteriorhodopsin-free mutant of *Halobacterium halobium*. *Biophys. J.*, Vol.44, No.1, pp. 59-64, ISSN 0006-3495

Heberle, J. (2000). Proton transfer reactions across bacteiorhodopsin and along the membrane. *Biochim. Biophys. Acta.*, Vol.1458, No.1, pp. 135-147, ISSN 0005-2728

Imasheva, E.S.; Balashov, S.P.; Wang, J.M.; Dioumaev, A.K. & Lanyi, J.K. (2004). Selectivity of retinal photoisomerization in proteorhodopsin is controlled by aspartic acid 227. *Biochemistry*, Vol.43, No.6, pp. 1648-1655, ISSN 0006-2960

Iwamoto, M.; Shimono, K.; Sumi, M.; Koyama, K. & Kamo, N. (1999). Light-induced proton uptake and release of *pharaonis* phoborhodopsin detected by a photoelectrochemical cell. *J. Phys. Chem.B.*, Vol.103, No.46, pp. 10311-10315, ISSN 1520-6106

Iwamoto, M.; Hasegawa, C.; Sudo, Y.; Shimono, K.; Araiso, T. & Kamo, N. (2004). Proton release and uptake of *pharaonis* phoborhodopsin (sensory rhodopsin II) reconstituted into phospholipid. *Biochemisty*, Vol.43, No.11, pp. 3195-3203, ISSN 0006-2960

Jung, K.H.; Trivedi, V.D. & Spudich, J.L. (2003). Demonstration of a sensory rhodopsin in eubacteria. *Mol. Microbiol.*, Vol.47, No.6, pp. 1513-1522, ISSN 0950-382X

Kamo, N.; Shimono, K.; Iwamoto, M. & Sudo Y. (2001). Photochemistry and photoinduced proton-transfer by *pharaonis* phoborhodopsin. *Biochemistry (Moscow)*, Vol.66, No.11, pp. 1277-1282, ISSN 0006-2979

Kikukawa, T.; Shimono, K.; Tamogami, J.; Miyauchi, S.; Kim, S.-Y.; Kimura-Someya, T.; Shirouzu, M.; Jung, K.-H.; Yokoyama, S. & Kamo, N. (2011). Photochemistry of *Acetabularia* rhodopsin II from a marine plant, *Acetabularia acetabulum*. *Biochemistry*, Vol.50, No.41, pp. 8888-8898, ISSN 0006-2960

Kitade, Y.; Furutani, Y.; Kamo, N & Kandori, H. (2009). Proton release group of *pharaonis* phoborhodopsin revealed by ATR-FTIR spectroscopy. *Biochemistry*, Vol.48, No.7, pp. 1595-1603, ISSN 0006-2960

Kitajima-Ihara, T.; Furutani, Y.; Suzuki, D.; Ihara, K.; Kandori, H.; Homma, M. & Sudo, Y. (2008). *Salinibacter* Sensory Rhodopsin Sensory rhodopsin I-like protein from a eubacterium. *J. Biol. Chem.*, Vol.283, No.35, pp. 23533–23541, ISSN 0021-9258

Klare, J.P.; Gordeliy, V.I.; Labahn, J.; Büldt, G.; Steinhoff, H.J. & Engelhard, M. (2004). The archaeal sensory rhodopsin II/transducer complex: a model for transmembrane signal transfer. *FEBS Lett.*, Vol.564, No.3, pp. 219-224, ISSN 0014-5793

Koyama, K.; Yamaguchi, N. & Miyasaka, T. (1994). Antibody-mediated bacteriorhodopsin orientation for molecular device architectures. *Science*, Vol.265, No.5173, pp. 762-765, ISSN 0036-8075

Koyama, K.; Miyasaka, T.; Needleman, R. & Lanyi, J.K. (1998a). Photoelectrochemical verification of proton-releasing groups in bacteriorhodopsin. *Photochem. Photobiol.*, Vol.68, No.3, pp. 400-406, ISSN 0031-8655

Koyama, K.; Sumi, M.; Kamo, N & Lanyi, J.K. (1998b). Photoelectric response of halorhodopsin from *Natronobacterium pharaonis*. *Bioelectrochem. Bioenerg.*, Vol.46, No.2, pp. 289-292, ISSN 0302-4598

Lakatos, M.; Lanyi, J.K.; Szakács, J. & Váró, G. (2003). The photochemical reaction cycle of proteorhodopsin at low pH. *Biophys.J.*, Vol.84, No.5, pp. 3252-3256, ISSN 0006-3495

Lanyi, J.K. (2004). Bacteriorhodopsin. *Annu. Rev. Physiol.*, Vol.66, pp. 665-88, ISSN 0066-4278

Lanyi, J.K. (2006). Proton transfers in the bacteriorhodopsin photocycle. *Biochim. Biophys. Acta.*, Vol.1757, No.8, pp. 1012-1018, ISSN 0005-2728

Lee, K.A.; Kim, S.Y.; Choi, A.R.; Kim, S.H.; Kim, S.J. & Jung, K.-H. Expression of membrane protein and photochemical properties of two *Acetabularia* rhodopsins, *Abstract of 14th International Conference on Retinal Proteins*, P-23, Santa Cruz, August 2-6, 2010

Luecke, H.; Schobert, B.; Stagno, J.; Imasheva, E.S.; Wang, J.M.; Balashov, S.P. & Lanyi, J.K. (2008). Crystallographic structure of xanthorhodopsin, the light-driven proton pump with a dual chromophore. *Proc. Natl. Acad. Sci. USA*, Vol.105, No. 43, pp. 16561-16565, ISSN 0027-8424

Marwan, W. & Oesterhelt, D. (1987). Signal formation in the halobacterial photophobic response mediated by a four-th retinal protein (P480). *J. Mol. Biol.*, Vol.195, No.2, pp. 333-342, ISSN 0022-2836

Matsuno-Yagi, A. & Mukohata, Y. (1977) Two possible roles of bacteriorhodopsin; A comparative study of strains of *Halobacterium halobium* differing in pigmentation. *Biochem. Biophys. Res. Commun.*, Vol.78, No.1, pp. 237-243, ISSN 0006-291X

Miranda, M.R.M.; Choi, A.R.; Shi, L.; Bezerra Jr., A.G.; Jung, K.-H. & Brown, L.S. (2009). The photocycle and proton translocation pathway in a cyanobacterial ion-pumping rhodopsin. *Biophys. J.*, Vol.96, No.4, pp. 1471-1481, ISSN 0006-3495

Miyasaka, T. & Koyama, K. (1991). Photoelectrochemical behavior of purple membrane Langmuir-Blodgett films at the electrode-electrolyte interface. *Chem. Lett.*, Vol.20, No.9, pp. 1645-1648, ISSN 0366-7022

Miyasaka, T., Koyama, K., Itoh, I. (1992). Quantum conversion and image detection by a bacteriorhodopsin-based artificial photoreceptor. *Science*, Vol.255, No.5042, pp. 342-344, ISSN 0036-8075

Miyazaki, M.; Hirayama, J.; Hayakawa, M. & Kamo, N. (1992). Flash photolysis study on *pharaonis* phoborhodopsin from a haloalkaliphilic bacterium (*Natronobacterium pharaonis*). *Biochim. Biophys. Acta.*, Vol.1140, pp. 22-29, ISSN 0005-2728

Mukohata Y., Ihara K., Tamura T. & Sugiyama Y. (1999). Halobacterial rhodopsins. *J. Biochem.*, Vol.125, No.4, pp. 649-657, ISSN 0021-924X

Nagel, G.; Ollig, D.; Fuhrmann, M.; Mustl, A.M.; Bamberg, E. & Hegemann, P. (2002). Channelrhodopsin-1: a light-gated proton channel in green algae. *Science*, Vol.296, No.5577, pp. 2395-2398, ISSN 0036-8075

Nakao, Y.; Kikukawa, T.; Shimono, K.; Tamogami, J.; Kimitsuki, N.; Nara, T.; Unno, M.; Ihara, K. & Kamo, N. (2011). Photochemistry of a putative new class of sensory rhodopsin (SRIII) coded by *xop2* of *Haloarcular marismortui*. *J. Photochem. Photobiol. B.*, Vol.102, No.1, pp. 45-54, ISSN 1011-1344

Oesterhelt, D. & Stoecknius, W. (1971). Rhodopsin-like protein from the purple membrane of *Halobacterium halobium*. *Nat. New. Biol.*, Vol.233, No.39, pp. 149-152, ISSN 0090-0028

Partha, R.; Krebs, R.; Caterino, T.L. & Braiman, M.S. (2005). Weakened coupling of conserved arginine to the proteorhodopsin chromophore and its counterion implies structural differences from bacteriorhodopsin. *Biochim. Biophys. Acta.*, Vol.1708, No.1, pp. 6-12, ISSN 0005-2728

Pebay-Peyroula, E.; Royant, A.; Landau, E.M.; Navarro, J. (2002). Structural basis for sensory rhodopsin function. *Biochim. Biophys. Acta.*, Vol.1565, No.2, pp. 196– 205, ISSN 0005-2728

Robertson, B. & Lukashev, E.P. (1995). Rapid pH change due to bacteriorhodopsin measured with a tin-oxide electrode. *Biophys. J.*, Vol.68, No.4, pp. 1507-1517, ISSN 0006-3495

Royant, A.; Nollert, P.; Edman, K.; Neutze, R.; Landau, E.M.; Pebay-Peyroula, E. & Navarro, J. (2001). X-ray structure of sensory rhodopsin II at 2.1-Å resolution. *Proc. Natl. Acad. Sci. USA*, Vol.98, No.18, pp. 10131-10136, ISSN 0027-8424

Sasaki, J. & Spudich, J.L. (1998). The transducer protein HtrII modulates the lifetimes of sensory rhodopsin II photointermediates. *Biophys. J.*, Vol.75, No.5, pp. 2435-2440, ISSN 0006-3495

Sasaki, J. & Spudich, J.L. (1999). Proton circulation during the photocycle of sensory rhodopsin II. *Biophys. J.*, Vol.77, No.4, pp. 2145-2152, ISSN 0006-3495

Sasaki, J. & Spudich, J.L. (2000). Proton transport by sensory rhodopsins and its modulation by transducer-binding. *Biochim. Biophys. Acta.*, Vol.1460, No.1, pp. 230-239, ISSN 0005-2728

Sato, M.; Kikukawa, T.; Araiso, T.; Okita, H.; Shimono, K.; Kamo, N., Demura, M. & Nitta, K. (2003). Role of Ser130 and Thr126 in chloride binding and photocycle of *pharaonis* halorhodopsin. *J. Biochem.*, Vol.134, No.1, pp. 151-158, ISSN 0021-924X

Schmies, G.; Engelhard, M.; Wood, P.G.; Nagel, G. & Bamberg, E. (2001). Electrophysiological characterization of specific interactions between bacterial sensory rhodopsins and their transducers. *Proc. Natl. Acad. Sci. USA*, Vol.98, No.4, pp. 1555–1559, ISSN 0027-8424

Schobert, B. & Lanyi, J.K. (1982). Halorhodopsin is a light-driven chloride pump. *J. Biol. Chem.*, Vol.257, No.17, pp. 306-313, ISSN 0021-9258

Sineshchekov, O.A.; Jung, K.-H. & Spudich, J.L. (2002). Two rhodopsins mediate phototaxis to low- and high-intensity light in *Chlamydomonas reinhardtii*. *Proc. Natl. Acad. Sci. USA*, Vol.99, No.13, pp. 8689-8694, ISSN 0027-8424

Spudich, E.N.; Sundberg, S.A.; Manor D. & Spudich J.L. (1986). Properties of a second sensory receptor protein in *Halobacterium halobium* phototaxis. *Proteins*, Vol.1, No.3, pp. 239-246, ISSN 0887-3585

Spudich, J.L. & Bogomolni, R.A. (1984). Mechanism of color discrimination by a bacterial sensory rhodopsin. *Nature*, Vol.312, No.5994, pp. 509-513, ISSN 0028-0836

Spudich, J.L.; Yang, C.-S.; Jung K.-H. & Spudich, E.N. (2000). Retinylidene proteins: Structures and functions from archaea to humans. *Annu. Rev. Cell Dev. Biol.*, Vol.16, pp. 365-392, ISSN 1081-0706

Spudich, J.L. & Luecke, H. (2002). Sensory rhodopsin II: functional insights from structure. *Curr Opin Struct Biol.*, Vol.12, No.4, pp. 540-546, ISSN 0959-440X

Spudich, J.L, & Jung, K.-H. (2005). Microbial rhodopsin: Phylogenetic and functional diversity, In: *Handbook of photosensory receptors*, Briggs, W. R. & Spudich, J.L., Ed., pp. 1-23, Wiley-VCH Verlag, ISBN 3-527-31019-3, Weinheim

Sudo, Y.; Iwamoto, M.; Shimono, K.; Sumi, M. & Kamo, N. (2001). Photo-induced proton transport of *pharaonis* phoborhodopsin (sensory rhodopsin II) is ceased by association with the transducer. *Biophys. J.*, Vol.80, No.2, pp. 916-922, ISSN 0006-3495

Suzuki, T.; Yamasaki, K.; Fujita, S.; Oda, K.; Iseki, M.; Yoshida, K.; Watanabe, M.; Daiyasu, H.; Toh, H.; Asamizu, E.; Tabata, S.; Miura, K.; Fukuzawa, H.; Nakamura, S. & Takahashia, T. (2003). Archaeal-type rhodopsins in *Chlamydomonas*: Model structure and intracellular localization. *Biochem. Biophys. Res. Commun.*, Vol.301, No.3, pp. 711-717, ISSN 0006-291X

Takahashi, T.; Tomioka, H.; Kamo, N. & Kobatake, Y. (1985). A photosystem other than P370 also mediates the negative phototaxis of *Halobacterium halobium*. *FEMS Microbiol. Lett.*, Vol.28, pp. 161-164, ISSN 0378-1097

Tamogami, J.; Kikukawa, T.; Miyauchi, S.; Muneyuki, E. & Kamo, N. (2009). A tin oxide transparent electrode provides the means for rapid time-resolved pH measurements: Application to photoinduced proton transfer of bacteriorhodopsin and proteorhodopsin. *Photochem. Photobiol.*, Vol.85, No.2, pp. 578-589, ISSN 0031-8655

Tamogami, J.; Kikukawa, T.; Ikeda, Y.; Takemura, A.; Demura, M. & Kamo, N. (2010). The photochemical reaction cycle and photoinduced proton transfer of sensory rhodopsin II (phoborhodopsin) from *Halobacterium salinarum. Biophys. J.*, Vol.98, No.7, pp. 1353-1363, ISSN 0006-3495

Tomioka, H.; Takahashi, T.; Kamo, N. & Kobatake, Y. (1986). Flash spectrophotometric identification of a fourth rhodopsin-like pigment in halobacterium halobium. *Biochem. Biophys. Res. Commun.*, Vol.139, pp. 389-395, ISSN 0006-291X

Tsunoda, S.P.; Ewers, D.; Gazzarrini, S.; Moroni, A.; Gradmann, D. & Hegeman, P. (2006). H⁺-pumping rhodopsin from the marine alga *Acetabularia. Biophys. J.*, Vol.91, No.4, pp. 1471-1479, ISSN 0006-3495

Váró, G.; Brown, L.S.; Needleman, R. & Lanyi, J.K. (1996). Proton transport by halorhodopsin. *Biochemistry*, Vol.35, No.21, pp. 6604-6611, ISSN 0006-2960

Váró, G. (2000). Analogies between halorhodopsin and bacteriorhodopsin. *Biochim. Biophys. Acta.*, Vol.1460, No.1, pp. 220-229, ISSN 0005-2728

Váró, G.; Brown, L.S.; Lakatos, M. & Lanyi, J.K. (2003). Characterization of the photochemical reaction cycle of proteorhodopsin. *Biophys. J.*, Vol.84, No.2, pp. 1202-1207, ISSN 0006-3495

Wada T.; Shimono, K.; Kikukawa, T.; Hato, M.; Shinya, N.; Kim, S. Y.; Kimura-Someya, T.; Shirouzu, M.; Tamogami, J.; Miyauchi, S.; Jung, K.-H.; Kamo, N. & Yokoyama, S. (2011). Crystal structure of the eukaryotic light-driven proton pumping rhodopsin, *Acetabularia* rhodopsin II, from marine alga. *J. Mol. Biol.*, Vol.411, No.5, pp. 986-998, ISSN 0022-2836

Wang, J.-P.; Song, L.; Yoo, S.-K. & El-Sayed, M.A. (1997). A comparison of the photoelectric current responses resulting from the proton pumping process of bacteriorhodopsin under pulsed and CW laser excitations. *J. Phys. Chem.B*, Vol.101, No.49, pp. 10599-10604, ISSN 1520-6106

Waschuk, S.A.; Bezerra, A.G.; Shi, Jr.L. & Brown, L.S. (2005). *Leptosphaeria* rhodopsin: Bacteriorhodopsin-like proton pump from a eukaryote. *Proc. Natl. Acad. Sci. USA*, Vol.102, No. 19, pp. 6879-6883, ISSN 0027-8424

Wolff, E.K.; Bogomolni, R.A.; Scherrer P.; Hess B. & Stoeckenius W. (1986). Color discrimination in halobacteria: Spectroscopic characterization of a second sensory receptor covering the blue-green region of the spectrum. *Proc. Natl. Acad. Sci. USA*, Vol.83, No.19, pp. 7272-7276, ISSN 0027-8424

Wu, J.; Ma, D.; Wang, Y.; Ming, M.; Balashov, S.P. & Ding, J. (2009). Efficient approach to determine the pKa of the proton release complex in the photocycle of retinal proteins. *J. Phys. Chem.B*, Vol.113, No.13, pp. 4482–4491, ISSN 1520-6106

Zimányi, L.; Váró, G.; Chang, M.; Ni, B.; Needleman, R. & Lanyi, J.K. (1992). Pathways of proton release in the bacteriorhodopsin photocycle. *Biochemistry*, Vol.31, No.36, pp. 8535-8543, ISSN 0006-2960

7

Function of Extrinsic Proteins in Stabilization of the Photosynthetic Oxygen-Evolving Complex

Yong Li[1], Yukihiro Kimura[1,2], Takashi Ohno[1] and Yasuo Yamauchi[1]
[1]Graduate School of Agricultural Science, Kobe University
[2]Organization of Advanced Science and Technology, Kobe University
Japan

1. Introduction

Oxygenic phototrophs convert photon energy into chemical energy through a series of light-induced electron transfer reactions initiated with charge separation of chlorophyll (Chl) special pairs located in the central part of photosystem I and II (PSI and PSII) (Fig. 1). The reducing power is transferred from PSII to PSI through cytochrome b_6f, and finally utilized for reduction of $NADP^+$ to assimilate CO_2. The oxidized equivalents accumulated on the PSII donor side are neutralized by substrate water molecules to release protons for driving ATP synthase and O_2 molecules as a by-product. This water oxidation takes place in the oxygen-evolving complex (OEC) of PSII [McEvoy & Brudvig, 2006, Renger & Renger, 2008]. The OEC assembly is largely similar between cyanobacteria and higher plants, except for a critical difference in the composition of extrinsic proteins [Roose *et al.*, 2007]. In cyanobacteria, PsbO, PsbV, and PsbU residing on the lumenal side of PSII play significant roles in the regulation and stabilization of the water oxidation machinery. Higher plants possess major nuclear gene-encoded extrinsic proteins named PsbO, PsbP, and PsbQ. PsbO is a common extrinsic protein highly conserved among the oxygenic phototrophs. PsbP and PsbQ are indigenous to plant PSII and have been proposed as the functional equivalents of PsbV and PsbU in bacterial PSII, having replaced them during the course of evolution from ancestral cyanobacteria to higher plants. These proteins play significant roles in the regulation and stabilization of the photosynthetic water oxidation [Roose *et al.*, 2007, Seidler, 1996, Williamson, 2008] although the details of their function(s) are still a matter of debate.

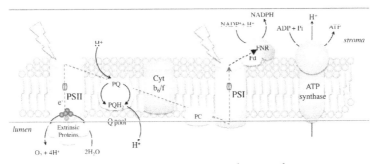

Fig. 1. Reaction scheme of photosynthesis in oxygenic phototrophs.

In this chapter, we describe the structural-functional roles of extrinsic proteins in the plant PSII. The effects of extrinsic proteins on the photosynthetic function of the Mn_4Ca cluster and the structural stability of the OEC core complex were investigated by spectroscopic and biochemical analyses. Based on the results presented here and reported previously, the structural and functional roles of extrinsic proteins in the regulation and stabilization of photosynthetic functions are discussed.

2. General structure and function of oxygenic phototrophs

2.1 Oxygen-evolving complex

Photosynthetic oxygen evolution occurs in the PSII OEC which is composed of a heterodimer of D1 (psbA) and D2 (psbD) proteins associated with two chlorophyll proteins (CP), CP47 (psbB) and CP43 (psbC), and involves a catalytic Mn_4Ca cluster located on the lumenal side of PSII (Fig. 2). These are highly conserved from cyanobacteria to higher plants to preserve the essential function of oxygenic phototrophs. The oxidized equivalents accumulated on the cluster and/or its ligands are reduced by electrons provided from a splitting reaction of substrate water molecules through a light-driven S-state cycle with five intermediate states S_n (n = 0 – 4), where n denotes the number of oxidizing equivalents stored. The OEC advances from the thermally stable S_1 state to the next oxidation state in a stepwise manner by absorbing each photon and attains the highest oxidation state S_4, followed by relaxation to the lowest oxidation state S_0 concurrent with a release of one oxygen molecule [Joliot et al., 1969, Kok et al., 1970]. Two water molecules are converted to one oxygen molecule by the OEC concurrent with release of four protons. Calcium and chloride ions are indispensable inorganic cofactors for playing functional and structural roles in the OEC [Debus, 1992, Yocum, 2008].

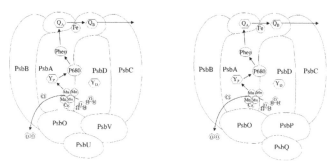

Fig. 2. OEC assemblies of cyanobacteria (left) and higher plants (right).

In the past decade, X-ray crystallography has revealed the structures of cyanobacterial PSII at resolutions of 3.8 Å to 2.9 Å [Ferreira et al., 2004, Guskov et al., 2009, Kamiya & Shen, 2003, Loll et al., 2005, Zouni et al., 2001]. A very recent structural model at the atomic resolution level has revealed details of the ligation structure of the Mn_4Ca cluster [Umena et al., 2011]. The Mn ions are bridged by several oxygen atoms and coordinated by water molecules as well as by Asp, Glu, Ala, and His residues from PsbA and/or PsbC proteins. In contrast, high-resolution structural analysis of higher plants has been delayed due to the instability of the membrane protein complex. The visualization of plant PSII structures has been limited to electron micrographs at low resolutions [Nield *et al.*, 2002, Nield & Barber, 2006]. Yet the findings to date strongly indicate that the structure and function of the PSII core assembly are almost identical to those of its prokaryotic counterparts, except for a critical difference in the composition of extrinsic proteins, which may provide valuable insights into the evolution of photosynthetic organisms [De Las Rivas *et al.*, 2004]. Based on the crystallographic structure of the OEC, several possible pathways for water, proton, and O_2 channels were proposed [Gabdulkhakov *et al.*, 2009, Guskov *et al.*, 2009]. However, photosynthetic water oxidation is a complex process that involves S-state cycling with five intermediate states, and therefore, the reaction mechanisms are not yet fully understood.

2.2 Extrinsic proteins in PSII

Higher plants possess gene-encoded extrinsic proteins, including PsbP, PsbQ, and PsbR, as well as PsbO, which commonly exists in all oxygenic phototrophs. These proteins play a key role in maintaining oxygen-evolving activity at physiological rates [Roose *et al.*, 2007, Williamson, 2008]. PsbO independently associates with the PSII core [Miyao & Murata, 1983, Miyao & Murata, 1989], and with PsbP through electrostatic interactions with PsbO [Miyao & Murata, 1983, Tohri *et al.*, 2004]. PsbQ requires both PsbO and PsbP for its binding [Miyao & Murata, 1983, Miyao & Murata, 1989]. In contrast, PsbO and PsbV independently bind to the PSII core, which lacks extrinsic proteins [Shen & Inoue, 1993], and the full binding of PsbU requires both PsbO and PsbV [Shen *et al.*, 1995, Shen & Inoue, 1993].

PsbP and PsbQ are thought to be the respective functional equivalents of PsbV and PsbU in the bacterial PSII [Enami *et al.*, 2005, Shen & Inoue, 1993], despite their low structural homology between PsbP(Q) and PsbV(U) [Balsera *et al.*, 2005, Ifuku *et al.*, 2004]. A phylogenetic study indicated that PsbP and PsbQ in the plant PSII were derived from PsbP and PsbQ homologues, respecively, in bacterial PSII [Thornton *et al.*, 2004], through intensive genetic modification during endosymbiosis and subsequent gene transfer to the host nucleus [De Las Rivas & Roman, 2005, Ifuku *et al.*, 2008, Ishihara *et al.*, 2007].

2.2.1 PsbO

PsbO is the most important protein for stabilization of the Mn_4Ca cluster, and therefore, it is common in all oxygenic phototrophs. The release of PsbO induces release of Mn ions from the cluster, resulting in the loss of O_2-evolving activity. The PsbO protein is common in every oxygenic phototroph but in varying proportions: one PsbO per PSII in cyanobacteria and two PsbO per PSII in higher plants [Williamson, 2008, Xu & Bricker, 1992]. High-resolution X-ray crystallographs of the PsbO protein associated with the PSII core are available for *Thermosynechococcus elongatus* [Ferreira *et al.*, 2004, Guskov *et al.*, 2009] and *Thermosynechococcus vulcanus* [Kawakami *et al.*, 2009, Umena *et al.*, 2011], in which PsbO is comprised of a β-barrel core with an extended α-helix domain. In contrast, the structural analysis of plant PsbO has been delayed and is limited to low-resolution images [Nield & Barber, 2006]. PsbO proteins are believed to play significant roles in protecting and stabilizing the catalytic center, however, none of the amino acid residue from PsbO serves as a direct ligand for the Mn_4Ca cluster.

Findings to date strongly indicate another significant role for PsbO: it is thought to modulate Ca^{2+} and Cl^- requirements for O_2 evolution [Seidler, 1996, Williamson, 2008]. However, the Ca^{2+}-binding properties of PsbO proteins are somewhat different between higher plants and cyanobacteria. It has been reported that plant PsbO induces structural changes upon the binding of Ca^{2+} [Heredia & De Las Rivas, 2003, Kruk et al., 2003], which is not the functional Ca^{2+} necessary for the water oxidation [Seidler & Rutherford, 1996]. In contrast, no significant Ca^{2+}-induced stuctural change was found in cyanobacterial PsbO [Loll et al., 2005], although it has been speculated that this protein serves as a low-affinity binding site for functional Ca^{2+} [Murray & Barber, 2006, Rutherford & Faller, 2001].

2.2.2 PsbP and PsbV

PsbP is also indispensable for the regulation and stabilization of PSII in higher plants [Ifuku et al., 2008]. Deletion of this protein disables the normal functions of the plant PSII [Ifuku et al., 2005]. This protein is related to the stability of the Mn_4Ca cluster as well as to the binding affinity of Ca^{2+} and Cl^- ions, which are essential cofactors for water oxidation reactions [Seidler, 1996]. Although it is unclear whether PsbP proteins directly interact with the Mn_4Ca cluster, the binding of this protein to the PSII core in the absence of Ca^{2+} is known to cause modification of its physicochemical properties, including redox potentials, magnetic structures, and ligation geometries of the Mn_4Ca cluster. It has also been speculated that PsbP is a metal-binding protein which reserves Mn^{2+} or Ca^{2+} to keep or deliver it to the apo-PSII [Bondarava et al., 2005].

PsbV is also thought to be involved with the binding of inorganic factors. PsbV-lacking mutants were unable to grow photoautotrophically in the absence of Ca^{2+} or Cl^- [Shen et al., 1998]. In contrast to PsbP, PsbV was capable of supporting water oxidation even in the absence of PsbO [Shen et al., 1995], suggesting that it serves to maintain a proper ion environment within the OEC for optimal oxygen-evolving activity [Nishiyama et al., 1994, Shen et al., 1995, Shen & Inoue, 1993, Shen et al., 1998, Shen et al., 1995]. Although none of residue from PsbV serves as a direct ligand for the Mn_4Ca cluster, it has been suggested that this protein participates in stabilizing PsbA through electrostatic interactions [Sugiura et al., 2010].

2.2.3 PsbQ and PsbU

The roles of PsbQ protein in photosynthetic functions are not yet fully understood. The PsbQ protein is not necessary for normal growth in higher plants [Ifuku et al., 2005, Yi et al., 2006]. However, this protein is required for photoautotrophic growth under low-light conditions [Yi et al., 2006] and it is involved in the binding of functional Cl ions [Balsera et al., 2005]. In cyanobacteria, PsbU-lacking mutants were capable of photoautotrophic growth in the absence of Ca^{2+} or Cl^-, but at reduced rates [Shen et al., 1998]. The oxygen-evolving ability was reduced by the removal of PsbU and restored in part by the addition of Cl^- but not Ca^{2+}, indicating that PsbU regulates Cl^- requirement [Inoue-Kashino et al., 2005, Shen et al., 1997]. Another of its functions is the suppression of light-induced D1 degradation [Inoue-Kashino et al., 2005]. Protection of the PSII core from reactive oxygen species (ROS) [Balint et al., 2006] has also been proposed.

3. Interaction of extrinsic proteins with the Mn₄Ca cluster in the OEC

3.1 PsbP and PsbQ

3.1.1 Depletion/reconstitution of extrinsic proteins and functional Ca^{2+}

In this study, we used two types of PSII membranes lacking functional Ca^{2+} with or without PsbP and PsbQ. The sample preparation method is shown in Fig. 3. Berthold-Babcock-Yocum (BBY)-type PS II membranes (untreated PSII, A) were prepared from spinach according to the method described previously [Ono et al., 2001]. The O_2-evolving activity was ~550 μmoles of O_2/mgChl/h. For depletion of Ca^{2+}, PsbP, and PsbQ, the membranes were suspended in medium A (2 M NaCl, 10 mM Mes/NaOH, and pH 6.5) at 0.5 mg of Chl per ml and gently stirred on ice under weak light (10 μmol/s/m²) for 30 min. Next, the following procedures were carried out in complete darkness or dim green light unless otherwise noted: EDTA was added to the suspension to achieve a final concentration of 1 mM, followed by 10-min incubation in the dark. The suspension was centrifuged and extensively washed with Chelex-treated medium B (400 mM sucrose, 20 mM NaCl, 20 mM Mes/NaOH, and pH 6.5) to yield PSII membranes depleted of Ca^{2+}, PsbP, and PsbQ (ExCa²⁺-depleted PSII, B). For depletion of PsbP and PsbQ proteins, PS II membranes were suspended in medium A at 0.5 mg of Chl per ml, and gently stirred on ice in darkness for 30 min. The extracted PsbP and PsbQ proteins were reconstituted into the NaCl/EDTA-treated PSII to obtain Ca²⁺-depleted PSII (C).

Alternatively, the PSII membranes were washed with medium C (400 mM sucrose, 20 mM NaCl, 0.1 mM Mes-NaOH, and pH 6.5) and then treated with medium D (400 mM sucrose, 20 mM NaCl, 40 mM citrate-NaOH , pH 3.0) at 2 mg of Chl per ml. After 5 min incubation on ice in darkness, the suspension was diluted with medium D (400 mM sucrose, 20 mM NaCl, 500 mM Mops-NaOH, pH 7.5), and incubated for 10 min to facilitate the rebinding of extrinsic proteins. Then, the sample was washed with medium E (400 mM sucrose, 20 mM NaCl, 40 mM Mes/NaOH, 0.5 mM EDTA, pH6.5) to obtain PSII membranes depleted of only Ca²⁺ (lowpH-treated PSII ,D). Finally, the resulting low-pH-treated PSII membranes were treated with medium A to produce PSII membranes depleted of both Ca²⁺ and extrinsic proteins (ExCa²⁺-depleted PSII, E).

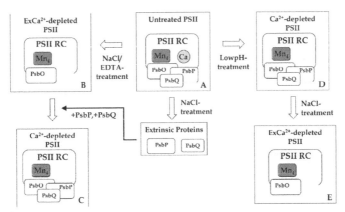

Fig. 3. Schematic representation for each sample preparation.

3.1.2 Effects of extrinsic proteins on the properties of the OEC

The PSII membranes prepared by the different methods were assessed by O_2-evolving activity and Fv/Fm values of chlorophyll fluorescence and the resulting data are summarized in Table 1. The O_2-evolving rate of the untreated PSII was decreased to 17% when PsbP, PsbQ and Ca^{2+} were depleted by the NaCl/EDTA treatments (B). The decreased activity was restored to 83% by reconstituting Ca^{2+}, as reported previously [Kimura & Ono, 2001, Ono et al., 2001]. However, the addition of PsbP and PsbQ to the ExCa^{2+}-depleted PSII in the absence of Ca^{2+} lowered the O_2-evolving rate to ~0% (C). Furthermore, the O_2-evolving activity was almost completely lost upon Ca^{2+} depletion by the low-pH treatment (D) but restored to 79% by adding Ca^{2+}. Notably, the lost activity was partially restored by the further depletion of PsbP and PsbQ to 25% (E). These results indicate that PsbP and PsbQ proteins completely suppress O_2 evolution in the absence of functional Ca^{2+}. Similar effects are also evident in the chlorophyll fluorescence measurements: the Fv/Fm values were much lower in the Ca^{2+}-depleted PSII (45%) than in the ExCa^{2+}-depleted PSII (64%), and both values were recoverd to ~80% after the supplementation with Ca^{2+}. In the Ca^{2+}-depleted PSII, the partial recovery to 68% was induced by the following depletion of the extrinsic proteins. Since Fv/Fm values are related to O_2-evolving activity, this strongly suggests that the functions of the OEC are disturbed by the extrinsic proteins in the absence of Ca^{2+}.

PSII preparation	Additives	O$_2$-evolving activity	Fv/Fm	FTIR S$_2$/S$_1$ carboxylate bands	Thermo-luminescence Q-band (°C)	S$_2$ EPR multiline signal
Untreated PSII (A)	No addition	100%	100%	Normal[b]	Normal[e,f]	Normal[f]
ExCa^{2+}-depleted PSII (B)	No addition	17%	64%	Normal[b]	Normal[b,d,e]	Normal[d]
	+Ca^{2+}	83%	79%	Normal[b]	Normal[b,d,e]	Normal[d]
	+PsbP, +PsbQ (C)	~0%	---[a]	---[a]	Abnormal[e]	Modified
Ca^{2+}-depleted PSII (D)	No addition	~0%	45%	Abnormal[c]	Abnormal[e,f]	Modified[f]
	+Ca^{2+}	79%	81%	Normal[c]	Normal[e,f]	Normal[f]
	-PsbP, -PsbQ (E)	25%	68%	---[a]	Normal[e]	Normal

[a]No data, [b][Kimura & Ono, 2001], [c][Noguchi et al., 1995], [d][Ono et al., 2001], [e][Ono et al., 1992], [f][Ono & Inoue, 1989].

Table 1. Effects of Ca^{2+} and extrinsic proteins (PsbP and PsbQ) on the properties of the OEC.

Next, the effects of extrinsic proteins and Ca^{2+} on the thermal stability of the OEC were examined. Fig. 4 shows the relative absorbance at 680 nm of the untreated control PSII (circle), ExCa^{2+}-depleted (triangle), and Ca^{2+}-depleted (square) PSII membranes during incubation at 50°C. The relative band intensity of the control PSII remained at ~85% after 64 min incubation, but was slightly decreased to ~75% in the ExCa^{2+}-depleted PSII and was markedly decreased to 50% in the Ca^{2+}-depleted PSII. This is consistent with the effects seen in the O_2-evolving activity and Fv/Fm values. These results strongly support the idea that PsbP and PsbQ lower the structural stability and disturb the normal functioning of the OEC in the absence of Ca^{2+}.

The present findings are largely in agreement with previous findings, as shown in Table 1. FT-IR spectroscopy provides valuable information on the structure and interactions within the OEC. The ligation geometry around the Mn_4Ca cluster is mostly similar between untreated and $ExCa^{2+}$-depleted PSII, at least in the S_1- and S_2-states [Kimura & Ono, 2001]. However, Ca^{2+}-depleted PSII exhibited marked deterioration in the carboxylate bands, which are thought to be from putative amino acid residues coordinating to the Mn_4Ca cluster [Noguchi et al., 1995]. Furthermore, the redox potential of the Mn_4Ca cluster has been reported to be abnormal when the extrinsic proteins bound to the PSII core in the absence of Ca^{2+}, as indicated by elevated peak temperatures of the thermoluminescence band for the

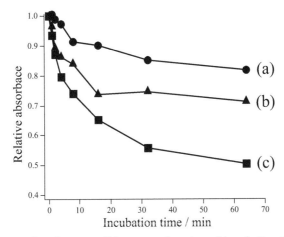

Fig. 4. Plots of relative absorbance at 680 nm as a function of incubation time at 50°C for (a) untreated, (b) $ExCa^{2+}$-depleted, and (c) Ca^{2+}-depleted PS II membranes.

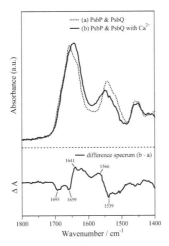

Fig. 5. ATR-FTIR spectra of isolated PsbP and PsbQ proteins in the absence (a, dotted line) and presence of Ca^{2+} (b, solid line). The difference spectrum obtained by subtracting spectrum a from spectrum b is shown in the lower panel.

$S_2Q_A^-$ recombination [Ono & Inoue, 1989, Ono et al., 1992]. Additional support for this view was obtained from electron paramagnetic resonance (EPR) studies which demonstrated abnormal magnetic structures of PSII lacking Ca^{2+} but retaining the extrinsic proteins as revealed by modified S_2-state multiline signals. These results are largely compatible with the present findings that the appropriate binding of extrinsic proteins in the presence of functional Ca^{2+} is required for the normal functioning of the OEC.

To understand function of these extrinsic proteins in the OEC, structural changes of the PsbP and PsbQ proteins induced by Ca^{2+} were observed by ATR-FTIR spectroscopy. Fig. 5 shows ATR-FTIR spectra of isolated PsbP and PsbQ (spectrum a) and those supplemented with Ca^{2+} (spectrum b). The control spectrum a exhibited characteristic bands for amide I (1700 – 1600 cm^{-1}) and amide II (1600 – 1500 cm^{-1}) vibrational modes from backbone polypeptides of the OEC. These bands were significantly modified when Ca^{2+} was added to the extrinsic proteins as can be clearly seen in the difference spectrum (lower part of Fig. 5). The IR bands at 1693, 1659 and 1539 cm^{-1} are decreased and new bands are visible at 1641 and 1566 cm^{-1}, strongly indicating that PsbP and/or PsbQ are metal-binding proteins that alter their secondary structures upon the binding of Ca^{2+}. Similar structural changes were evident in the spectrum of the purified PsbP protein (data not shown). Although high-resolution crystallographic studies have revealed the structure of the PsbP protein in *Nicotiana tabacum* [Ifuku et al., 2004], this protein lacks the N-terminal region which are thought to contain the Ca^{2+}-binding site, and therefore, the relationship between PsbP and Ca^{2+} remains unclear [Ifuku & Sato, 2002]. However, the authors of a previous study hypothesized that PsbP acts to reserve Mn^{2+} or Ca^{2+} ions [Bondarava et al., 2005]. These results strongly support the idea that the PsbP protein is a metal-binding protein that directly and/or indirectly interacts with the catalytic center of the OEC in the absence of sufficient Ca^{2+}.

3.2 PsbO

The most important physiological role of PsbO is to stabilize the binding of the Mn_4Ca cluster, which is essential for oxygen-evolving activity [Debus, 2001]. The PsbO protein can be dissociated from the PSII by a variety of chemical treatments including washing with alkaline Tris buffer, a high concentration of $CaCl_2$, and chaotropic agents [Enami et al., 1994, Ghanotakis & Yocum, 1990]. In particular, Lys residue-modifying chemicals such as *N*-succinimidyl propionate and 2,4,6-trinitrobenzene sulfonic acid caused release of PsbO from PSII and loss of oxygen-evolving activity [Miura et al., 1997], suggesting that the positive charge of Lys is important for the electrostatic interaction between PsbO and PSII. Alternatively, the release of PsbO can be caused by thermal denaturation. However, PsbO itself is a thermostable protein [Lydakis-Simantiris et al., 1999], and therefore, other factors might also be responsible for the release of PsbO as described later in this chapter.

Several spectroscopic studies using isolated PsbO reported different Ca^{2+}-binding properties between higher plants and cyanobacteria. It has been suggested that plant PsbO can bind Ca^{2+}, which induces slight changes in secondary structure from a β-sheet to a loop or nonordered structure, and facilitated the association of PsbO with the PSII core [Heredia & De Las Rivas, 2003, Kruk et al., 2003]. However, an EPR study indicated that the functional Ca^{2+} ion was not involved in the binding to PsbO [Seidler & Rutherford, 1996]. In cyanobacteria, PsbO does not bind Ca^{2+}, at least before the protein associates with the PSII core, since no significant conformational change upon the Ca^{2+}-binding was induced in isolated PsbO [Loll *et al.*, 2005]. In contrast, the low-affinity Ca^{2+}-binding site in PsbO located at the luminal exit of the proton channel has been suggested to be responsible for water oxidation [Murray & Barber, 2006, Rutherford & Faller, 2001]. These results strongly indicate that the structural-functional role of PsbO is not identical between higher plants and cyanobacteria. Interestingly, thermal stability was enhanced when plant PsbO proteins were replaced with thermally stable homologues from thermophilic *Phormidium laminosum* [Pueyo *et al.*, 2002]. Therefore, slight variation in the primary structure and/or the protein folding pattern is possibly responsible for the difference in thermal stability of PsbO between higher plant and thermophilic cyanobacteria.

4. Protective role of extrinsic proteins in regulation and stabilization of photosynthetic functions

4.1 PsbP and PsbQ

The present study revealed that PsbP significantly affects the structure and function of the Mn_4Ca cluster in the OEC only in the absence of sufficient Ca^{2+} in the OEC. This result is compatible with the previous analyses that involved FT-IR, thermoluminescence, and EPR spectoroscopies [Kimura & Ono, 2001, Noguchi *et al.*, 1995, Ono *et al.*, 2001, Ono & Inoue, 1989, Ono *et al.*, 1992]. In addition, it has been reported that PsbP has Ca^{2+}-binding sites in the N-terminal region [Ifuku & Sato, 2002] and functions as a reserver of Mn^{2+} or Ca^{2+} ions to supply them as needed by the impaired OEC [Bondarava *et al.*, 2005]. Therefore, it is possible that the PsbP completely eliminates functional Ca^{2+} or interacts with the Mn_4Ca cluster directly and/or indirectly to inhibit the O_2-evolving activity and modify the ligation geometry, redox potentials and magmetic structures of the Mn_4Ca cluster.

It has been suggested that normal functioning of PSII requires 15 highly conserved residues in the N-terminal region of the PsbP protein as well as the PsbQ protein for retention of functional Ca^{2+} [Ifuku *et al.*, 2005]. A recent FTIR study indicated that the PsbP protein, but not the PsbQ protein, has an effect on S_2/S_1 conformational changes of the intrisic polypeptide backbone around the Mn_4Ca cluster through the N-terminal region of the PsbP [Tomita *et al.*, 2009]. In addition, little change was found in characteristic carboxylate stretching modes from putative amino acid ligands for the Mn_4Ca cluster in the presence of Ca^{2+} when PsbP and PsbQ were depleted by NaCl washing, or all the extrinsic proteins were eliminated by $CaCl_2$ washing [Tomita *et al.*, 2009]. Based on these results, it is possible that the PsbP protein interacts with intrinsic proteins, which may be closely related to the Mn_4Ca cluster, and preserves the OEC functions appropriately in the presence of Ca^{2+}, but modifies the properties of the cluster directly and/or indirectly through intrinsic proteins in the absence of Ca^{2+}.

It is intriguing to note that PsbV in cyanobacteria exhibits functional similarity with PsbP in higher plants, although their primary and 3D crystallographic structures are largely different [Ifuku *et al.*, 2004, Kerfeld *et al.*, 2003]. The apparent inconsistency in the structural-functional consequence may reflect the fact that PsbP and PsbV in plant and cyanobacterial PSII are not involved in specific interactions between the protein and the Mn_4Ca cluster, but serve to maintain indispensable inorganic cofactors in the proximity of the cluster and to protect it from invasion. Additionally, PsbQ and PsbU also play a key role for tuning O_2-

evolving activity and enahncing structural stability through the interaction with PsbP and PsbV, respectively [Nishiyama et al., 1997, Nishiyama et al., 1999].

4.2 PsbO

In higher plants, PSII is much more susceptible to high temperatures than PSI [Berry & Bjorkman, 1980]. The thermal stability of the PSII core is closely related to the acquisition of cellular thermal tolerance in oxyphototrophs. The thermosensitivity of oxygen evolution in higher plants has been studied through simple experiments using PSII particles or isolated thylakoid membranes. Previous in-vivo and in-vitro studies have estimated the heat-labile properties of the OEC [Berry & Bjorkman, 1980, Havaux & Tardy, 1996, Mamedov et al., 1993]. These studies demonstrated that the release of PsbO occurs first, followed by liberation of two of the four Mn ions from the Mn_4Ca cluster of the OEC [Enami et al., 1998, Enami et al., 1994, Nash et al., 1985], and finally by the loss of oxygen evolution at high temperatures [Enami et al., 1994, Yamane et al., 1998].

Another form of damage to the physiological function of the PSII can be caused by reactive oxygen species (ROS) generated under high light conditions. The D1 proteins are degraded by the ROS species and inhibited in their ability to repair the photodamaged PSII by suppressing the synthesis of D1 proteins [Murata et al., 2007]. The ROS species are thought to arise from heat-induced inactivation of a water-oxidizing manganese complex and through lipid peroxidation [Yamashita et al., 2008]. On the other hand, saturation of polyunsaturated fatty acids (PUFAs) contributes to the acquisition of heat tolerance of photosynthesis by altering physicochemical properties [Alfonso et al., 2001, Murakami et al., 2000, Thomas et al., 1986]. The increased saturation of PUFAs raises the temperature at which lipids phase-separate into non-bilayer structures, providing the proper assembly and dynamics of PSII tolerant to higher temperatures [Alfonso et al., 2004].

Recently, we published biochemical evidence that the biological effect of reactive carbonyls such as malondialdehyde (MDA) and acrolein is greatly enhanced under heat-stressed conditions. [Yamauchi & Sugimoto, 2010]. PsbO is one of the proteins most frequently modified by MDA, which is an end-product of peroxidized polyunsaturated fatty acids. Detailed biochemical experiments indicated that the modification of PsbO by MDA affects its binding to the PSII complex and causes inactivation of the OEC (a schematic diagram is shown in Fig. 6). Purified PsbO and PSII membranes, from which extrinsic proteins had been eliminated, of the oxygen-evolving complex (PSIIΔOEE) of spinach were separately treated with MDA. The binding was diminished when both PsbO and PSIIΔOEE were modified, but when only PsbO or PSIIΔOEE was treated, the binding was not impaired. In an experiment using thylakoid membranes, the release of PsbO from PSII and a corresponding loss of oxygen-evolving activity were observed when thylakoid membranes were treated with MDA at 40°C but not at 25°C. In spinach leaves treated at 40°C under light, the maximum efficiency of PSII photochemistry (Fv/Fm ratio of chlorophyll fluorescence) and oxygen-evolving activity decreased. Simultaneously, the MDA content of the heat-stressed leaves increased, and PsbO and PSII core proteins (including 47 kDa and 43 kDa chlorophyll-binding proteins) were modified by MDA. In contrast, these changes were less profound when these experiments were performed at 40°C in the dark. Thus, MDA modification of PSII proteins likely causes the release of PsbO from PSII, an effect that is particuarly marked in heat and oxidative conditions.

First, ROS attack trienoic fatty acids in thylakoid membranes, resulting in the generation of MDA. MDA attaches to critical Lys residues of PsbO and PsbB (CP47) for the interaction between PsbO and PSII in a temperature-dependent manner. When both sides of PsbO and PSII are modified by MDA, PsbO is released from PSII. Finally, the Mn_4Ca cluster is spontaneously released from PSII, causing loss of oxygen-evolving activity.

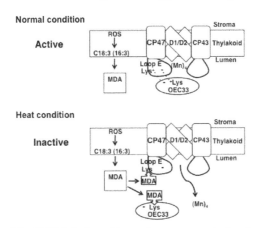

Fig. 6. A schematic model of MDA-induced loss of oxygen evolution in heat-stressed spinach PSII complexes.

5. Conclusion

In this article, we focused on the structural and functional roles of extrinsic proteins in the plant PSII. Since PSII is an integrated pigment-protein complex embedded in plant membranes, the structures and interactions of these extrinsic proteins in the membrane interface are of significance for protecting the RC. This involves protecting the Mn_4Ca cluster from exogenous invasion and/or alteration of physiological conditions. However, based on the results presented here and reported previously, we consider it very likely that the extrinsic protein itself is also responsible for the deterioration of the normal functioning of the OEC under inappropriate conditions. Further studies on the plant PSII, including high-resolution crystallographic strudies, will be required for understanding the functions of extrinsic proteins in the structural stability and the water oxidation chemistry in PSII.

6. Acknowledgment

This research was supported by a grant from the Hyogo Science and Technology Association of Japan.

7. References

Alfonso, M.; Collados, R.; Yruela, I. & Picorel, R. (2004). Photoinhibition and recovery in a herbicide-resistant mutant from Glycine max (L.) Merr. cell cultures deficient in fatty acid unsaturation, *Planta*, Vol. 219, No.3, 428-439.

Alfonso, M.; Yruela, I.; Almarcegui, S.; Torrado, E.; Perez, M. A. & Picorel, R. (2001). Unusual tolerance to high temperatures in a new herbicide-resistant D1 mutant from Glycine max (L.) Merr. cell cultures deficient in fatty acid desaturation, *Planta*, Vol. 212, No.4, 573-582.

Balint, I.; Bhattacharya, J.; Perelman, A.; Schatz, D.; Moskovitz, Y.; Keren, N. & Schwarz, R. (2006). Inactivation of the extrinsic subunit of photosystem II, PsbU, in Synechococcus PCC 7942 results in elevated resistance to oxidative stress, *FEBS Lett.*, Vol. 580, No.8, 2117-2122.

Balsera, M.; Arellano, J. B.; Revuelta, J. L.; de las Rivas, J. & Hermoso, J. A. (2005). The 1.49 angstrom resolution crystal structure of PsbQ from photosystem II of Spinacia oleracea reveals a PPII structure in the N-terminal region, *J. Mol. Biol.*, Vol. 350, No.5, 1051-1060.

Berry, J. & Bjorkman, O. (1980). Photosynthetic Response and Adaptation to Temperature in Higher-Plants, *Annu. Rev. Plant Phys.*, Vol. 31, 491-543.

Bondarava, N.; Beyer, P. & Krieger-Liszkay, A. (2005). Function of the 23 kDa extrinsic protein of photosystem II as a manganese binding protein and its role in photoactivation, *Biochim. Biophys. Acta*, Vol. 1708, No.1, 63-70.

De Las Rivas, J.; Balsera, M. & Barber, J. (2004). Evolution of oxygenic photosynthesis: genome-wide analysis of the OEC extrinsic proteins, *Trends in Plant Sci.*, Vol. 9, No.1, 18-25.

De Las Rivas, J. & Roman, A. (2005). Structure and evolution of the extrinsic proteins that stabilize the oxygen-evolving engine, *Photochem. Photobiol. Sci.*, Vol. 4, No.12, 1003-1010.

Debus, R. J. (1992). The Manganese and Calcium-Ions of Photosynthetic Oxygen Evolution, *Biochim. Biophys. Acta*, Vol. 1102, No.3, 269-352.

Debus, R. J. (2001). Amino acid residues that modulate the properties of tyrosine Y_Z and the manganese cluster in the water oxidizing complex of photosystem II, *Biochim. Biophys. Acta*, Vol. 1503, No.1-2, 164-186.

Enami, I.; Kamo, M.; Ohta, H.; Takahashi, S.; Miura, T.; Kusayanagi, M.; Tanabe, S.; Kamei, A.; Motoki, A.; Hirano, M.; Tomo, T. & Satoh, K. (1998). Intramolecular cross-linking of the extrinsic 33-kDa protein leads to loss of oxygen evolution but not its ability of binding to photosystem II and stabilization of the manganese cluster, *J. Biol. Chem.*, Vol. 273, No.8, 4629-4634.

Enami, I.; Kitamura, M.; Tomo, T.; Isokawa, Y.; Ohta, H. & Katoh, S. (1994). Is the Primary Cause of Thermal Inactivation of Oxygen Evolution in Spinach PS-II Membranes Release of the Extrinsic 33 Kda Protein or of Mn, *Biochim. Biophys. Acta*, Vol. 1186, No.1-2, 52-58.

Enami, I.; Suzuki, T.; Tada, O.; Nakada, Y.; Nakamura, K.; Tohri, A.; Ohta, H.; Inoue, I. & Shen, J. R. (2005). Distribution of the extrinsic proteins as a potential marker for the evolution of photosynthetic oxygen-evolving photosystem II, *FEBS J.*, Vol. 272, No.19, 5020-5030.

Ferreira, K. N.; Iverson, T. M.; Maghlaoui, K.; Barber, J. & Iwata, S. (2004). Architecture of the photosynthetic oxygen-evolving center, *Science*, Vol. 303, No.5665, 1831-1838.

Gabdulkhakov, A.; Guskov, A.; Broser, M.; Kern, J.; Muh, F.; Saenger, W. & Zouni, A. (2009). Probing the Accessibility of the Mn_4Ca Cluster in Photosystem II: Channels Calculation, Noble Gas Derivatization, and Cocrystallization with DMSO, *Structure*, Vol. 17, No.9, 1223-1234.

Ghanotakis, D. F. & Yocum, C. F. (1990). Photosystem-II and the Oxygen-Evolving Complex, *Annu. Rev. Plant Phys.*, Vol. 41, 255-276.

Guskov, A.; Kern, J.; Gabdulkhakov, A.; Broser, M.; Zouni, A. & Saenger, W. (2009). Cyanobacterial photosystem II at 2.9-angstrom resolution and the role of quinones, lipids, channels and chloride, *Nature Struc. & Mol. Biol.*, Vol. 16, No.3, 334-342.

Havaux, M. & Tardy, F. (1996). Temperature-dependent adjustment of the thermal stability of photosystem II in vivo: Possible involvement of xanthophyll-cycle pigments, *Planta*, Vol. 198, No.3, 324-333.

Heredia, P. & De Las Rivas, J. (2003). Calcium-dependent conformational change and thermal stability of the isolated PsbO protein detected by FTIR Spectroscopy, *Biochemistry*, Vol. 42, No.40, 11831-11838.

Ifuku, K.; Ishihara, S.; Shimamoto, R.; Ido, K. & Sato, F. (2008). Structure, function, and evolution of the PsbP protein family in higher plants, *Photosynth. Res.*, Vol. 98, No.1-3, 427-437.

Ifuku, K.; Nakatsu, T.; Kato, H. & Sato, F. (2004). Crystal structure of the PsbP protein of photosystem II from Nicotiana tabacum, *EMBO Reports*, Vol. 5, No.4, 362-367.

Ifuku, K.; Nakatsu, T.; Shimamoto, R.; Yamamoto, Y.; Ishihara, S.; Kato, H. & Sato, F. (2005). Structure and function of the PsbP protein of Photosystem II from higher plants, *Photosynth. Res.*, Vol. 84, No.1-3, 251-255.

Ifuku, K. & Sato, F. (2002). A truncated mutant of the extrinsic 23-kDa protein that absolutely requires the extrinsic 17-kDa protein for Ca^{2+} retention in photosystem II, *Plant Cell Physiol.*, Vol. 43, No.10, 1244-1249.

Ifuku, K.; Yamamoto, Y.; Ono, T.; Ishihara, S. & Sato, F. (2005). PsbP protein, but not PsbQ protein, is essential for the regulation and stabilization of photosystem II in higher plants, *Plant Physiol.*, Vol. 139, No.3, 1175-1184.

Inoue-Kashino, N.; Kashino, Y.; Satoh, K.; Terashima, I. & Pakrasi, H. B. (2005). PsbU provides a stable architecture for the oxygen-evolving system in cyanobacterial photosystem II, *Biochemistry*, Vol. 44, No.36, 12214-12228.

Ishihara, S.; Takabayashi, A.; Ido, K.; Endo, T.; Ifuku, K. & Sato, F. (2007). Distinct functions for the two PsbP-like proteins PPL1 and PPL2 in the chloroplast thylakoid lumen of Arabidopsis, *Plant Physiol.*, Vol. 145, No.3, 668-679.

Joliot, P.; Barbieri, G. & Chabaud, R. (1969). A New Model of Photochemical Centers in System-2, *Photochem. Photobiol.*, Vol. 10, No.5, 309-329.

Kamiya, N. & Shen, J. R. (2003). Crystal structure of oxygen-evolving photosystem II from Thermosynechococcus vulcanus at 3.7-angstrom resolution, *Proc. Natl. Acad. Sci. U. S. A.*, Vol. 100, No.1, 98-103.

Kawakami, K.; Umena, Y.; Kamiya, N. & Shen, J. R. (2009). Location of chloride and its possible functions in oxygen-evolving photosystem II revealed by X-ray crystallography, *Proc. Natl. Acad. Sci. U. S. A.*, Vol. 106, No.21, 8567-8572.

Kerfeld, C. A.; Sawaya, M. R.; Bottin, H.; Tran, K. T.; Sugiura, M.; Cascio, D.; Desbois, A.; Yeates, T. O.; Kirilovsky, D. & Boussac, A. (2003). Structural and EPR characterization of the soluble form of cytochrome c-550 and of the psbV2 gene product from the cyanobacterium Thermosynechococcus elongatus, *Plant Cell Physiol.*, Vol. 44, No.7, 697-706.

Kimura, Y. & Ono, T. (2001). Chelator-induced disappearance of carboxylate stretching vibrational modes in S_2/S_1 FTIR spectrum in oxygen-evolving complex of photosystem II, *Biochemistry*, Vol. 40, No.46, 14061-14068.

Kok, B.; Forbush, B. & McGloin, M. (1970). Cooperation of charges in photosynthetic O_2 evolution-I. A linear four step mechanism, *Photochem. Photobiol.*, Vol. 11, No.6, 457-475.

Kruk, J.; Burda, K.; Jemiola-Rzeminska, M. & Strzalka, K. (2003). The 33 kDa protein of photosystem II is a low-affinity calcium- and lanthanide-binding protein, *Biochemistry*, Vol. 42, No.50, 14862-14867.

Loll, B.; Gerold, G.; Slowik, D.; Voelter, W.; Jung, C.; Saenger, W. & Irrgang, K. D. (2005). Thermostability and Ca^{2+} binding properties of wild type and heterologously expressed PsbO protein from cyanobacterial photosystem II, *Biochemistry*, Vol. 44, No.12, 4691-4698.

Loll, B.; Kern, J.; Saenger, W.; Zouni, A. & Biesiadka, J. (2005). Towards complete cofactor arrangement in the 3.0 angstrom resolution structure of photosystem II, *Nature*, Vol. 438, No.7070, 1040-1044.

Lydakis-Simantiris, N.; Hutchison, R. S.; Betts, S. D.; Barry, B. A. & Yocum, C. F. (1999). Manganese stabilizing protein of photosystem II is a thermostable, natively unfolded polypeptide, *Biochemistry*, Vol. 38, No.1, 404-414.

Mamedov, M.; Hayashi, H. & Murata, N. (1993). Effects of Glycinebetaine and Unsaturation of Membrane-Lipids on Heat-Stability of Photosynthetic Electron-Transport and Phosphorylation Reactions in Synechocystis PCC6803, *Biochim. Biophys. Acta*, Vol. 1142, No.1-2, 1-5.

McEvoy, J. P. & Brudvig, G. W. (2006). Water-splitting chemistry of photosystem II, *Chem. Rev.*, Vol. 106, No.11, 4455-4483.

Miura, T.; Shen, J. R.; Takahashi, S.; Kamo, M.; Nakamura, E.; Ohta, H.; Kamei, A.; Inoue, Y.; Domae, N.; Takio, R.; Nakazato, K.; Inoue, Y. & Enami, I. (1997). Identification of domains on the extrinsic 33-kDa protein possibly involved in electrostatic interaction with photosystem II complex by means of chemical modification, *J. Biol. Chem.*, Vol. 272, No.6, 3788-3798.

Miyao, M. & Murata, N. (1983). Partial Disintegration and Reconstitution of the Photosynthetic Oxygen Evolution System - Binding of 24 Kilodalton and 18 Kilodalton Polypeptides, *Biochim. Biophys. Acta*, Vol. 725, No.1, 87-93.

Miyao, M. & Murata, N. (1983). Partial Reconstitution of the Photosynthetic Oxygen Evolution System by Rebinding of the 33-KDa Polypeptide, *FEBS Lett.*, Vol. 164, No.2, 375-378.

Miyao, M. & Murata, N. (1989). The Mode of Binding of 3 Extrinsic Proteins of 33-Kda, 23-Kda and 18-Kda in the Photosystem-Ii Complex of Spinach, *Biochim. Biophys. Acta*, Vol. 977, No.3, 315-321.

Murakami, Y.; Tsuyama, M.; Kobayashi, Y.; Kodama, H. & Iba, K. (2000). Trienoic fatty acids and plant tolerance of high temperature, *Science*, Vol. 287, No.5452, 476-479.

Murata, N.; Takahashi, S.; Nishiyama, Y. & Allakhverdiev, S. I. (2007). Photoinhibition of photosystem II under environmental stress, *Biochim. Biophys. Acta*, Vol. 1767, No.6, 414-421.

Murray, J. W. & Barber, J. (2006). Identification of a calcium-binding site in the PsbO protein of photosystem II, *Biochemistry*, Vol. 45, No.13, 4128-4130.

Nash, D.; Miyao, M. & Murata, N. (1985). Heat Inactivation of Oxygen Evolution in Photosystem-Ii Particles and Its Acceleration by Chloride Depletion and Exogenous Manganese, *Biochim. Biophys. Acta*, Vol. 807, No.2, 127-133.

Nield, J.; Balsera, M.; De Las Rivas, J. & Barber, J. (2002). Three-dimensional electron cryo-microscopy study of the extrinsic domains of the oxygen-evolving complex of spinach - Assignment of the PsbO protein, *J. Biol. Chem.*, Vol. 277, No.17, 15006-15012.

Nield, J. & Barber, J. (2006). Refinement of the structural model for the Photosystem II supercomplex of higher plants, *Biochim. Biophys. Acta*, Vol. 1757, No.5-6, 353-361.

Nishiyama, Y.; Hayashi, H.; Watanabe, T. & Murata, N. (1994). Photosynthetic Oxygen Evolution Is Stabilized by Cytochrome C(550) against Heat Inactivation in Synechococcus sp. PCC-7002, *Plant Physiol.*, Vol. 105, No.4, 1313-1319.

Nishiyama, Y.; Los, D. A.; Hayashi, H. & Murata, N. (1997). Thermal protection of the oxygen-evolving machinery by PsbU, an extrinsic protein of photosystem II, in Synechococcus species PCC 7002, *Plant Physiol.*, Vol. 115, No.4, 1473-1480.

Nishiyama, Y.; Los, D. A. & Murata, N. (1999). PsbU, a protein associated with photosystem II, is required for the acquisition of cellular thermotolerance in Synechococcus species PCC 7002, *Plant Physiol.*, Vol. 120, No.1, 301-308.

Noguchi, T.; Ono, T. & Inoue, Y. (1995). Direct-Detection of a Carboxylate Bridge between Mn and Ca^{2+} in the Photosynthetic Oxygen-Evolving Center by Means of Fourier-Transform Infrared-Spectroscopy, *Biochim. Biophys. Acta*, Vol. 1228, No.2-3, 189-200.

Ono, T.; Rompel, A.; Mino, H. & Chiba, N. (2001). Ca^{2+} function in photosynthetic oxygen evolution studied by alkali metal cations substitution, *Biophys. J.*, Vol. 81, No.4, 1831-1840.

Ono, T. A. & Inoue, Y. (1989). Roles of Ca^{2+} in O_2 Evolution in Higher-Plant Photosystem-Ii - Effects of Replacement of Ca^{2+} Site by Other Cations, *Arch. Biochem. Biophys.*, Vol. 275, No.2, 440-448.

Ono, T. A.; Izawa, S. & Inoue, Y. (1992). Structural and Functional Modulation of the Manganese Cluster in Ca^{2+}-Depleted Photosystem-Ii Induced by Binding of the 24-Kilodalton Extrinsic Protein, *Biochemistry*, Vol. 31, No.33, 7648-7655.

Pueyo, J. J.; Alfonso, M.; Andres, C. & Picorel, R. (2002). Increased tolerance to thermal inactivation of oxygen evolution in spinach Photosystem II membranes by substitution of the extrinsic 33-kDa protein by its homologue from a thermophilic cyanobacterium, *Biochim. Biophys. Acta*, Vol. 1554, No.1-2, 29-35.

Renger, G. & Renger, T. (2008). Photosystem II: The machinery of photosynthetic water splitting, *Photosynth. Res.*, Vol. 98, No.1-3, 53-80.

Roose, J. L.; Wegener, K. M. & Pakrasi, H. B. (2007). The extrinsic proteins of photosystem II, *Photosynth. Res.*, Vol. 92, No.3, 369-387.

Rutherford, A. W. & Faller, P. (2001). The heart of photosynthesis in glorious 3D, *Trends Biochem. Sci.*, Vol. 26, No.6, 341-344.

Seidler, A. (1996). The extrinsic polypeptides of Photosystem II, *Biochim. Biophys. Acta*, Vol. 1277, No.1-2, 35-60.

Seidler, A. & Rutherford, A. W. (1996). The role of the extrinsic 33 kDa protein in Ca^{2+} binding in photosystem II, *Biochemistry*, Vol. 35, No.37, 12104-12110.

Shen, J. R.; Burnap, R. L. & Inoue, Y. (1995). An Independent Role of Cytochrome C-550 in Cyanobacterial Photosystem-Ii as Revealed by Double-Deletion Mutagenesis of the Psbo and Psbv Genes in Synechocystis Sp Pcc-6803, *Biochemistry*, Vol. 34, No.39, 12661-12668.

Shen, J. R.; Ikeuchi, M. & Inoue, Y. (1997). Analysis of the psbU gene encoding the 12-kDa extrinsic protein of photosystem II and studies on its role by deletion mutagenesis in Synechocystis sp. PCC 6803, *J. Biol. Chem.*, Vol. 272, No.28, 17821-17826.

Shen, J. R. & Inoue, Y. (1993). Binding and functional properties of two new extrinsic components, cytochrome c-550 and a 12-kDa protein, in cyanobacterial photosystem II, *Biochemistry*, Vol. 32, No.7, 1825-1832.

Shen, J. R.; Qian, M.; Inoue, Y. & Burnap, R. L. (1998). Functional characterization of Synechocystis sp. PCC 6803 delta psbU and delta psbV mutants reveals important roles of cytochrome c-550 in cyanobacterial oxygen evolution, *Biochemistry*, Vol. 37, No.6, 1551-1558.

Shen, J. R.; Vermaas, W. & Inoue, Y. (1995). The Role of Cytochrome C-550 as Studied through Reverse Genetics and Mutant Characterization in Synechocystis Sp Pcc-6803, *J. Biol. Chem.*, Vol. 270, No.12, 6901-6907.

Sugiura, M.; Iwai, E.; Hayashi, H. & Boussac, A. (2010). Differences in the Interactions between the Subunits of Photosystem II Dependent on D1 Protein Variants in the Thermophilic Cyanobacterium Thermosynechococcus elongatus, *J. Biol. Chem.*, Vol. 285, No.39, 30008-30018.

Thomas, P. G.; Dominy, P. J.; Vigh, L.; Mansourian, A. R.; Quinn, P. J. & Williams, W. P. (1986). Increased Thermal-Stability of Pigment-Protein Complexes of Pea Thylakoids Following Catalytic-Hydrogenation of Membrane-Lipids, *Biochim. Biophys. Acta*, Vol. 849, No.1, 131-140.

Thornton, L. E.; Ohkawa, H.; Roose, J. L.; Kashino, Y.; Keren, N. & Pakrasi, H. B. (2004). Homologs of plant PsbP and PsbQ proteins are necessary for regulation of photosystem II activity in the cyanobacterium Synechopystis 6803, *Plant Cell*, Vol. 16, No.8, 2164-2175.

Tohri, A.; Dohmae, N.; Suzuki, T.; Ohta, H.; Inoue, Y. & Enami, I. (2004). Identification of domains on the extrinsic 23 kDa protein possibly involved in electrostatic interaction with the extrinsic 33 kDa protein in spinach photosystem II, *Eur. J. Biochem.*, Vol. 271, No.5, 962-971.

Tomita, M.; Ifuku, K.; Sato, F. & Noguchi, T. (2009). FTIR Evidence That the PsbP Extrinsic Protein Induces Protein Conformational Changes around the Oxygen-Evolving Mn Cluster in Photosystem II, *Biochemistry*, Vol. 48, No.27, 6318-6325.

Umena, Y.; Kawakami, K.; Shen, J. R. & Kamiya, N. (2011). Crystal structure of oxygen-evolving photosystem II at a resolution of 1.9 angstrom, *Nature*, Vol. 473, No.7345, 55-U65.

Williamson, A. K. (2008). Structural and functional aspects of the MSP (PsbO) and study of its differences in thermophilic versus mesophilic organisms, *Photosynth. Res.*, Vol. 98, No.1-3, 365-389.

Xu, Q. & Bricker, T. M. (1992). Structural Organization of Proteins on the Oxidizing Side of Photosystem-II - 2 Molecules of the 33-Kda Manganese-Stabilizing Proteins Per Reaction Center, *J. Biol. Chem.*, Vol. 267, No.36, 25816-25821.

Yamane, Y.; Kashino, Y.; Koike, H. & Satoh, K. (1998). Effects of high temperatures on the photosynthetic systems in spinach: Oxygen-evolving activities, fluorescence characteristics and the denaturation process, *Photosynth. Res.*, Vol. 57, No.1, 51-59.

Yamashita, A.; Nijo, N.; Pospisil, P.; Morita, N.; Takenaka, D.; Aminaka, R. & Yamamoto, Y. (2008). Quality control of photosystem II: reactive oxygen species are responsible for the damage to photosystem II under moderate heat stress, *J. Biol. Chem.*, Vol. 283, No.42, 28380-28391.

Yamauchi, Y. & Sugimoto, Y. (2010). Effect of protein modification by malondialdehyde on the interaction between the oxygen-evolving complex 33 kDa protein and photosystem II core proteins, *Planta*, Vol. 231, No.5, 1077-1088.

Yi, X. P.; Hargett, S. R.; Frankel, L. K. & Bricker, T. M. (2006). The PsbQ protein is required in Arabidopsis for photosystem II assembly/stability and photoautotrophy under low light conditions, *J. Biol. Chem.*, Vol. 281, No.36, 26260-26267.

Yocum, C. F. (2008). The calcium and chloride requirements of the O_2 evolving complex, *Coordin. Chem. Rev.*, Vol. 252, No.3-4, 296-305.

Zouni, A.; Witt, H. T.; Kern, J.; Fromme, P.; Krauss, N.; Saenger, W. & Orth, P. (2001). Crystal structure of photosystem II from Synechococcus elongatus at 3.8 angstrom resolution, *Nature*, Vol. 409, No.6821, 739-743.

Part 4

Computational Aspects of Photochemistry

8

Computational Modelling of the Steps Involved in Photodynamic Therapy

L. Therese Bergendahl and Martin J. Paterson

Department of Chemistry, Heriot Watt University, Edinburgh, Scotland

1. Introduction

Photodynamic therapy (PDT) is a branch of phototherapy that has seen a surge of interest in the last few decades, due to its potential in the treatment of various cancers, infections and heart disease.(Bonnett, 2000) This chapter aims to give an overview of the various photochemical steps involved in PDT as a cancer therapy, and in particular the challenges and insight gained from their theoretical description. After a brief review of PDT in general, in a biological and chemical context, the photochemical steps involved will be discussed, detailing the computational techniques required to model these chemical pathways theoretically. We will detail the methodologies that can currently be applied, as well as their limitations of use at present, and areas requiring further development.

1.1 Review of PDT

Photodynamic therapy is perhaps best introduced by considering the photochemical pathways illustrated in the example Jablonski-type diagram shown below, which describes the various processes that would take place within the vicinity of a target cell.

The photochemical processes involved in PDT starts with irradiation (**A**). The irradiation can take place externally, normally from a laser source, as well as internally, through the use of optical fibers, and targets tissue containing a *photosensitiser* molecule (**B**). The sensitiser is functionalised so that it will accumulate in the cells that are targeted, for example in a cancer lesion or an area of diseased skin cells. It is also designed so that it will effectively absorb light of a specific wavelength. The absorption process (**C**) takes place either through the absorption of one photon per sensitiser molecule, *one-photon absorption* (OPA), or, if a light source with high fluency is used (e.g., from an intense laser beam), two photons per molecule (TPA). In the Jablonski diagram the eigenstates are separated according to their energy as well as spin multiplicity, where S denotes a singlet state and T a triplet state. These multiplicities, and their relative order, are typical for many organic chromophores that are used as photosensitisers, and the transitions taking place between them are governed by the various section rules for electronic transitions (e.g. non-parity conserving transitions for OPA and parity conserving for TPA). After very fast non-radiative relaxation processes, collectively termed *internal conversion* (IC) (**D**), the absorption results in the generation of the sensitiser in its first excited singlet state, S_1. The sensitiser molecule must

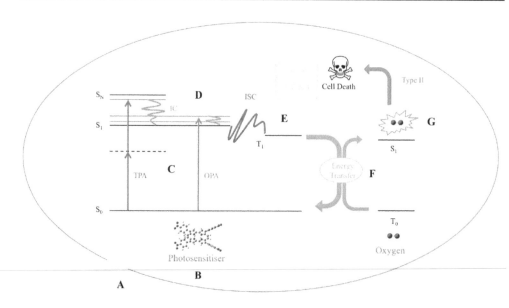

Scheme 1. Overview of the photochemical pathways taking place in photodynamic therapy, described through the energy flow starting from irradiation of the photosensitiser molecule and ultimately ending in cell death.

then be able to undergo efficient *intersystem crossing* (ISC) (**E**) that generates the first excited triplet state of the molecule, T_1. The sensitiser has to be designed so that ISC takes place with high quantum yield. Any radiative processes, where the molecule will return to its ground state after emitting a photon, are looked upon as loss processes in PDT and should be not effective deactivation channels in an ideal photosensitiser molecule. The same applies to any intra-molecular reactive channels in the singlet or triplet manifolds, which generate new photoproducts. The comparably long-lived triplet state can react in two ways when confined in a biological environment: Type I reactions, which involves the creation of radicals via electron or hydrogen transfer reactions, or Type II reactions, where an *energy transfer process* (**F**) takes place from the triplet state of the photosensitiser, and generates the cytotoxic *singlet oxygen* species (**G**) from ground state oxygen which is always abundant in a biological environment. During the energy transfer process the photosensitiser is simultaneously brought back to its singlet ground state where it can, in principle, take part in further sensitisation cycles. The singlet oxygen created is exceptionally reactive, and is therefore very short-lived in a biological medium where it has been shown to be toxic to essentially any biomolecule and organelle within the cell. Both pathways ultimately lead to oxidative degradation of the biomolecules within the cell, but in most cases Type II mechanisms are dominant, and will be the pathway considered in this chapter. Only in environments with low oxygen levels are Type I mechanisms observed, which means it will mainly have to be considered for the targeting of hypoxic parts of fast-growing tumours. (Bonnett, 2000; MacDonald & Dougherty, 2001; Missailidis, 2008)

The knowledge of the benefits of sunlight on the general health have been traced back over 5000 years with references from Egypt and India as well as ancient Greek texts mentioning Herodetus and Hippocrates recommending "heliotherapy" to their fellow Greeks and documenting the benefits of sunlight on bone growth. (Bonnett, 2000) The development of PDT as we know it today has been in progress since the early 1900s, an example being the 1903 Nobel Prize in Physiology or Medicine awarded to the Danish scientist Niels Rydberg Finsen.(Nobelprize.org, 2011) Finsens work was the first of its kind where light was used actively and directly as a treatment of disease. The first descriptions of the use of a chemical photosensitiser compound in light therapy are also from the early 1900s through the observations of Raab as well as Jasionek and Von Trappen.(Ethirajan, Chen et al., 2011) Raabs work involved the investigation of bacterial cultures and the effect of an added porphyrin dye. He observed completely unexpected cell death in the experiments taking place during the daytime, but none during his evening experiments, which lead him to draw conclusions about the light activation of his dye molecules and their therapeutic outcome. (Allison, Downie et al., 2004) One of the most notable examples from this period is the 1913 experiment by the German scientist Meyer, who injected himself with 200 milligrams of *haematoporphyrin*, a carbonyl functionalised porphyrin derivative (Fig. 1). After sunlight exposure he suffered extreme swelling and experienced photosensitivity for several months after the experiment. (Dolphin, Sternberg et al., 1998) A more recent discovery in the development of PDT is that a heamatoporphyrin derivative has tumour localising properties as well as being phototoxic. This discovery prompted a spur of development between 1960s and 1980s which ultimately lead to the first clinically approved drug Photofrin®, based on the haematoporphyrin derivative, as well as being the starting point for the development of PDT as a promising tool in cancer treatment. (Hasan, Celli et al., 2010)

The current use of PDT in cancer treatment is typically palliative and for very advanced stages of bladder, lung and oesophageal cancers. Its only success as a general treatment is in the case of age-related macular degeneration, which is a leading cause of blindness in the western world, and this still represents the only area where PDT is the first treatment of choice. (Schmidt-Erfurth, 2000) The main limitation in the general application of PDT as an anticancer therapy is the poor level of tissue penetration of the light used. Skin cancers are the only type that can be treated with relative ease, whilst for lung and oesophagus tumours the administration of light normally requires the use of an endoscope. A further problem arising if the light dose can be administrated is the targeting of the photosensitiser compound. Poor targeting leads to general photosensitivity in patients for weeks after the treatment, and even high strength light bulbs can be problematic. General side effects from the treatment of internal cancer are nausea and fever, caused by swelling and inflammation of the nearby tissue.(Davids & Kleemann, 2011)

Despite these side effects and even though developments in PDT have not yet resulted in its general use as an anti-cancer therapy, it is still an exceptionally attractive goal and research continues today along many promising avenues. The attractiveness of PDT as a clinical therapy comes from several aspects that could in principle elevate it above current modes of treatment. It is a non-invasive therapy that has no toxicity in the absence of a light source, and compared to surgery it has excellent outcome with respect to scarring and other cosmetic considerations. As the technique generally requires light in the ultraviolet to infrared region the radiation damage is minimal, compared to other forms of radiotherapy, and it has also been suggested that PDT can trigger and help mobilise the immune system

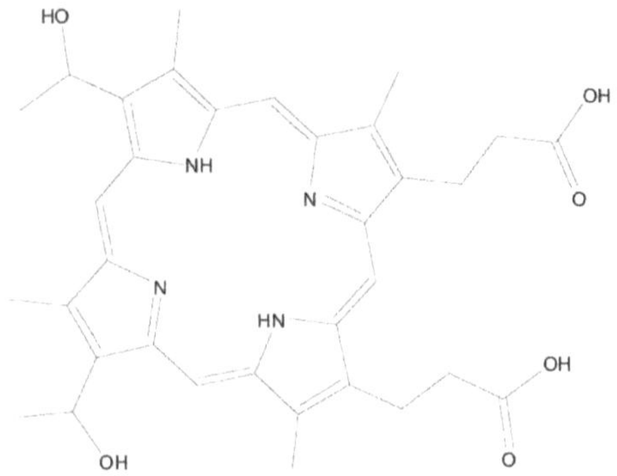

Fig. 1. Haematoporphyrin. The PDT drug Photofrin® consists of oligomers haematoporphyrin, of varying size.

against the tumour.(Brackett & Gollnick, 2011) A further major advantage PDT has, in comparison to various forms of chemotherapy, is the inability of the body to develop resistance to the treatment as there is not necessarily a specific target biomolecule or organelle involved in the mechanism of its cytotoxicity. These facts have stimulated continued research as we detail below. One of the most important aspects is a deeper understanding of the spectroscopic and mechanistic features as shown in Fig. 1. First principles computation has only recently matured to the point where it can be routinely applied to some of these problems, and still other challenges remain. Nevertheless such studies are now regarded as invaluable in rationalising the existing classes of photosensitizers, and can now be used in the design aspect of photosensitisers to have a set of desired molecular properties (Arnbjerg, Jimenez-Banzo et al., 2007; Arnbjerg, Paterson et al., 2007; Bergendahl & Paterson, 2011; Johnsen, Paterson et al., 2008).

Any molecule that show promising behaviour for the use as a photosensitiser will have to be further optimized extensively in order to consider targeting within the human body, toxicity, pharmacokinetics, ease of total synthesis of the molecule, and so on. With computational chemistry having reached this stage, systematic studies of molecular structure for PDT properties can hopefully aid in the design of drug molecules for application in this field.

2. Computational techniques

In order to fully explain the fate of an electronically excited species theoretically the Jablonski diagram above needs to be expanded, and the energy of each state needs to be related to the *molecular structure* of the system under investigation. This is done by the expression of the energy as a function of the degrees of freedom in the molecule. For any molecular system, apart from simple diatomic molecules, this will result in a multidimensional *potential energy surface* (PES). This concept, where the motions of the atomic nuclei are treated separately from the motion of the electrons, is the result of one of the fundamental approximations employed in molecular quantum mechanics: the *Born-Oppenheimer* (BO) *Approximation*. Conceptually it is based upon the idea that the relatively large velocity and low mass of the electrons in comparison to the nuclei means their motion can be decoupled and treated separately, as the dynamics of these species will take place on vastly differing timescales. The BO approximation can be used with good results in the theoretical description of most areas of chemistry, and the topology of a PES can be used routinely to analyse the thermodynamics and kinetics of most thermal reactions. We will see however that this approximation fails when describing certain photochemical pathways where two PESs can be looked upon as crossing each other, as in IC or ISC when discussing PDT, a problem that needs to be addressed for an accurate theoretical description. Thus, such treatment of so-called vibronically coupled states is one of the most challenging areas, but also one of the most important in Fig. 1 above, as this determines the exchange of the initial photo-energy between the electronic and nuclear degrees of freedom.

2.1 Modelling of the electronic ground state

When attempting to describe theoretically the PES of the ground state of a large molecular system from first principles it is convenient to divide the solution to the electronic Schrödinger equation, the total electronic energy, E_{el}, into various part:

$$E_{el} = E_T + E_V + E_J + E_X + E_C \tag{1}$$

where E_T is the kinetic energy of the electrons, E_V is the coulomb interaction between the electrons and the nuclei and the coulomb term E_J describes the energy the electrons would have if they moved independently and if each electron also repelled themselves (i.e., the repulsion of the "smeared out" charge clouds of each electron). The final two terms correct for the false parts of the latter, E_X, the exchange term, corrects for the fact that electrons with the same spin will avoid each other strongly as a result of the Pauli principle, whilst E_C, the correlation term, corrects for the smaller correlation of the movement of electrons with opposite spin.(Gill, 1998) Most computer codes will calculate an approximate solution to these parts of the total electronic energy separately, and the way in which they deal with each of them is what distinguishes various methods from each other. That, and of course the computational cost of each method (in terms of CPU time).

2.1.1 Wavefunction methods

The exact solution to the electronic Schrödinger equation is an intimidating task for any system with more than one electron due to the complexity of dealing with interacting electrons. The first approximation generally applied is therefore to make a crude assumption that the total many-electron wavefunction, Φ, can be constructed as a product of functions, ϕ, which describes one electron at the time, i.e. *molecular orbitals*. This is the approach used in the *Hartree-Fock method*, which is the best approximation available when using one-electron molecular orbitals to describe a single *electronic configuration*, i.e., the orbital occupancy of the system. To account for electron spin and indistinguishability the orbitals are extended to *spin-orbitals* and described as a *Slater determinant*, which satisfies the Pauli principle for electrons. The total many-electron Hamiltonian is then used with these one-electron orbitals, and the resulting energy is optimised variationally. Crucially however, the resulting operator after this treatment is expressed in terms of the molecular orbitals obtained as its eigenfunction, creating a non-linear problem that needs to be solved iteratively through a so called *self-consistent field* (SCF) calculation. (Cramer, 2004; Jensen,

2007; Szabo & Ostlund, 1996) It should be noted that in the HF method the last term in eqn. 1 above is completely unaccounted for, and thus this MO method is generally described as an *uncorrelated* method. The electronic configuration described by the HF method need not be that of the electronic ground state but if a configuration of a different symmetry is constructed then the variational nature of the method can be used to determine the optimum molecular orbitals for that configuration. Thus, for an excited state that can be qualitatively described by a single electron configuration, of a different symmetry to the ground state, then this method is capable of describing such a state. In this so called Δ-SCF approach two problems arise, (i) most excited electronic states are poorly represented by a single configuration, and (ii) this method is unable to describe excited states of the same symmetry as a lower lying state due to variational collapse.

One key part in carrying out any computation is the choice of *basis functions* used to describe the one electron orbitals, ϕ, through an expansion of the molecular orbitals via known atom-centred functions (i.e., the well known LCAO expansion). It is a dilemma to choose functions that accurately describe the electron distribution in a given region of space, but at the same time are not so numerous as to make the computational cost too steep. Strategies that are employed to increase the quality of a given basis set are to include more functions of each angular momentum type that are used, as the addition of *diffuse* functions. (Papajak, Zheng et al., 2011) Diffuse functions have a larger radial component than the valence functions and can therefore more accurately describe wavefunctions at a larger distance from the nucleus, which is useful when considering properties of excited states, as well as properties such as polarisabilities, or when dealing with negatively charged species. A veritable plethora of basis sets exists for excited state computations.

The HF method exactly accounts for the fact that electrons with the same spin will avoid each other strongly, so-called *electron exchange*, through the anti-symmetry of the Slater determinant. However it completely ignores the smaller correction from the movement of electrons with the same spin, this is termed *electron correlation*. This missing electron correlation energy is normally only a fraction of the total energy, but it is often crucial for the correct description of a system as it varies greatly with molecular geometry. This is especially true for excited states, where there might be more than one electronic configuration needed to describe the true bonding situation in the system, so called *static*, or *non-dynamic* correlation. Thus, static correlation describes the cases where a qualitatively correct wavefunction requires several electronic configurations (i.e., a multiconfigurational wavefunction). This can either be due to spatial quasi-degeneracies (i.e., small HOMO-LUMO gaps) or to spin-symmetry (*vide infra*). The correlation that takes the instantaneous Coloumb repulsion between electron pairs into account, rather than the mean field SCF orbital description, is termed *dynamic* correlation. This can either be recovered from single or multiconfigurational reference states.

The problem of accounting for the missing electron correlation energy is treated differently in various methods, but the common aspect of all approaches is the understanding that electron-electron interactions can be described more accurately if more than one configuration is used to express the final many electron wavefunction, Φ. (Jensen, 2007) Conceptually the most straightforward solution to the correlation problem is to generate more electronic configurations by promoting electrons in occupied orbitals to virtual, unoccupied, orbitals, so that the system is described by a linear combination of all excited configurations. This is termed *the Configuration Interaction* (CI) method:

$$\Psi^{CI} = a_0 \phi^{HF} + \overset{\substack{Single \\ Excitations}}{\underset{a_i^S}{\sum}} a_i^S \phi^S + \overset{\substack{Double \\ Excitations}}{\underset{a_i^D}{\sum}} a_i^D \phi^D + ... + \overset{\substack{N-fold \\ Excitations}}{\underset{a_i^N}{\sum}} a_i^N \phi^N \qquad (2)$$

If the optimisation of the CI expansion coefficients (also known as amplitudes), a_i, were to be carried out over all possible configurations the CI wavefunction would be the exact wavefunction, and hence give an exact solution to the time-independent Schrödinger equation within that given one-electron basis set. This is an extraordinarily difficult task as the number of possible configuration will grows factorially with system size, and the full CI wavefunction normally needs to be truncated in order for it to be routinely used. Even though a truncated CI wavefunction will accurately account for a large fraction of the missing correlation energy, there is one unfortunate drawback. Truncated CI expansions are not size extensive, which means that as the system size gets larger a proportionally smaller amount of the correlation energy is returned at a given truncation level. A related issue is that the CI method is also not size consistent, i.e. the energy is not proportional to the number of non-interacting units comprising the system. This is a significant flaw when calculating for example dissociation energies, as the energy of two dissociated molecular fragments will not be equal to the energy of the two fragments calculated separately, as some multiple excitations are omitted.(Piela, 2007)

A method related to CI is the *Coupled Cluster* (CC) method. It is arguably the most sophisticated and accurate method for retrieving some of the dynamic correlation energy in use today. The *cluster operator* $\hat{T}$ is related to the exact wave function through a reference determinant, usually the HF determinant, through the coupled cluster ansatz;

$$\Psi = e^{\hat{T}} \phi = \left(1 + \hat{T} + \frac{\hat{T}^2}{2!} + \frac{\hat{T}^3}{3!} + ... \right) \phi \qquad (3)$$

where the CC wavefunction is generated through an exponential expansion of the cluster operator that generates the substituted configurations from ϕ in the same way as in CI theory.

It is assumed that the cluster operator is a sum of excitations operators for all possible excitations from occupied to unoccupied orbitals;

$$\hat{T} = \hat{T}_1 + \hat{T}_2 + ... + \hat{T}_N \qquad (4)$$

where $\hat{T}_1$ is an operator for single excitations and so on. If all possible excitations are considered the result will be the same as for a full CI treatment, and again this full treatment is too computationally expensive to be practical. Yet again a truncation of the wavefunction must take place, and this is where the difference to CI becomes evident. Unlike CI methods the excitation operators will automatically include important so-called disconnected terms, so truncating at $\hat{T}_2$ in the cluster operator automatically includes $\hat{T}_2^2$, $\hat{T}_2^3$ up to $\hat{T}_2^N$, from the Taylor expansion of $e^{\hat{T}}$, which results in not only a more accurate description of electron correlation but also generates wavefunctions that are both size extensive and size consistent.(Piela, 2007) The disconnected terms like $\hat{T}_2^2$ can be understood to account for two simultaneous pair-wise interactions away from the HF-SCF mean-field description. The one disadvantage is that the CC method is not variational (the exponential ansatz unfortunately does not generate truncated left and right states in an expectation value of the energy).(Crawford & Schaefer, 2000) However this is not as important as the advantages gained and therefore CC theory occupies the prime position in correlated wavefunction theory. A final advantage the CC approach shares with CI theory is that as the excitation operator is expanded one can systematically improve the properties of the molecule calculated from the resulting wavefunction (with these methods for ground state properties, but see below for excited state properties). However, even with truncation, the computational cost of CC computations are very high and the size of the molecules that can be investigated are limited, and specialist techniques, such as the use of significant computer resources as well as the development of non-iterative versions of the methods, need to be employed if larger systems are to be investigated.(Kowalski, Krishnamoorthy et al., 2010) In order to routinely be able to probe photosensitiser molecules the size of substituted porphyrinic macrocyclic systems different approaches needs also to be considered, for example through the use of Density Functional Theory (DFT), although for the core chromophores of various photosensitisers coupled cluster theory has set important benchmarks against which lower cost methods can be calibrated.

2.1.2 Density based methods

In DFT an alternative starting point for the calculation of the total electronic energy is employed through the use of the total electron density, ρ. The main idea of DFT is that the total energy can be completely determined by the total electron density of the system through a *functional* that connects it with the energy. (Koch & Holthausen, 2001) The advantage to using the density, rather than the molecular wavefunction, is that the problem will not depend on the number of electrons present in the system, and therefore avoid becoming increasingly complicated as the system increases in size (i.e., the energy ultimately depends on only 3 spatial coordinates rather than $3n$ for an n-electron wavefunction). In order to develop both an accurate and computationally tractable scheme W. Kohn and L.J. Sham developed a fictitious model system where the electrons do not interact at all so that they are described exactly by a set of molecular orbitals. Crucially, they realised that the density for this non-interacting system can be exactly the same as that of the same system with interacting electrons, and this was therefore used to develop the so-called Kohn-Sham expression for the ground state electronic energy (where each term is obtained from an expression involving molecular orbitals):

$$E = E_T^{KS} + E_V + E_J + E_{XC}^{KS} \qquad (5)$$

As the expression deals with non-interacting electrons, the kinetic energy term is not complete since the exact electronic kinetic energy cannot be written as a functional of the density, and its correction is introduced in addition to the exchange and correlation terms, which are all then grouped together in the E_{XC}^{KS} term. All other terms can be evaluated exactly, and so the main problem of DFT is reduced to the estimation of the exchange correlation energy. The way in which various DFT *functionals* retrieve the exchange correlation energy varies, and is what sets them apart. A number of strategies exist and the improvement of various functionals is generally derived from both theoretical foundations by expanding the parameter set linking the energy with the density (via gradients etc.), and the fitting of such parameters to experimental data.

The Kohn-Sham equation is derived from the Shrödinger equation for the non-interacting system and its solution yields the so-called Kohn-Sham orbitals, φi (r), which are defined such that:

$$\rho(r) = \sum_{i=1}^{N} |\phi_i(r)|^2 \qquad (6)$$

These orbitals are then expanded in a basis set, and arranged in a single Slater determinant, both of which are completely analogous to HF MO theory. When finding the energy of a specific molecular structure using DFT the Kohn-Sham equation is solved self consistently using an electron density that initially is a guess in order to calculate E_{XC}^{KS}, and hence the energy of the optimised system.(Koch & Holthausen, 2001)

2.1.3 Multiconfiguration methods

The problem with correlation that we have dealt with so far takes the instantaneous Coloumb repulsion between opposite spin electrons into account, the *dynamic correlation* (since the Kohn-Sham non-interacting system has a single configuration wavefunction). For some systems however there is also a further complication in that more than one electron configuration might be needed to fully describe the bonding situation in the system. This is especially true for most *open-shell* systems, of which the oxygen molecule is a prime example, as well as in the description of almost all excited states of closed-shell systems. This effect is the *static* or *non-dynamic correlation* discussed above, and is clearly something we need to be able to model if we are to accurately describe all photochemical pathways involved in PDT. In the cases where static correlation is important the use of one single determinant to describe a many electron wavefunction, with the correct spin, is not always possible. There has however been some single-reference methods developed recently that can describe some multiconfigurational situations, for example the so-called *spin-flip* approach. In this approach a high-spin triplet reference state is chosen, $M_s = 1$. The target state, e.g., the $M_s = 0$ state of an open-shell singlet state, can then be described as a spin-flipping excitation from the reference state.(Shao, Head-Gordon et al., 2003) This is a promising avenue of attack for the application of DFT to open-shell systems. A more common approach in order to deal with static correlation however, is to turn to a strategy of taking a linear combination of a number of electron configurations (as in CI theory described above), and define a *multiconfigurational* wavefunction. This wavefunction is then optimised variationally by an SCF procedure in which all expansion coefficients are variables. That means that the expansion coefficients for the basis functions (the LCAO parameters), which describe the molecular orbitals, as well as the configurational expansion coefficients will be optimised, which adds flexibility to the wavefunction. Thus the optimum orbitals for the given CI-type wavefunctions are obtained, unlike in standard CI theory where the reference orbitals are fixed. The most popular variant of MCSCF theory is the *complete active space* (CAS) SCF approach. The popularity stems mainly from the fact that the choice of which electron configurations that are important to include in the MC wavefunction can be difficult, and this is automatically taken care of in CASSCF. In the CASSCF approach an active space is selected, that includes specific orbitals, important to describe the chemistry one is interested in, e.g., for benzene photochemistry an obvious choice is the six π-electrons and six π-orbitals.(Li, Mendive-Tapia et al., 2010) In aromatic organic molecules in general, with or without heteroatoms, the frontier molecular orbitals, FMOs, (the Highest Occupied Molecular Orbitals, HOMOs, and the Lowest Unoccupied Molecular Orbitals, LUMOs) are crucial to describe any photochemical mechanisms, and needs therefore to be described accurately. Orbitals of less importance, that are expected to stay either fully occupied or un-occupied in all determinants, are kept unchanged, and thus no configurations arising from changes to these doubly occupied or empty orbitals are included. The remaining orbitals are considered to be *active*, and a full CI is carried out on this subset. CASSCF is exceptionally flexible and can be used to compute essentially any part of the PES of excited as well as ground states, including surface crossings, which are described later.

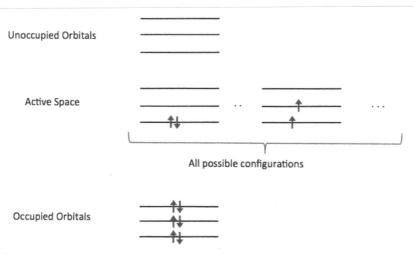

Fig. 2. Overview of the CASSCF strategy, where a full CI is carried out in the active space, whilst the occupied and unoccupied orbitals are kept constant in all configurations.

2.2 Modelling of excited states

The single reference methods mentioned so far are all constructed for the description of the ground state potential energy surface of a molecular system, in particular around regions describing stable configurations of the molecule (i.e., PES minima). When dealing with photochemical events obviously more than one PES needs to be considered.

There are many approaches to the theoretical modelling of the properties of a molecular system, such as excitation energies, one example being *state specific methods*. In these a trial wavefunction is chosen so that it corresponds to the state we are interested in, which could be an excited state rather than the ground state. (Nicolaides, 2011; 2011) This is the idea behind the Δ-SCF method discussed above. More powerful are methods that can construct arbitrary excited state wavefunctions of any symmetry. These are most often obtained as higher roots in CI-type eigenvalue problems, and as such CI methods are the method of choice in constructing both qualitative and quantitative excited state wavefunctions. For example the simplest qualitative correct excited state wavefunctions can often be represented by the CIS approach. Also in this manner CASSCF can generate excited wavefunctions of the same accuracy as for ground electronic states. It should be noted that such excited states invariably always require such a multiconfigurational wavefunction at most points on a photochemical reaction path.(Malmqvist and Roos, 1989; Olsen, 2011) Dynamic correlation effects can then be added to such multiconfigurational wavefunctions by using the underlying multi-reference configurations for further CI (MR-CI), or perturbation theory treatments (e.g., CASPT2).(Pulay, 2011)

2.2.1 One- and two-photon absorption

One method for the determination of excited state information that has been very successful and popular in recent years is the so-called *response function method*. The reason for its popularity is that it can be adapted to suit DFT, MCSCF as well as CC methods. Thus, most molecular properties, such as excitation energies and transition moments can be obtained from a response function calculated from a ground state quantity (the wavefunction in MCSCF and CC theory, and the density in DFT). In response theory any molecular property is expressed in terms of a response function, as a property of a molecule can be defined as the response of an expectation value of an operator to a perturbation. (Bast, Ekstrom et al., 2011) When dealing with perturbations in the form of an external field interacting with a molecular system, as when dealing with absorption of light, a time-dependent approach needs to be considered. The time-dependent perturbation term is expressed by an interaction operator, which is added to the Hamiltonian for the problem.

The linear response function for operator P with respect to perturbation Q can be written as:

$$\langle\langle P;Q\rangle\rangle_{\omega_b} = -\hat{X}^{PQ}\sum_j\left(\frac{\langle\psi_i|P|\psi_j\rangle\langle\psi_j|Q|\psi_i\rangle}{\omega_b-\omega_j}\right) \quad (7)$$

An example is the linear response of a system subject to a homogenous electric field of frequency ω_b , where the perturbation operator then will refer to the electric dipole operator, μ .This equation is identical to the dipole-dipole *polarizability* tensor at frequency ω_b, i.e. the frequency-dependent polarizability. This has *poles* (i.e. where the denominator is equal to zero) at the excitation energies of the system. The *residues* of the response function (i.e. the numerator at the poles) give the transition moment (related to the probability of a transition between states). The poles of the response functions are obtained as solutions to an eigenvalue equation, with the eigenvectors giving the contribution from orbital excitations to each excited state. This method is very efficient since there is no need to construct all the excited states wavefunctions explicitly as in the sum-over-states expression in the equation above. All the information regarding the excited state is contained in the response function, so there is only a need for one well defined wavefunction or density, such as a ground state wavefunction from DFT, MCSCF or CC, or density from the Kohn-Sham orbitals in DFT, in order to access information on the excited states. Thus the response function approach allows us to access excited state information from nominally ground state methods and take advantage of what they bring (i.e., systematic accuracy in CC theory, low cost in DFT)

As previously mentioned, there is also a possibility of a two-photon absorption process for the excitation of the photosensitiser, as per Scheme 2. This has numerous benefits for the application to PDT, such as access to the so-called tissue transparency window (i.e., avoiding absorbance from water, melanin and haemoglobin) and increased spatial resolution for the treatment, thus the accurate modelling of this non-linear absorption process is of great importance in the development of ideal photosensitiser molecules.

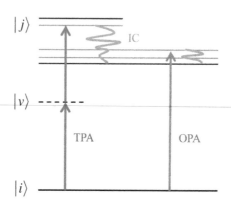

Scheme 2. Overview of the TPA process, normally described as taking place through a virtual state, v.

The use of a state specific method for the computation of TPA is inaccurate, as a few state model would need to be constructed, so the general strategy is the use of response theory through higher order terms of the response function. For TPA the *quadratic* response function needs to be considered under two dipole perturbations, one of frequency ω_b and one of frequency ω_c. The energy representation of the generalised quadratic response function is then described as:

$$\langle\langle P; Q, R \rangle\rangle_{\omega_b \omega_c} = -\sum \hat{X}^{PQR} \sum_{v, j \neq i} \left[\frac{\langle \psi_i | P | \psi_v \rangle \langle \psi_v | \bar{Q} | \psi_j \rangle \langle \psi_j | R | \psi_i \rangle}{\left(\omega_v - \omega_b + \omega_c \right)\left(\omega_j - \omega_c \right)} \right] \tag{8}$$

When the operators P, Q and R are dipole operators then this expression is identical to the first dipole *hyperpolarisability* tensor at frequencies ω_b and ω_c. This frequency-dependent hyperpolarisability can describe non-linear phenomena involving two photons of frequencies ω_b and ω_c respectively, as well as TPA information, where two identical photons are absorbed simultaneously. (Paterson, Christiansen et al., 2006) The residues of the quadratic response function can be shown to give the conventional expression for the two-photon absorption *transition matrix element*, M, between states i and j through a virtual state (written as linear combination of all real eigenstates v):

$$M_{\alpha\beta} = \sum_v \frac{\langle i|\mu_\alpha|v\rangle\langle v|\mu_\beta|j\rangle}{\omega_v - \omega} + \frac{\langle i|\mu_\beta|v\rangle\langle v|\mu_\alpha|j\rangle}{\omega_v - \omega} \tag{9}$$

where ω the frequency of the irradiating light, and ω_v is the excitation frequency of state v. The labels α and β refers to the Cartesian coordinates x, y and z of the TPA tensor. The transition matrix element is then used to calculate the rotationally averaged two-photon absorption cross section, δ.

Even though the formal energy expressions for the quadratic response function appears to involve a tedious sum over all states of the molecule to describe the virtual state v above, as mentioned previously the actual computational machinery of response theory calculates the properties directly without constructing the states explicitly which saves both time and computational effort. Additionally, one does not need to make any *a priori* estimate of the importance of certain states to the TPA tensor. Not only that, this technique has also been proven to be exceptionally accurate in the prediction of TPA spectra. When response theory is used with a CC ground state wavefunction quantitative agreement with experimental results can be achieved, and for a DFT ground state density very good agreement is also often obtained.(Arnbjerg, Jimenez-Banzo et al., 2007; Hattig, Christiansen et al., 1998) When it comes to DFT however, the choice of functional is crucial for the application of quadratic response theory, as it has to be able to model excitations to and from the virtual state, which essentially means it has to be considerably more robust than the functionals routinely used together with linear response theory for one-photon absorption. If this requirement is satisfied quadratic response DFT has been shown to be very accurate, and even comparable to high-level CC calculations for TPA properties.(Arnbjerg, Paterson et al., 2007; Johnsen, Paterson et al., 2008; Paterson, M. J., Christiansen et al., 2006) This is very encouraging, considering the size of the photosensitisers we are interested in modelling for the description of TPA in PDT (*vide infra*). The improvement of non-linear absorbing qualities of photosensitiser molecules can involve many complicated synthetic routes(Drobizhev, Karotki et al., 2002; Drobizhev, Meng et al., 2006) As the benefits of the use of TPA in PDT are plentiful, a theoretical input will be invaluable, as it can give important information on molecular structure-TPA relationships which are crucial for the development of TPA PDT on a daily basis. This has for example been demonstrated in an experimental and theoretical study, where the *porphycene* macrocycle was shown to be a promising photosensitiser system for TPA PDT.(Arnbjerg, Jimenez-Banzo et al., 2007)

We have carried out further computational TPA studies of this porphycene molecule and shown that the non-linear absorption can be effected to a somewhat unexpected degree by the substitution of heteroatoms in the macrocyclic core.(Bergendahl & Paterson, 2011) The structural changes afforded very effective absorption in the so-called Soret-region of the TPA spectra (an unusual feature for the porphyrin photosensitisers so far investigated for TPA PDT) without having a large impact on the one-photon absorption spectra. We investigated a set of dioxaporphycenes, where the pyrrolic nitrogen atoms were substituted for oxygen atoms systematically as per Scheme 3.

After optimising the geometries of the molecules, the OPA and TPA was calculated using linear and quadratic DFT response theory respectively, with the CAM-B3LYP functional and

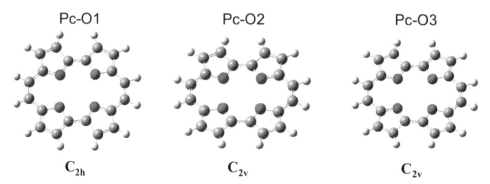

Scheme 3. The optimised geometries of the three dioxaporphycenes investigated, with resulting symmetries. N atoms in blue, and O atoms in red.

6-31G* Pople-type basis set. Previous work has shown that the CAM-B3LYP functional is very accurate at modelling TPA, as it has correct asymptotic behaviour in the exchange-correlation functional, which gives a good description of absorption to and from the virtual state.(Paterson, Christiansen et al., 2006; Rudberg, Salek et al., 2005)

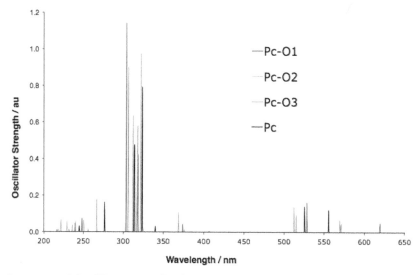

Fig. 3. Summary of the OPA spectra for the dioxaporphycenes investigated. Note the absorption band at 300 - 350 nm, the *Soret*-band, and at 520 – 580 nm, the *Q*-band, which are characteristic bands for the porphyrin family

The OPA results are summarised in the simulated absorption spectra in Figure 3, and show clear Soret- (~ 300-350 nm) and Q-bands (~ 500 - 600 nm) for all species. As expected, there are no major effects observed on the spectral profile upon substitution of the core heteroatoms, compared with porphycene. The TPA spectra are presented in Figure 4. It is important to note that the absorption wavelength indicated is the total for the two-photon process, i.e. the actual excitation takes place at twice the wavelength. The spectra all show a main absorption in the Soret-region (~ 250 – 270 nm), and there is a significant disparity between the calculated TPA cross-sections for the four isomers. These two facts can be explained in terms of one-photon resonance enhancement of the absorption. This phenomenon is best discussed in terms of the sum-over-states expression in the equation for the transition matrix element M, as seen previously, where the magnitude of M depends on the difference between the excitation frequency of the v-th state and the frequency of the total excitation, from the magnitude of the denominator in the expression. As the virtual state can be described as a linear combination of all real eigenstates in the system, a TPA via a virtual state that can be described mainly in terms of a real state will be significantly enhanced. Thus, an allowed OPA Q-band transition accidentally degenerate at half the Soret excitation energy will greatly enhance the Soret TPA transition. These results suggest that the tuning of TPA does not have to be carried out through large structural changes, and small electronic factors can be used to improve the magnitude of the TPA cross-section, hence broadening the scope for the development of TPA photosensitisers in PDT.

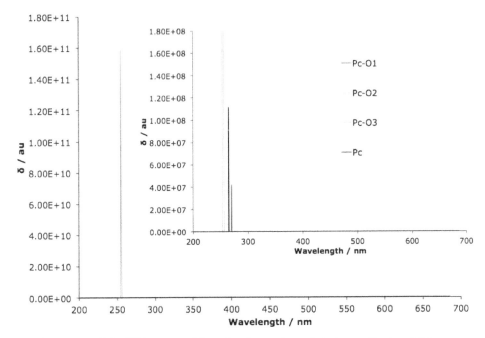

Fig. 4. Summary of the TPA spectra. Note the change in scale between the two TPA cross-section axes.

2.2.3 Photochemistry on excited state potential surfaces

Even though the information we gain from response theory can accurately predict the electronic spectra for both one and two photon absorption processes, it is insufficient to give a full overview of the whole reaction path. This is true when we are modelling reactive photochemistry in general, and the paths involved in PDT in particular where IC and ISC plays a significant role in the understanding of the reaction path. The modelling of reaction paths which cross multiple surfaces have been studied both experimentally and theoretically for number of years, and the understanding of the pathways involved in the photochemistry of small molecules have expanded in great detail.(Zewail, 2000; 2000)

The non-radiative IC pathways have been known to exist for a very long time, an early description being *Kashas rule*, which states that radiative transitions and ISC will take place from the lowest energy excited level of that spin multiplicity due to the very short time-scale of the IC processes between upper electronic levels.(Kasha, 1950) More recently however, the description and understanding of these events have experienced a dramatic increase through the development of electronic structure methods to theoretically model these events. It is known that IC events are normally dominated by so called *conical intersections*, where two, or more, surfaces cross each other. Conical intersections are mathematical objects that have been known since the 1930s, and many excellent reviews exist on the topic. (Matsika & Krause, 2011; Worth & Cederbaum, 2004; Yarkony, 1998) They were well known early on to exist in high symmetry species, and manifest through phenomena like the Jahn-Teller effect, showing that molecules in orbitally degenerate states undergo non-totally symmetric distortions to remove the degeneracy.(Paterson, Bearpark et al., 2005) However, advances in electronic structure theory in the last 20 years or so have highlighted that accidental conical intersections, not requiring high symmetry, are exceptionally common in polyatomic systems, and especially in heteroaromatic molecules, which is of interest from a photosensitiser point of view.

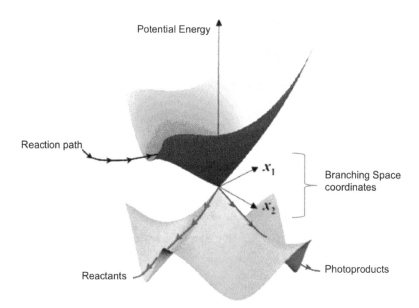

Fig. 5. When the energies of the crossing surfaces are plotted agains two coordinates making up the so-called *branching space*, the surfaces meet in a characteristic double cone shape forming a funnel along the reaction path.

The experimental evidence for the presence of conical intersections often require advanced set-ups and generate complex data sets, such that a theoretical input is often required for full characterisation.(Hadden, Wells et al., 2011; Oliver, King et al., 2011) In non-radiative processes, where a system moves from potential energy surface to another, this population transfer is most efficient at points of degeneracy (i.e., conical intersection seams). Furthermore, the timescale for such transitions is frequently found to be ultrafast (i.e., femtosecond) as a single molecular vibration can cause a change in electronic state. Clearly non-adiabatic terms are very important, as the dynamical processes involve the coupling of nuclear and electronic motion to be taken into account if the full reaction path is to be understood.(Worth & Cederbaum, 2004)

Modern computational codes can straightforwardly locate stationary points on a surface, but in the case of two surfaces crossing, the actual crossing point is not a stationary point of either of the surfaces. The intersection seams are of dimension 3N-8 for an N-atom molecule (every point on the seam being a conical intersection geometry). Modern optimisation algorithms optimise to points of zero gradient in this so-called intersection space, generating minimum energy crossing points.(Paterson, Bearpark et al., 2004; Yarkony, 2005) These act as photochemical funnels for the reaction in the upper electronic state. It is important to note however that this minimum energy crossing point might not be the only one of importance when it comes to the generalisation of various reaction paths, but a good characterisation of the full seem is crucial.(Paterson, Robb et al., 2005) The procedure that is needed to find the minimum energy crossing point is actually fairly straight forward, at least in principle, and involves computing both the energy and its gradient on both surfaces, as well as the non-adiabatic coupling, which essentially describes the extent to which the *electronic* states are coupled by *nuclear* motion (i.e., a break-down of the BO approximation discussed earlier). Multiconfigurational methods are the most common, and appropriate, for the calculations of the IC pathways, as they can handle the multi state character needed near a crossing point. However, as the ubiquitous nature of conical intersections, and their importance in the description of photochemistry, in molecules with biological importance, such as photosensitiser molecules, becomes more evident there is a need to develop methods that can handle systems of a larger size. The use of TD-DFT, which is successful for excited states in the vertical region, in the characterisation of crossing points have been investigated, and has been shown to be of some use for the location of minimum energy crossing points between excited levels (i.e., not with the ground electronic state) They can not yet describe excited state surfaces accurately in general, so care must be taken in their application to problems in reactive photochemistry as opposed to spectroscopy. (Dreuw & Head-Gordon, 2005)

When it comes to the description of surface crossings of larger molecules, alternative multiconfigurational approaches are required. This can be carried out in two general ways: the truncation of the computational method we want to use, or the truncation of the molecule to a representative model system. Bearpark et al, investigated a collection of photochemical reactions involving multi-state processes using a CASSCF starting point.(Bearpark, Ogliaro et al., 2007) They primarily considered so called *hybrid methods* where the molecular system is partitioned so that one area, such as the reaction centre or chromophore, is treated with a more accurate, and computationally more expensive, model.

In addition they investigated a conventional method, where the whole system is treated at the same level of theory, although biased towards a particular region of the molecule. At the moment the practical limit of the use of CASSCF lies in the region of 14 active electrons in 14 active orbitals. Methods that can combine a quantum mechanical method (QM) with a Molecular Mechanics method (MM) (which determines molecular properties through the use of empirically built force fields) are generally referred to as hybrid methods. Hybrid methods, such as ONIOM investigated in this study, rely on the fact that some parts of the molecule are more crucial in the representation of the chemical reaction we want to study, whilst the remainder are viewed as playing a minor role. The ONIOM version of QM/MM method has the advantage of being very general, as in principle more layers of levels of theory can be applied. This method was found to be appropriate for excited state potential energy surfaces, provided the chromophore could be accurately separated from the full molecule. This highlights the importance of these models in the investigations of chromophores in for example a protein or solvent environment. When it comes to molecules where the chromophore is essentially the entire system however, these methods were not always appropriate. This is a key factor when dealing with photosensitisers for use in PDT, as they normally are of a large structure incorporating many aromatic features, such as porphyrins, where the whole system is crucial for the modelling of its photochemistry. For molecules such as this the strategy would be to turn to models where the method itself, rather than the molecular structure, has been truncated, such as the RASSCF approach.(Merchan, Orti et al., 1994; 1994) The RASSCF method is analogous to CASSCF where the active space has been divided into smaller parts, as to reduce the total number of electron configurations generated. The active space is separated according to maximum and minimum occupancy criteria, where the RAS 1 space only allows a limited number of vacancies, holes, and the RAS 3 space only allows a limited number of electrons. In the remaining RAS 2 space a full CI is carried out, as in the general CASSCF approach. In this study it was found that the usefulness of the various methods studied was very system-dependent. However, providing the partitioning of the system, or selections of orbitals in the RASSCF case, were adequate the calculations lead to small errors when modelling the excited state surfaces, when compared to CASSCF, on a 14-electron system.(Bearpark, Ogliaro et al., 2007)

We are currently investigating the use of RASSCF for analysis of surface crossings in molecules of the size of common photosensitiser molecules. The RAS space chosen here highlights the difficulty in the use of multiconfigurational methods for large molecules, as essentially the whole molecule forms a crucial part of the chromophore (Fig 7). The selection of the orbitals to be included in RAS-2 has to be founded on photochemical background of the problem. In the case of porphyrin derivatives it is well known that the two dominating absorption bands in the spectrum (the Soret- and Q-bands) arise from excitation from the HOMO and the HOMO -1 to the LUMO and the LUMO+1. The use of these FMOs to describe the absorption characteristics of porphyrin systems is termed the *Gouterman Four Orbital* (GFO) model and it is obvious that an accurate description (i.e., full CI in the RAS 2 space) of these orbitals will be crucial for an accurate RASSCF calculation (Fig 7). A methodical description of a 26-electron 24-orbital system like these is already a challenging task, and in order for these methods to be useful for the investigations into photosensitiser molecules with various specialist properties, like large TPA cross section, advances in both electronic structure codes and computational performance will be crucial and lead to important insight for these types of investigations.

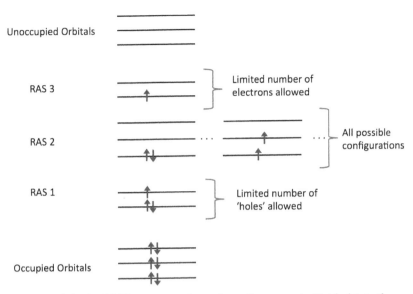

Fig. 6. Overview of the RASSCF strategy, where the active space is divided into three parts, RAS 1-3.

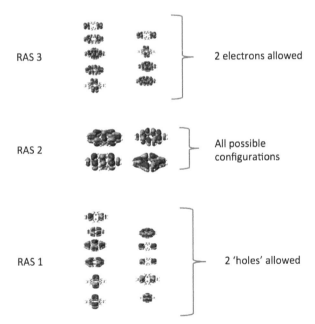

Fig. 7. Overview of the orbitals chosen in the description of the porphyrin chromophore, where the whole π-system have to be accounted for. RAS 2 includes the frontier molecular orbitals, FMOs, of the porphyrin macrocycle, as these have been shown to be vital for an accurate description of the absorption in molecules of this type.

In the computation of the crossing of states with different spin symmetry, ISC, it is necessary to consider spin-orbit coupling, the coupling of the net spin and orbital angular momenta of the electrons. This can be viewed as a small amount of spin being mixed between two states, which lead to intensity from 'spin-allowed' transitions getting transferred to formally 'spin-forbidden' transitions, hence facilitating the increased probability of the latter. The process of ISC can be modelled through multiconfigurational methods, as described above for an IC process. It should be noted that for weak spin-orbit coupling in photosensitisers the geometries at which the singlet and triplet surfaces cross are the dominant features that need to be understood. These crossings are 3N-7 dimensional seams in contrast to conical intersections due to the non-adiabatic coupling being zero by symmetry. Standard conical intersection algorithms can be routinely applied to find minimum energy points on these ISC seams. The use of RASSCF to model this is now possible given recent advances in the RASSCF method, in particular the use of gradient driven RASSCF.

There has also been considerable development in the use of DFT to model a change in spin in the ground state of a molecule, mainly spurred from the interest in the modelling of chemical reactions involving a change of spin in transition metal compounds, which can be of interest from the point of view of photosensitiser molecules.(Poli & Harvey, 2003) Even though DFT has had most success as a ground state method, the modelling of crossings between spin-states can be carried out effectively, as the event can be viewed as a single surface reaction. This is done by considering the reactants and products of the photochemical reaction as being located on two diabatic surfaces that have significant mixing in the area of the transition state. If the surfaces of the two spin-states are described accurately by DFT, the method can be used with good results to predict the energy of the crossing point between them. Of course this only applies to the crossings between the lowest energy states of each multiplicity. The main issue with DFT however is still the choice of functional to apply to the problem, as there is no direct recipe even for what class of functional will preform better in these cases. (Reiher, Salomon et al., 2001)

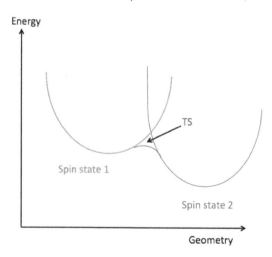

Fig. 8. Two diabatic surfaces, each desribing a spin state, crossing in an area which can be viewed as the adiabatic TS.

3. Photosensitisers in PDT

As in the development of any drug molecule there are many aspects to consider in the advancement of an ultimate photosensitiser molecule, such as pharmacokinetics, targeting and cytotoxicity. However, in the case of a photosensitiser molecule, there is also the absorption qualities to consider and, in particular, the location of the wavelength of maximum absorption in relation to the of tissue transparency window.

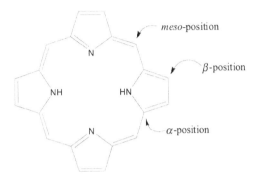

Fig. 9. The Porphyrin free base, with crucial structural sites indicated.

All modern photosensitisers used in practice today are based on the porphyrin macrocyclic structure, and PDT with porphyrin derivatives have been under development since the 1960s (Fig 9).(Peng & Moan, 2003) In the body, the most recognisable porphyrin derivative that exists naturally is coordinated to Fe in the haem structure, present in the haemoglobin protein. The unchelated porphyrin structure is known as protoporphyrin IX (PpIX), which has been shown to be an effective photosensitiser.(Kennedy, Pottier et al., 1990) The strategy for PpIX treatment is to administer a synthetic version of δ-Aminolaevulic acid, the biosynthetic precursor to PpIX, which leads to a build-up of PpIX in the targeted area. In Europe, PpIX is approved for use in PDT of actinic keratosis and basal cell carcinoma.(Morton, McKenna et al., 2008) The most common photosensitiser in use today, however, is still the heamatoporphyrin derivative (HpD), which consists of a mixture of oligomeric porphyrin chains. In its purified form it is known as Photofrin®, which was the first PDT sensiter to be approved for clinical use.(Elsaie, Choudhary et al., 2009) Photofrin or HpD is approved in Europe, the USA, Canada and Japan and is used mainly for cervical, bladder and gastric cancers.

Although Photofrin®, and other porphyrin derivatives, show excellent singlet oxygen quantum yields the longest wavelength at which they are excited is at 630 nm.(Dolphin, Sternberg et al., 1998) This maximum absorption is within the transparency window of human tissue as desired but not at an optimum position. Light of a longer wavelength would penetrate deeper and the clinical use of the photosensitiser would be expanded. This has lead to extended research in ways to manipulate the wavelength of maximum absorption in porphyrin related molecules and has lead to substituted porphyrins being the most studied class of sensiter molecules.(Peng & Moan, 2003) One strategy to alter the porphyrin structure is to add substituents in the meso-position. Meso-substituted porphyrins were investigated by Oliviera and co-workers, and they could demonstrate that unsymmetrical substitutions on the porphyrin molecule had neglible effect on the absorbance as well as on the fluorescence spectrum. The effect on the singlet oxygen quantum yield was also neglible. These were significant results, as they points out a porphyrinic site that is available for potential modification with drug targeting in mind, without necessarily affecting the chromophore itself.(Oliveira, Licsandru et al., 2009) However, most known porphyrins do not absorb at longer wavelengths than Photofrin, so it is likely that interest as well as importance of pure porphyrins for use a photosensitiser agents in PDT will decrease in the future, whilst concentration is being shifted towards porphyrin *derivatives*, so called *second generation* photosensitisers.

Substituting one N atom in the porphyrin core for a heavier S or Se atom, in order to facilitate larger spin-orbit coupling, have little effect on the absorption spectrum, apart from a shift in the absorption maximum to slightly longer wavelengths (~665 nm), (Rath, Sankar et al., 2005) (Detty, Hilmey et al., 2002) however without any effect in the generation of singlet oxygen. Substitution of a further N atom with S or Se moves the absorption maximum to longer wavelengths, but shows a negative effect in the generation of singlet oxygen.(Detty, Gibson et al., 2004) Even though the singlet oxygen generation is decreased in the structures where two core heteroatoms have been substituted, they have been proven to be more effective than meso-substituted pure porphyrin cores in in vivo studies.(Detty, Gibson et al., 2004) The avenue for the use of core-substituted macrocycles in TPA PDT is still in its early stages of investigation, but does show some promise.(Bergendahl & Paterson, 2011)

Porphyrin isomers that has one of their β-carbons and pyrrolic nitrogen atoms switched are termed confused porphyrins and they show a maximum absorption at longer wavelengths than porphyrin (~ 730 nm).(Furuta, Asano et al., 1994) Even though these types of compounds have potential as photosensitisers, with moderate singlet oxygen yield, their difficult total synthesis is a major obstacle for their development as photosensitiser agents in PDT.

A molecular structure related to the porphyrin core can be found in the porphycene compounds, a further reduced porphyrin isomer. The porphycene core structure have been shown to experience a large absorption at slightly longer wavelengths than porphyrins (~ 633 nm) which is thought to be due to lower molecular symmetry, and in vitro studies prove a modest singlet oxygen quantum yield.(Stockert, Canete et al., 2007) Porphycenes have also been shown in in vivo studies to exhibit selective uptake in membrane related organelles, which was proven to ultimately lead to apoptosis of the cells after application of radiation. (Stockert, Canete et al., 2007). The importance of the porphycene system has been suggested to lie mainly in the application of TPA PDT, as early theoretical studies suggests its superiority at effective TPA compared to many related porphyrin derivatives.

As the main focus of the investigations into effective photosensitiser system had been dominated by synthetic developments of porphyrin related systems, this is also the avenue that recent theoretical efforts have gone down. Hence the theoretical methods mentioned in this chapter have been discussed with systems like these in mind. It is however important to keep in mind that non-porphyrin photosensitisers also are being investigated, including various dye molecules and systems on the nano-scale such as semiconductor quantum dots and fullerenes.(Bakalova, Ohba et al., 2004; Mroz, Tegos et al., 2007) As the field of theoretical chemistry develops, with improvements in algorithms as well as advances in computational hardware, the input into PDT research has potential of becoming very important in the future.

4. Oxygen in PDT

The species we have so far referred to as 'singlet oxygen' in our discussion on PDT is the first excited state of molecular oxygen ($a^1\Delta_g$). Reactions involving interactions from this state play a key role in many photodegradation processes in photochemistry as well as photobiology and various applications exists, aside from PDT, examples including mechanistic cell death studies, bleaching processes and general organic synthesis applications. (Schweitzer & Schmidt, 2003)

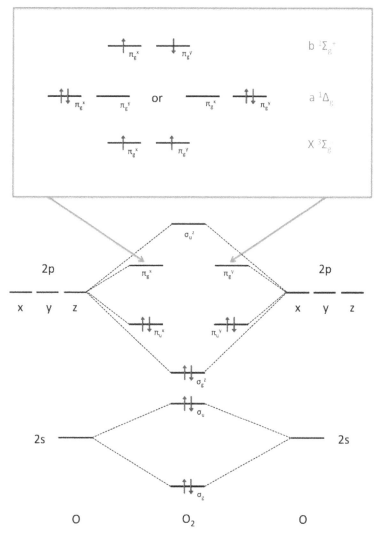

Fig. 10. Molecular orbital diagram for oxygen. The occupation of the antibonding π-orbitals, which give rise to the various excited states of O_2, is indicated.
Adapted from (Paterson, M. J., Christiansen et al., 2006)

The a $^1\Delta_g \leftarrow X^3\Sigma_g^-$ transition is arguably the most forbidden electronic transition in nature, with violation of not only the spin selection rule, but parity and orbital selection rules as well, and in fact all the radiative processes of the singlet states of the O_2 molecule are forbidden to a various degree. This leads to an extraordinary metastability of these states in the gas phase, with lifetimes of 45 min for the $^1\Delta_g$ state and 7.1 sec for the $^1\Sigma_g^+$ state.(Kasha & Brabham, 1979) Crucially, however, the degree that these radiative transitions are forbidden decreases rapidly as soon as the molecule is perturbed by other atoms or molecules. The nature of this perturbation is generally thought of as a weakly bound collision complex, M-O_2, where M could be a number of species, including solvents or O_2 itself. In the case of sensitisation $O_2(X^3\Sigma_g^-)$ forms a collision complex with an excited state of a photosensitiser molecule, as will be described more later. The weakness of this interaction presents problems for computational modelling. Firstly, many possible conformations may be possible and some dynamical sampling will be required. There are at present no good force-fields to model this with MD. Secondly, methods such as DFT do not generally describe such non-covalent interactions very well, although empirically corrected functionals describing such dispersion effects have recently been developed.

As illustrated in the molecular orbital diagram in Figure 10, 14 electrons (including the $1s^2$ pairs) of the O_2 molecule can be paired up in the lowest lying molecular orbitals. This leaves two remaining electrons that have to occupy a doubly degenerate set of orbitals (π_g). There are several possible ways to accomplish this. This is a basic and simple view of the electronic structure of O_2, but it highlights some important aspects of the system, such as the fact that the differences in the three lowest lying states in O_2 are only a result of the pattern in which the final two electrons occupy two orthogonal and degenerate molecular orbitals. It also highlights the fact that the photochemically important $a^1\Delta_g$ and $X^3\Sigma_g^-$ states are both open-shell states, i..e, states involving partially occupied orbitals. This open-shell character is crucial to consider if we are to attempt to accurately model O_2, and M-O_2 collision complexes computationally.(Schmidt & Schweitzer, 2003) (Paterson, Christiansen et al., 2006) When modelling the oxygen molecule, and its role in PDT, the important states are in fact all of open-shell character to some extent, and often require several electronic configurations to describe them. This open-shell character gives rise to physical phenomena that a computational method needs to be able to account for. One example is so called *spin-polarisation*, which accounts for the fact that up and down spin electrons will respond differently to an excess of spin density, which in turn affects the spatial distribution of up and down spin electrons, meaning that it is no longer equal for each as is the case for a closed shell wavefunction. The wavefunction should also be constructed so that spin angular momentum is accounted for properly, i.e., the wavefunction should be an eigenfunction of the spin angular momentum operators describing the total spin (S) and the projection of spin along one direction (the operators $\hat{S}^2$ and $\hat{S}_z$ with eigenvalues S(S+1) and M_s respectively). This ensures that the wavefunction describes a pure spin state, without being *contaminated* by terms from other spin states that describe additional unpaired electrons in the model, a problem referred to as spin-contamination. (Jensen, 2007) High-spin open-shell states such as $X^3\Sigma_g^-$ can be treated qualitatively using a single-configurational model, while low-spin open-shell states (including $a^1\Delta_g$) cannot. If one wishes to use a single configurational model then both the polarisation and contamination cannot be simultaneously accounted for. One can either use a restricted approach in which spin up and spin down electrons are maximally paired and share a common spatial component, or one can use an unrestricted approach in which each up and down spin electron have different spatial orbitals. In the restricted approach the single configuration is automatically an eigenfunction of $\hat{S}_z$ with an eigenvalue corresponding to half the excess number of spin up electrons (using the convention of the open-shell electrons having spin up), and the eigenvalue of $\hat{S}^2$, corresponding to a pure value of S, given by the sum of spin up quantum numbers for all open-shell electrons. Since these open-shell electrons see the same spatial distribution of the paired closed shell spin up and spin down electrons there is no polarisation leading to non-physical effects such as no hyperfine coupling of excess spin-density to nuclear spin. On the other hand using an unrestricted approach allows for spin-polarization as the spin up and spin down orbitals can have different amplitudes in different regions of space. Such unrestricted wavefunctions mix in states of higher S (but common M_s). In general the higher this spin-contamination (i.e., how far above S(S+1) the expectation value $<\hat{S}^2>_U$) the worse the description of the state, and properties, including molecular geometries, are less accurate. Thus, one needs to go beyond a single configuration to describe these high-spin open-shell states more accurately, and such methods are the only starting point for the low-spin open-shell states. For a fuller description see the excellent review by Bally et al. which has a chemical discussion of all these aspects.(Bally & Borden, 2007) Multi-configurational methods, as described in 2.1.3, are therefore essential for a balanced treatment of all the states in the oxygen molecule, and this includes when oxygen weakly interacts with a sensitizer, the topic covered in the next section.

4.1 Photosensitisation

In the discussion of the various aspects of the photochemistry that underpins the pathways involved in PDT, we have now arrived at the crucial stage where the reactive singlet oxygen species is generated through photosensitisation by the excited state of the irradiated photosensitiser molecule. The direct absorption of pure gaseous O_2 is very weak due to the nature of the electronic transitions, as discussed above, and in order to describe the generation of singlet oxygen through sensitisation, the nature of the perturbed O_2 molecule needs to be considered.(Badger, Wright et al., 1965) As mentioned previously, the perturbation is normally viewed as taking place through a collision complex between oxygen and the perturbing molecule, in our case the photosensitiser molecule, PS. This complex, which we will denote as PS-O_2, and a selection of the lowest lying electronic states in the complex, is described in figure 11.

One of the first attempts to describe the mechanisms behind photosensitisation originated from observations involving heteroaromatic solvent molecules as perturbers of the O_2 molecule. The fact that oxygen interacts with heteroaromatic organic molecules was first inferred in the 1950s when it was shown that samples where oxygen had been dissolved gave rise to an absorption band at longer wavelengths than the aromatic compounds alone. The first suggestion as to the nature of this interaction, and its precise mechanism, was presented in a seminal paper in the 1960s when Mulliken and Tsubomura suggested it to be due to absorption by a charge transfer complex (denoted ^{1,3}(PS$^+$O$_2^-$) in Figure 11) originating from the interaction between oxygen and the aromatic perturbing molecule, where the oxygen has the role as electron acceptor. The interaction in these complexes were considered

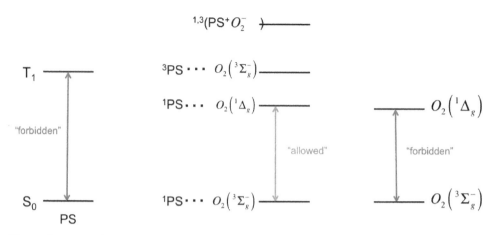

Fig. 11. Diagram illustrating some states of the PS-O2 complex. As new wavefunctions are needed to describe the complex, transitions that were forbidden in the isolated molecules can now become allowed.

to be very small, as the added electron in O_2^- will vacate one of the already occupied π_g orbitals, which have a very small overlap with the heteroaromatic n-donor orbitals.(Tsubomura, 1960) The mechanism of the enhancement of a transition in O_2 through a collision complex has been suggested to be due to a 'borrowing' of intensity from the spin allowed transition. The degree of enhancement in these cases will however depend on the nature as well as the orientation of the collider, as the degenerate anti-bonding $\pi_{g,x}$ and $\pi_{g,y}$ will be perturbed differently depending on the nature and the *orientation* of the valence MOs of the colliding sensitiser molecule. In the case of a photosensitiser the size of a porphyrin derivative, this is a crucial point in the modelling of the charge transfer event.

Several more possible mechanisms were suggested through the years, where the main theme was the enhanced intersystem crossing of the photosensitiser, where the states are mixing either directly or through charge transfer (CT) states of the complex. Kawaoka however later suggested that there might not be a need to populate the actual CT states of the PS-O2 complex in order to generate singlet oxygen, as its states were mixed to a varying degree.(Kawaoka, 1966) It was suggested that CT character will in fact have a negative effect on the production of $O_2(a^1\Delta_g)$. This holds true for intramolecular CT within the photosensitiser, which facilitates coupling of an excited state to a ground state, as well as for intermolecular CT. The latter is normally described as the degree of mixing of the CT states with the excited or ground state of the exciplex. (Kawaoka, 1966) If the PS-O2 complex has large CT character, radicals are easily formed, which has a negative effect on the singlet O_2 yield. Also, the CT state will facilitate coupling between excited PS-O2 states and the ground triplet state $^3(^1PS-^3O2)$. A further mechanism has also been suggested as it has been shown that $O_2(^1\Delta_g)$ can be produced through a *direct* cooperative absorption by the PS-O2 $(^3\Sigma_g^-)$ complex. (Pagani, Salice et al., 2010; Scurlock & Ogilby, 1988) This highlights the fact that many uncertainties still exist with regards to the understanding of the actual mechanism of the charge transfer pathway taking in the photosensitisation process.

It has become clear that O_2 quenches excited states of organic molecules of both multiplicities, with varying rates depending on the environment. The quenching of the S_1 state does not generally contribute to the generation of singlet oxygen, but leads to the formation of T_1 of the sensitiser, through O_2 enhanced ISC, or the generation of the sensitiser singlet ground state, O_2 enhanced IC, with regeneration of the $O_2(^1\Sigma_g^+)$.(Schmidt & Schweitzer, 2003) With respect to PDT applications, quenching of the S_1 of the sensitiser can therefore be looked upon largely as a loss process.

In terms of modelling the interaction between the oxygen and the sensitizer one is faced with several problems. Firstly the weak nature of interaction complex, and the possibility of many geometrical conformers contributing in the condensed phase has to be considered. Secondly, and the open-shell nature of the collision complex requires the same techniques as isolated O_2 requires, i.e., multi-configurational methods. Given the often large size of the sensitiser this presents many challenges to theory. In fact it is fair to say that accurate multi-state, multi-configurational investigations of this system have not been attempted to date. Modern methods (for example as discussed in section 2.2.3) can in principle be applied to the full interaction complex. These when combined with molecular dynamics simulations, and sampling, offer up the possibility that this final crucial step in PDT can be modelled as accurately as the other steps and thus hopefully lead to predictive use of modelling in the design of new PDT drugs.

5. Conclusions

We have introduced Photodynamic Therapy in terms of the photochemical pathways taking place within a target biological system. We presented an outline of the challenges associated with the modelling photosensitiser molecules, focusing on the issues related to electron correlation problem, in terms of methods using the electronic wavefunction as well as methods based on the electronic density of the system. Further we have highlighted the computational advances that have made it possible to accurately describe both linear and non-linear absorption processes in molecules the size of common photosensitisers. An accurate description of these processes can be crucial in the development of photosensitisers with absorption maximum within the optical window of tissue penetration. The modelling of ultrafast processes taking place from an excited state of a molecule is challenging and we have summarised the issues associated with the description of these processes in large molecules. Finally, we have discussed the difficulty in an accurate theoretical description of the energy transfer step involving molecular oxygen present in the target cell, mainly in terms of the challenges associated with the modelling of open-shell systems.

Despite several drawbacks, including poor light penetration of cancer tissue and poor targeting of the photosensitiser systems, PDT remains a very attractive goal, and research continues through many avenues. The search for the ideal photosensitiser, which will make PDT a realised anti-cancer treatment for general use is very much under way and the input that can be gained from computational studies cannot be ignored, as it has the potential to increase the understanding of the entire photochemical pathways involved. Computational techniques have now matured so that molecules the size of standard photosensitisers can be investigated routinely for their absorption properties. This is also true for non-linear absorption properties, which means that this approach to PDT can be developed and hence potentially improve the scope of the technique, and be an important tool both when rationalising the existing classes of photosensitizers and the design of future photosensitisers with desired molecular properties.

Even though significant challenges clearly still remain, the insight a theoretical input can give to the understanding of the fundamental photochemical mechanisms has potential to greatly accelerate the development of PDT as an everyday technique.

6. Acknowledgements

We thank the European Research Council (ERC) for funding under the European Union's Seventh Framework Programme (FP7/2007-2013)/ERC Grant No. 258990.

7. References

Allison, R. R., G. H. Downie, et al. (2004). Photosensitizers in clinical PDT. *Photodiagnosis and Photodynamic Therapy*,1: 27-42.

Arnbjerg, J., A. Jimenez-Banzo, et al. (2007). Two-photon absorption in tetraphenylporphycenes: Are porphycenes better candidates than porphyrins for providing optimal optical properties for two-photon photodynamic therapy? *Journal of the American Chemical Society*,129, 16: 5188-5199.

Arnbjerg, J., M. J. Paterson, et al. (2007). One- and two-photon photosensitized singlet oxygen production: Characterization of aromatic ketones as sensitizer standards. *Journal of Physical Chemistry A*,111, 26: 5756-5767.

Badger, R. M., A. C. Wright, et al. (1965). Absolute Intensities of the Discrete and Continuous Absorption Bands of Oxygen Gas at 1.26 and 1.065 μ and the Radiative Lifetime of the 1Dg State of Oxygen. *The Journal of Chemical Physics*,43: 4345-4351.

Bakalova, R., H. Ohba, et al. (2004). Quantum dots as photosensitizers? *Nature Biotechnology*,22, 11: 1360-1361.

Bally, T. and W. T. Borden (2007). Calculations on Open-Shell Molecules: A Beginner's Guide. *Reviews in Computational Chemsitry*,13.

Bast, R., U. Ekstrom, et al. (2011). The ab initio calculation of molecular electric, magnetic and geometric properties. *Physical Chemistry Chemical Physics*,13, 7: 2627-2651.

Bearpark, M. J., F. Ogliaro, et al. (2007). CASSCF calculations for photoinduced processes in large molecules: Choosing when to use the RASSCF, ONIOM and MMVB approximations. *Journal of Photochemistry and Photobiology A-Chemistry*,190, 2-3: 207-227.

Bergendahl, L. T. and M. J. Paterson (2012). Two-Photon Absorption in Porphycenic Macrocycles: The effect of Tuning The Core Aromatic Electronic Structure. *Chemical Communications*, 48: 1544-1546

Bonnett, R. (2000). *Chemical Aspects of Photodynamic Therapy*. Amsterdam, Gordon and Breach Science Publishers.

Brackett, C. M. and S. O. Gollnick (2011). Photodynamic therapy enhancement of anti-tumor immunity. *Photochemical & Photobiological Sciences*,10, 5: 649-652.

Cramer, C. J. (2004). *Essentials of Computaional Chemistry - Theories and Models*. Chichester, John Wiley & Sons Ltd.

Crawford, T. D. and H. F. Schaefer (2000). An introduction to coupled cluster theory for computational chemists. *Reviews in Computational Chemistry, Vol 14*,14: 33-136.

Davids, L. M. and B. Kleemann (2011). Combating melanoma: The use of photodynamic therapy as a novel, adjuvant therapeutic tool. *Cancer Treatment Reviews*,37, 6: 465-475.

Detty, M. R., S. L. Gibson, et al. (2004). Current clinical and preclinical photosensitizers for use in photodynamic therapy. *Journal of Medicinal Chemistry*,47, 16: 3897-3915.

Detty, M. R., D. G. Hilmey, et al. (2002). Water-soluble, core-modified porphyrins as novel, longer-wavelength-absorbing sensitizers for photodynamic therapy. II. Effects of core heteroatoms and meso-substituents on biological activity. *Journal of Medicinal Chemistry*,45, 2: 449-461.

Dolphin, D., E. D. Sternberg, et al. (1998). Porphyrin-based photosensitizers for use in photodynamic therapy. *Tetrahedron*,54, 17: 4151-4202.

Dreuw, A. and M. Head-Gordon (2005). Single-reference ab initio methods for the calculation of excited states of large molecules. *Chemical Reviews*,105, 11: 4009-4037.

Drobizhev, M., A. Karotki, et al. (2002). Resonance enhancement of two-photon absorption in porphyrins. *Chemical Physics Letters*,355, 1-2: 175-182.

Drobizhev, M., F. Q. Meng, et al. (2006). Strong two-photon absorption in new asymmetrically substituted porphyrins: Interference between charge-transfer and intermediate-resonance pathways. *Journal of Physical Chemistry B*,110, 20: 9802-9814.

Elsaie, M. L., S. Choudhary, et al. (2009). Photodynamic therapy in dermatology: a review. *Lasers in Medical Science*,24, 6: 971-980.

Ethirajan, M., Y. H. Chen, et al. (2011). The role of porphyrin chemistry in tumor imaging and photodynamic therapy. *Chemical Society Reviews*,40, 1: 340-362.

Furuta, H., T. Asano, et al. (1994). N-Confused Porphyrin - a New Isomer of Tetraphenylporphyrin. *Journal of the American Chemical Society*,116, 2: 767-768.

Gill, P. M. W. (1998). Density Functional Theory (DFT), Hartree-Fock (HF), and the Self-consistent Field. *Encyclopedia of Computational Chemistry*. P. v. R. Schleyer. Chichester, John Wiley. **1**: 678-689.

Hadden, D. J., K. L. Wells, et al. (2011). Time resolved velocity map imaging of H-atom elimination from photoexcited imidazole and its methyl substituted derivatives. *Physical Chemistry Chemical Physics*,13, 21: 10342-10349.

Hasan, T., J. P. Celli, et al. (2010). Imaging and Photodynamic Therapy: Mechanisms, Monitoring, and Optimization. *Chemical Reviews*,110, 5: 2795-2838.

Hattig, C., O. Christiansen, et al. (1998). Multiphoton transition moments and absorption cross sections in coupled cluster response theory employing variational transition moment functionals. *Journal of Chemical Physics*,108, 20: 8331-8354.

Jensen, F. (2007). *Introduction to Computational Chemistry*. Chichester, John Wiley and Sons.

Johnsen, M., M. J. Paterson, et al. (2008). Effects of conjugation length and resonance enhancement on two-photon absorption in phenylene-vinylene oligomers. *Physical Chemistry Chemical Physics*,10, 8: 1177-1191.

Kasha, M. (1950). Characterization of Electronic Transitions in Complex Molecules. *Discussions of the Faraday Society*,9: 14-19.

Kasha, M. and D. E. Brabham (1979). Singlet Oxygen Electronic Structure and Photosensitization, In *Singlet Oxygen*. H. H. Wasserman. London, Academic Press Ltd.

Kawaoka, H., Khan, A.U., Kearns, D.R. (1966). Role of Singlet Excited States of Molecular Oxygen in the Quenching of Organic Triplet States. *The Journal of Chemisl Physics*,46, 5: 1842-1853.

Kennedy, J. C., R. H. Pottier, et al. (1990). Photodynamic Therapy with Endogenous Protoporphyrin .9. Basic Principles and Present Clinical-Experience. *Journal of Photochemistry and Photobiology B-Biology*,6, 1-2: 143-148.

Koch, W. and M. C. Holthausen (2001). *A Chemist's Guide to Density Functional Theory*. Chichester, Wiley-VCH.

Kowalski, K., S. Krishnamoorthy, et al. (2010). Active-space completely-renormalized equation-of-motion coupled-cluster formalism: Excited-state studies of green fluorescent protein, free-base porphyrin, and oligoporphyrin dimer. *Journal of Chemical Physics*,132, 15.

Li, Q. S., D. Mendive-Tapia, et al. (2010). A global picture of the S(1)/S(0) conical intersection seam of benzene. *Chemical Physics*,377, 1-3: 60-65.

MacDonald, I. J. and T. J. Dougherty (2001). Basic principles of photodynamic therapy. *Journal of Porphyrins and Phthalocyanines*,5, 2: 105-129.

Malmqvist, P. A. and B. O. Roos (1989). The Casscf State Interaction Method. *Chemical Physics Letters*,155, 2: 189-194.

Matsika, S. and P. Krause (2011). Nonadiabatic Events and Conical Intersections. *Annual Review of Physical Chemistry, Vol 62*,62: 621-643.

Merchan, M., E. Orti, et al. (1994). Ground-State Free-Base Porphin - C2v or D2h Symmetry - a Theoretical Contribution. *Chemical Physics Letters*,221, 1-2: 136-144.

Merchan, M., E. Orti, et al. (1994). Theoretical Determination of the Electronic-Spectrum of Free-Base Porphin. *Chemical Physics Letters*,226, 1-2: 27-36.

Missailidis, S. (2008). *Anticancer Therapeutics*. Chichester, John Wiley & Sons, Ltd.

Morton, C. A., K. E. McKenna, et al. (2008). Guidelines for topical photodynamic therapy: update. *British Journal of Dermatology*,159, 6: 1245-1266.

Mroz, P., G. P. Tegos, et al. (2007). Photodynamic therapy with fullerenes. *Photochemical & Photobiological Sciences*,6, 11: 1139-1149.

Nicolaides, C. A. (2011). State- and Property-Specific Quantum Chemistry. *Advances in Quantum Chemistry, Vol 62*,62: 35-103.

Nicolaides, C. A. (2011). State- and Property-Specific Quantum Chemistry: Basic Characteristics, and Sample Applications to Atomic, Molecular, and Metallic Ground and Excited States of Beryllium. *International Journal of Quantum Chemistry*,111, 13: 3347-3361.

Nobelprize.org. (2011). "The Nobel Prize in Physiology or Medicine 1903." Retrieved 9 Sept 2011, Available from:
http://www.nobelprize.org/nobel_prizes/medicine/laureates/1903/%3E.

Oliveira, A. S., D. Licsandru, et al. (2009). A Singlet Oxygen Photogeneration and Luminescence Study of Unsymmetrically Substituted Mesoporphyrinic Compounds. *International Journal of Photoenergy*.

Oliver, T. A. A., G. A. King, et al. (2011). Position matters: competing O-H and N-H photodissociation pathways in hydroxy- and methoxy-substituted indoles. *Physical Chemistry Chemical Physics*,13, 32: 14646-14662.

Olsen, J. (2011). The CASSCF Method: A Perspective and Commentary. *International Journal of Quantum Chemistry*,111, 13: 3267-3272.

Pagani, G. A., P. Salice, et al. (2010). Photophysics of Squaraine Dyes: Role of Charge-Transfer in Singlet Oxygen Production and Removal. *Journal of Physical Chemistry A*,114, 7: 2518-2525.

Papajak, E., J. J. Zheng, et al. (2011). Perspectives on Basis Sets Beautiful: Seasonal Plantings of Diffuse Basis Functions. *Journal of Chemical Theory and Computation*,7, 10: 3027-3034.

Paterson, M. J., M. J. Bearpark, et al. (2004). The curvature of the conical intersection seam: An approximate second-order analysis. *Journal of Chemical Physics*,121, 23: 11562-11571.

Paterson, M. J., M. J. Bearpark, et al. (2005). Conical intersections: A perspective on the computation of spectroscopic Jahn-Teller parameters and the degenerate 'intersection space'. *Physical Chemistry Chemical Physics*,7, 10: 2100-2115.

Paterson, M. J., O. Christiansen, et al. (2006). Invited review - Overview of theoretical and computational methods applied to the oxygen-organic molecule photosystem. *Photochemistry and Photobiology*,82, 5: 1136-1160.

Paterson, M. J., O. Christiansen, et al. (2006). Benchmarking two-photon absorption with CC3 quadratic response theory, and comparison with density-functional response theory. *Journal of Chemical Physics*,124, 5: 054322.

Paterson, M. J., M. A. Robb, et al. (2005). Mechanism of an Exceptional Class of Photostabilizers: A Seam of Conical Interseaction Parallel to Excited State Intramolecular Proton Transfer (ESIPT) in o-Hydroxyphenyl-(1,3,5)-triazine. *Journal of Physical Chemistry A*,109: 7527-7537.

Peng, Q. and J. Moan (2003). An outline of the hundred-year history of PDT. *Anticancer Research*,23, 5A: 3591-3600.

Piela, L. (2007). *Ideas of Quantum Chemistry*. Oxford, Elsevier.

Poli, R. and J. N. Harvey (2003). Spin forbidden chemical reactions of transition metal compounds. New ideas and new computational challenges. *Chemical Society Reviews*,32, 1: 1-8.

Pulay, P. (2011). A Perspective on the CASPT2 Method. *International Journal of Quantum Chemistry*,111, 13: 3273-3279.

Rath, H., J. Sankar, et al. (2005). Core-modified expanded porphyrins with large third-order nonlinear optical response. *Journal of the American Chemical Society*,127, 33: 11608-11609.

Reiher, M., O. Salomon, et al. (2001). Reparameterization of hybrid functionals based on energy differences of states of different multiplicity. *Theoretical Chemistry Accounts*,107, 1: 48-55.

Rudberg, E., P. Salek, et al. (2005). Calculations of two-photon charge-transfer excitations using Coulomb-attenuated density-functional theory. *Journal of Chemical Physics*,123, 18: -.

Schmidt, R. and C. Schweitzer (2003). Physical mechanisms of generation and deactivation of singlet oxygen. *Chemical Reviews*,103, 5: 1685-1757.

Schmidt-Erfurth (2000). Photodynamic therapy with verteporfin for choroidal neovascularization caused by age-related macular degeneration: Results of retreatments in a phase I and 2 study (vol 117, 1177, 1999). *Archives of Ophthalmology*,118, 4: 488-488.

Schweitzer, C. and R. Schmidt (2003). Physical mechanisms of generation and deactivation of singlet oxygen. *Chemical Reviews*,103, 5: 1685-1757.

Scurlock, R. D. and P. R. Ogilby (1988). Spectroscopic Evidence for the Formation of Singlet Molecular-Oxygen (1-Delta-6-O-2) Upon Irradiation of a Solvent Oxygen (3-Sigma-G-O(-)-2) Cooperative Absorption-Band. *Journal of the American Chemical Society*,110, 2: 640-641.

Shao, Y. H., M. Head-Gordon, et al. (2003). The spin-flip approach within time-dependent density functional theory: Theory and applications to diradicals. *Journal of Chemical Physics*,118, 11: 4807-4818.

Stockert, J. C., M. Canete, et al. (2007). Porphycenes: Facts and prospects in photodynamic therapy of cancer. *Current Medicinal Chemistry*,14, 9: 997-1026.

Szabo, A. and N. S. Ostlund (1996). *Modern Quantum Chemistr - Introduction to Advanced Electronic Structure Theory*. Mineola, Dover Publications, INC.

Tsubomura, T. M., S. (1960). Molecular Complexes and their Spectra. XII. Ultraviolet Absorption Spectra Caused by the Interaction of Oxygen with Organic Molecules. *Journal of the American Chemical Society*,82: 5966-5974.

Worth, G. A. and L. S. Cederbaum (2004). BEYOND BORN-OPPENHEIMER: Molecular Dynamics Through a Conical Intersection. *Annual Review of Physical Chemistry*,55: 127-158.

Yarkony, D. R. (1998). Conical intersections: Diabolical and often misunderstood. *Accounts of Chemical Research*,31, 8: 511-518.

Yarkony, D. R. (2005). Escape from the double cone: Optimized descriptions of the seam space using gateway modes. *Journal of Chemical Physics*,123, 13.

Zewail, A. H. (2000). Femtochemistry: Atomic-scale dynamics of the chemical bond. *Journal of Physical Chemistry A*,104, 24: 5660-5694.

Zewail, A. H. (2000). Femtochemistry. Past, present, and future. *Pure and Applied Chemistry*, 72, 12: 2219-2231.

Part 5

Applications of Photochemistry

9

High Power Discharge Lamps and Their Photochemical Applications: An Evaluation of Pulsed Radiation

Lotfi Bouslimi[1,2], Mongi Stambouli[1], Ezzedine Ben Braiek[1],
Georges Zissis[3] and Jean Pascal Cambronne[3]
[1]Ecole Supérieure des Sciences et Techniques de Tunis,
[2]Institut Supérieur de Pêche et d'Aquaculture de Bizerte
[3]Université de Toulouse; UPS, INPT;
LAPLACE (Laboratoire Plasma et Conversion d'Energie)
[1,2]Tunis
[3]France

1. Introduction

The photochemical applications of the ultraviolet (UV) radiation develop with rate accelerated so much in the field of general public technologies as in lighting, descriptive, and imagery, and too of the advanced technologies (treatment and engraving of surfaces, air, water and agro-alimentary treatment). The radiation sources used are generally high, medium and low pressure gas discharge lamps.

In the past decades, gas discharge lamps have gained widespread use in industrial applications. Due to their unique design properties concerning spectral, electrical and geometrical features, all types of gas discharge lamps can been found in technical applications. Mercury based lamps are the workhorses in many applications upgraded by their relatives, the metal halide versions. The low and medium pressure mercury lamps are usually used as sources of UV radiation. Low pressure mercury lamps are used extensively for disinfection of drinking water, packing material and air. Medium pressure lamps are applied in printing industry to dry inks and cure adhesives, in waste water treatment plants to reduce the total organic compounds (TOC) and as a competing technology to low pressure versions in germicidal applications. Metal halide doped versions of medium and high pressure mercury lamps open the possibility to adjust spectral output to specific requirements.

The control of the spectral distribution of energy is considered as the main parameter affecting the system flexibility and the product quality. However, even though the lamp characteristics have an important impact on the spectral distribution of radiation, the power supply characteristics cannot be neglected. Indeed, the temporal characteristics of the system are controlled mainly by the used power supply.

Indeed, in the case of the high pressure lamps, the significant interactions between particles, it is difficult, with traditional power supply (electromagnetic ballasts) to move the energy distribution of the electronic cloud compared to Local Thermodynamic Equilibrium (LTE). However, by using short pulses of current one can hope to obtain such a result and to modify of this fact the distribution of the atomic excitation and the spectral distribution of the radiation (mainly visible and ultraviolet). Former works showed that the form of the current wave imposed on the lamp could be selected so as to improve the production of the radiation (Chalek, 1981; Brates, 1987; Chammam et al., 2005; Mrabet et al., 2006; Bouslimi et al., 2009a, 2009c). It remains, for these sources, to optimize the parameters of excitation (form, amplitude, frequency, duration of pulses) according to those of the discharge (natural of the mixture gas, energy spectral distribution).

Current UV radiation technology is dominated by two techniques, the continuous radiation and pulsed-radiation. The first technique provides a lower-level constant-flux UV radiation. The second technique provides radiation doses through flashing a source lamp. The effect of this pulsing technique is to provide short pulses of higher energy into the system.

The technology of the electronic pulsed supply is a field of studies relatively new related to the development of the generators, switches and electric applications of high energy, with weak durations and face of fast rise. These pulsed operations created by current or voltage pulses produce a pulsed light rich in UV.

The pulsed light system rich in UV radiation from 100 to 400 nm seems to be a promising alternative for the decontamination of the foodstuffs, and the sterilization of packing. Its effectiveness is now fully proven in experiments for decontamination on the surface of the products. Recent studies show the effectiveness of this treatment on products in powder form in fine layer. Bacteria in vegetative form, the ascosporous of moulds, the viruses and the parasites are destroyed by this instantaneous contribution of energy. Many works was completed on the biological effects of the UV carried out an excellent bibliographical analysis on this subject (Mimouni, 2004; Dunn et al., 1997a, 1997b; Dunn et al., 1990; Jagger, 1967; Fine, 2004).

The UV radiation as a disinfection technique has been also proven in multiple industrial applications, especially in the water treatment. Applications for water UV treatment are numerous: Potabilization of water, waste water treatment, treatment of seawater for aquaculture and shellfish culture. An historic perspective on UV disinfection has been published in several review articles (Groocock, 1984; Schenck, 1981; USEPA, 1996; Wolfe, 1990; Zoeteman et al., 1982).

The general objective of this work consists in studying the effect of the current pulses, provided by a feeding system (prototype) designed in our laboratory, on the spectral radiant flux emitted by two types of lamps: high and medium pressure mercury vapour lamps. The first is used mainly for screen printing, copying, and light curing adhesives and varnishes, and the second is germicidal gas-discharge lamps, intended particularly for water treatment. The spectral results obtained by two mode of current, highlight and evaluate the effectiveness of the pulsed current on the radiation production in the ultraviolet and the visible part of the spectrum.

In the remainder of this chapter, we present in the second section an overview of the ultraviolet applications. In the third section, we explore some special lamps for technical applications and their power suppliers. The experimental results of time-dependent electrical and spectral measurements carried out on high and medium pressure mercury lamps operated in pulsed current, are compared with the square wave operation for the same consumption in section 4. The paper is finally summarized with some conclusions in section 5.

2. Overview of UV-lamp applications

2.1 Ultraviolet radiation

Like visible light, Ultraviolet light (UV) is a classification of electromagnetic radiation having a wavelength bandwidth between 100 and 400nm, between the X-ray portion of the spectrum and the visible portion (Fig. 1). UV radiation is subdivided into four wavebands, which we use for a wide range of applications. These four subgroups within the UV spectrum are located in the 100nm - 380nm waveband (Meulemans, 1986):

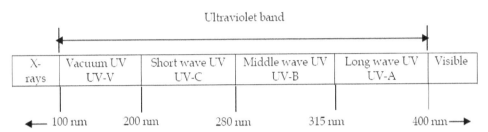

Fig. 1. Ultraviolet bandwidth

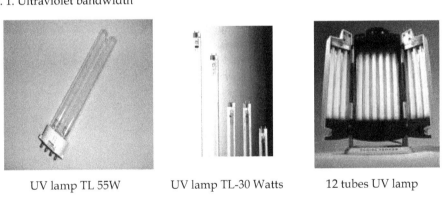

UV lamp TL 55W   UV lamp TL-30 Watts   12 tubes UV lamp

Fig. 2. Ultraviolet lamps for water disinfection

- UVA (380-315nm) is used for curing UV adhesives and plastics. It is also used for fluorescent inspection purposes.
- UVB (315-280nm) is the most energetic region of natural sunlight and is used in conjunction with UVA light for artificial accelerated aging of materials.
- UVC (280-200nm) is used for rapid drying of UV inks and lacquers. It is also used for sterilization of surfaces, air and water
- VUV (vacuum-UV, 200-100nm) can only be used in a vacuum and is therefore of minor importance.

Practical application of UV disinfection relies on the germicidal ability of UVC and UVB and depends on artificial sources of UV. The most common sources of UV are commercially available low and medium pressure mercury arc lamps (Fig. 2).

2.2 UV applications in photochemistry

Photochemistry is the study of the action of light on chemical reactions. In a more precise, it includes works whose purpose is to determine the nature of the reactive excited states of molecules obtained by absorption of light, to study the deactivation process of these states, especially those that lead to products different reagents and irradiated to establish the mechanisms by which rearrangements occurring intra-and intermolecular initiated by radiation (Hecht, 1920).

The chemical reactions induced by light indirectly as a result of electronic energy transfer are an area of study and implementation has long been known (F. Weigert, 1907) and highly developed now. In general, photo-chemical processes are part of different modes of deactivation of molecules previously made in their metastable excited states by absorption of a photon.

Ideally, a photochemical process is performed by irradiating the sample with monochromatic light, since the reaction may depend on the excitation wavelength. Most of polychromatic light sources; the wavelength is selected or required by filters or by a monochromator (Hecht, 1920). Light sources are now almost always discharge tubes containing either xenon or mercury vapor alone or in carefully selected impurities. Some of these lamps are very powerful and they can consume tens of kilowatts of electricity.

The first application of photochemistry was the isomerization of benzene in the liquid state: Under the influence of radiation from a mercury vapor lamp (253.7 nm), the isomerization of benzene in liquid product benzvalene and fulvene, while in the field of wavelength range 166-200 nm, irradiation produces more benzene told Dewar (Fig.3).

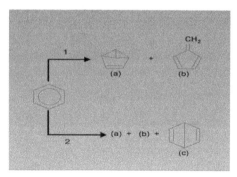

Fig. 3. Isomerization of benzene in liquid

Today industrial, photochemistry has made its biggest breakthrough in the field of setting polymers on different surfaces, such as the "drying" of printing inks and the manufacture of electronic circuits. The notion of quantum efficiency is very important in photochemistry. For that performance is great each application needs a specific wavelength. Although its optimal wavelengths are known. We find that the sources most frequently used are medium pressure mercury lamps (possibly doped with iron iodide) and xenon lamps.

Photochemistry is also used in the curing (polymerization) of specially formulated printing inks and coatings. Since it was originally introduced in the 1960's, UV curing has been widely adopted in many industries including automotive, telecommunications, electronics, graphic arts, converting and metal, glass and plastic decorating.

Ultraviolet curing (commonly known as UV curing) is a photochemical process in which high-intensity ultraviolet light is used to instantly cure or "dry" inks, coatings or adhesives. Offering many advantages over traditional drying methods; UV curing has been shown to increase production speed, reduce reject rates, improve scratch and solvent resistance, and facilitate superior bonding.

Using light instead of heat, the UV curing process is based on a photochemical reaction. Liquid monomers and oligomers are mixed with a small percent of photoinitiators, and then exposed to UV energy. In a few seconds, the products - inks, coatings or adhesives instantly harden.

UV curable inks and coatings were first used as a better alternative to solvent-based products. Conventional heat- and air-drying works by solvent evaporation. This process shrinks the initial application of coatings by more than 50% and creates environmental pollutants. In UV curing, there is no solvent to evaporate, no environmental pollutants, no loss of coating thickness, and no loss of volume. This results in higher productivity in less time, with a reduction in waste, energy use and pollutant emissions.

UV-VIS spectroscopy is one of the main applications of photochemistry. It allows us to determine the concentration of a molecule in a sample, and sometimes, it can aid in identifying an unknown molecule. The molecule being tested must absorb light in the ultraviolet (about 200 to 400nm) or the visible (about 400 to 700nm) range in order to be detected by this equipment. A light beam containing multiple wavelengths gets passed through a small container holding your sample, and the computer records which wavelength(s) were absorbed, and at which intensity.

Another emerging UV application is the photocatalysis. This is the photoactivation of a surface covered with TiO_2 (sometimes doped) which is causing a hyper-hydrophilicity.This process leads us to make self-cleaning surfaces using the following procedure in figure 4.

Usually, this type of process uses UV-C (<180 nm). Xenon lamps with low-pressure radiation at 172nm are well positioned for this application. However, today by doping the layer of TiO_2, we get a photoactivatable produce materials with wavelengths larger located in the UV-A (380 nm). Lamps or dielectric barrier to Xe_2 or Xe-halide combinations seem to be the most promising sources.

2.3 Mechanism of UV disinfection

The UV disinfection process corresponds to the inactivation of microorganisms, following a modification of their genetic information: the UV affect the DNA double helix, as well as RNA, cells, blocking all biochemical processes used for their reproduction. The maximum efficiency of UV disinfection depends on the energy emitted (with peaks near 200 and

260nm), more precisely; it corresponds to output energy of 253.7 nm (absorption peak of UV radiation by micro-organisms) (Wright & Cairns, 1998; Sonntag et al., 1992).

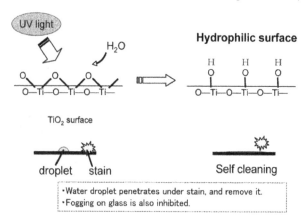

Fig. 4. The process of self-cleaning surfaces

Absorbed UV promotes the formation of bonds between adjacent nucleotides, creating double molecules or dimmers (Jagger, 1967). While the formation of thymine-thymine dimers are the most common, cytosine-cytosine, cytosine-thymine, and uracil dimerization also occur. Formation of a sufficient number of dimmers within a microbe prevents it from replicating its DNA and RNA, thereby preventing it from reproducing. Due to the wavelength dependence of DNA UV absorption, UV inactivation of microbes is also a function of wavelength. Figure 5 presents the germicidal action spectra for the UV

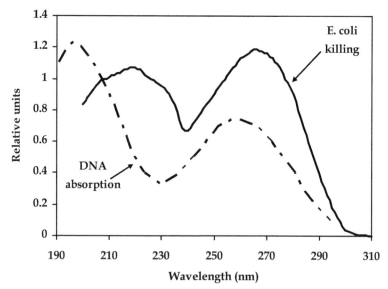

Fig. 5. Comparison of the action spectrum for E. coli inactivation to the absorption spectrum of nucleic acids (Wright & Cairns, 1998)

inactivation of E. coli. The action spectra of E. coli peaks at wavelengths near 265nm and near 220nm. It is convenient that the 254nm output of a low pressure lamp coincides well with the inactivation peak near 265nm (Wright & Cairns, 1998).

2.4 Biological effects of UV radiation

The effects of ultraviolet radiation on living organisms are due to its photochemical action. The best known are the erythema or "sunburn", for which the area of activity is between 320 and 280 nm (with a maximum at 297 nm), and "tan", which involves training, migration and oxidation of melanin, and whose field of activity is wider towards longer wavelengths, which allows you to tan without the risk of rash using products such as filters stopping the radiation of shorter lengths waveform.

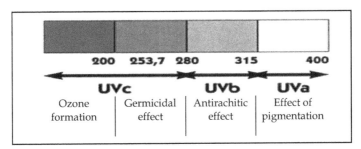

Fig. 6. UV-region spectrum of Sun

In terms of medical treatment, in addition to its use in some diseases of the skin, ultraviolet was mainly used for the treatment of rachitis; its action has the effect of the conversion of vitamin D sterols: direct radiation (sterols present in the skin), irradiation of food containing these elements, or for the direct synthesis of vitamin D.

As for dermatological applications in the fight against diseases such as vitiligo and psoriasis, there are two types of treatment. The first consists in irradiating the skin with a UV-A radiation at 308 nm, which inhibits locally the patient's immune system by calming for period more or less limited effects of the disease. For this application, dermatologists now use lasers. However, dielectric barrier lamps using a mixture of Xe-Cl2 begin to appear. The advantages of systems using these lamps are numerous: they are easy to handle, they require less maintenance, they are lighter and can be portable, compared to a laser. They produce a lower UV power and thus limit the risk of burns. The second method of treatment is called "PUVA". PUVA therapy is a method that combines a photosensitizing drug (in the series of psoralen) administered orally and irradiation of the skin lesions to be treated by long ultraviolet (UVA). The comparison of the effectiveness of each of psoralens used does not show clear-cut superiority of one or the other of them. Their general tolerance appears to be satisfactory, except for minor digestive problems. The tolerances are checked blood and liver in each case respectively by the blood counts, blood count and assay of transaminases. Elevated levels of these enzymes involves discontinuation of treatment is followed by a rapid normalization. It has been shown that the presence of radiation, psoralens are all capable of combining with the pyrimidine bases of DNA chains. This can lead to very different metabolic changes, which would explain why the PUVA appears to have contradictory effects. However, it might be a good alternative to chemotherapy against some types of skin cancer because UV radiations associated with psoralens have the power to destroy the offending cells.

The effects of UV on microorganisms depend on the doses, ranging from the reduction of vital processes (cell division, cell motility, synthesis of nucleic acid) to the destruction of organisms. The germicidal action, observed during exposure to UVC radiation type, is most effective when the wavelength is between 250 and 260 nm (253.7 nm). At this level, the UVC damage the nucleic acids of microorganisms, causing the amount of energy following implementation (afigfoessel.fr):

- A bacteriostatic effect in the case of low radiation level of the cell. In this case it continues to live while unable to reproduce.
- A bactericidal effect in the case of a significant radiation at the cellular level. In this case it is destroyed.

The germicidal action has received applications where mercury vapour lamps are used (253.7 nm): surface sterilization of food products or pharmaceuticals in their packaging, disinfection of objects, air and water (difficult because of the absorption if the water is not pure).

Today we do not really know the answer of microorganisms to UV radiation, but, empirically, we know what is the ultraviolet dose required to kill different microorganisms to a certain percentage (usually for the treatment of water, this is between 90 and 99%).

For water treatment (potable and tertiary) where the rate of destruction of microorganisms required no more than 2log (99%), now the most commonly lamps used are UV lamps, low pressure (using amalgam thereby obtaining high power of about 100 W / lamp) and HID lamps "medium pressure" (pure mercury lamps with power ratings up to 5 kW, see more in some cases). Some systems based on the phenomenon of photo-catalysis are emerging in the market. Regardless of the application cited above photo-biological lamps used do not produce the optimal wavelength.

2.5 Other UV applications

The following table summarizes some other UV wavelength applications "optimal"

Xe2 (172 nm)	KrCl (222 nm)	XeBr (282 nm)
<ul><li>Cleaning of surfaces</li><li>Photochemical vapour deposition</li><li>Modification of structure and composition of surfaces</li><li>Activation of surfaces</li><li>UV matting</li><li>Ozone generation</li></ul>	<ul><li>Photolysis of to hydrogen peroxide</li><li>Inactivation of microorganism</li><li>UV curing for printing processes</li><li>Photochemical vapour deposition</li></ul>	<ul><li>Inactivation of microorganisms</li><li>UV curing for printing processes</li></ul>

Table 1. Various UV wavelength applications

In view of that, practical application of UV disinfection depends on artificial sources of UV and their mode of electrical power supplier. The most common sources of UV are commercially available low and medium pressure mercury arc lamps. The power suppliers (named ballasts) for mostly lamps may be characterized as either electromagnetic or electronic ballasts (O'Brian et al., 1995; Phillips, 1983).

3. Lamps for technical applications and their power suppliers

In addition to low, medium and high pressure mercury discharge lamps, mercury short arc lamps with high operating pressures are found wherever high brightness and good imaging is required, for example in steppers for micro-lithography or as ultra-high pressure types in projectors. Besides the huge field of specialty lighting (stage-studio-TV, floodlights, effect-lighting and car headlights) with special focus on the response function of human eyes, these lamps are also used in reprographic machines, photo-chemistry, medical applications and by the tanning industry. Thus covering the whole field from pure industrial use to directly consumer related applications.

Pure rare gas fillings are used in flash-lamps for pumping the active medium of solid state lasers, whereas long arc xenon lamps satisfy the request for simulating solar radiation in chambers to test the radiation resistance of textiles and colours. Highly stable deuterium-lamps are operating in UV spectrometer and analytical instruments (HPLC, LC) as a source for broadband UV-radiation between 150 and 300nm.

In addition to the above mentioned lamp types, excimer lamps have gained increased interest during the last decade due to their quasi monochromatic spectrum. Intense and efficient UV generations of these lamps have revealed their potentials in the application field of surface modification, cleaning, curing and disinfection.

Discharge lamps are a source of light in which light is produced by the radiant energy generated from a gas discharge. A typical mercury arc lamp consists of a hermetically sealed tube of UV -transmitting vitreous silica or quartz with electrodes at both ends (Phillips, 1983). The tube is filled with a small amount of mercury and an inert gas, usually argon. Argon is present to aid lamp starting, extend electrode life, and reduce thermal losses. Argon does not contribute to the spectral output of the lamp. Most gas discharge lamps are operated in series with a current-limiting device. This auxiliary, commonly called ballast, limits the current to the value for which each lamp is designed. It also provides the required starting and operating voltages.

Ballasts are classified into two major types: electromagnetic ballasts and high-frequency electronic ballasts. The conventional ballast, made of a simple electromagnetic coil, has many significant disadvantages, such as large size, heavy weight, including low-frequency humming, low efficiency, poor power regulation, and high sensibility to voltage changes, etc. Since the electronic ballast can overcome these drawbacks. The high operating frequency allows to the ballast to be smaller and lighter-weight than the electromagnetic ballast. Unfortunately, there is a serious problem of acoustic resonance when the lamps operate in certain frequency range; this phenomenon is even severe for low-wattage lamps. These types of ballasts is more widely developed and used in many applications (Bouslimi et al., 2009b).

The structure of the electronic pulsed power supply developed in our laboratory presents several advantages in this domain. The main advantage of the proposed topology is to provide to the lamp a various shapes of current (square wave, rectangular and pulses) with optimization of the excitation parameter (form, amplitude, frequency, number and duration of pulses).

4. Experimental results

We present in this section, the effect of the current pulses, provided by the feeding system designed in our laboratory, on the ultraviolet and visible spectral flux emitted by two types of lamps: high and medium pressure mercury vapour lamps. In order to highlight and evaluate the effectiveness of the pulsed current on the radiation production, we give a comparison of spectral results obtained by two mode of excitation, rectangular and pulsed current.

4.1 Structure of the pulsed power supply

The bloc diagram of the lamp circuitry is shown in figure 7. The lamp is supplied mainly through an inverter connected with two electrical separate sources: the first source provides a rectangular wave current and the second provides a pulsed current.

The rectangular wave operation is achieved using a (DC) constant current source (S1) in conjunction with an electronic full bridge IGBTs inverter and an active protection system that allows protecting the IGBTs and the drivers against the over-voltage at the time of starting and the hot restarting of the lamp or by an unexpected opening of the circuit.

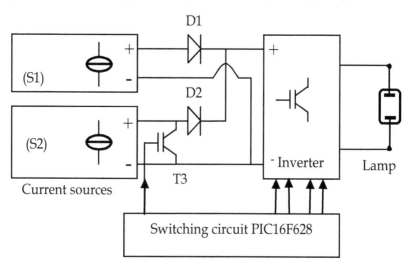

Fig. 7. Bloc diagram of the pulsed power supply

The pulsed operation is achieved by the second (DC) current source (S2) switched by a pulse switching circuit (transistor T3). The control signals for the pulse switching circuit and full wave bridge are ensured by a microcontroller (PIC16F628). The microcontroller provides flexibility in the integration of the two current source AC/DC converters with the full-bridge inverter. It allows obtaining a low-frequency rectangular wave with one or more pulses superimposed on each half cycle. The current in each source is controlled by a regulating circuit. D1 and D2 are anti-return fast diodes (Bouslimi et al., 2008, 2009a, 2009b).

This power supply allows as more studying the energy effectiveness and the photometric behaviour of various gas-discharge lamps (low, medium and high pressure), and this with an aim of evaluating the visible and ultraviolet radiation and of comparing it with the continuous radiation for the same consumption by the discharge lamps.

We also note that the proposed current pulsed power supply can be more exploited in photochemical applications exactly for the treatment of water whose needs a variation of the amplitude and the duration of the pulse (UV dose) according to the virus and bacteria lifespan (Severin et al, 1984). It cans also feeding power lamps going until 3kW.

4.2 The radiation produced by high pressure lamp in pulsed operation

4.2.1 Lamp characteristics

The main characteristics of filling, geometrical and electric of the discharge lamp used in this investigation are consigned in the table below.

Characteristics	Rating values
Diameter (mm)	18.2
Inter-electrode length (mm)	72
Total mercury mass (mg)	70
Argon pressure at the ambient temperature (torr)	10
Power (W)	400
I arc (A)	3.2
V arc (V)	140

Table 2. Characteristics of the studied lamp

The lamp operates vertically through a current inverter and all the measurements have been done in a steady state after the flux and circuit stabilization. Below, we present the results of our electrical and spectral measurements.

4.2.2 Spectral results (Bouslimi et al., 2009b)

Relative average spectral flux was recorded for a rectangular current and a pulsed current. In these two modes, the power provided to the lamp was the same one. Thus, it is possible to evaluate the influence of the current pulses on the radiation production effectiveness in the ultraviolet part and the visible part of the spectrum. Theses results are illustrated in figures 8 and 9.

If you look at the figures above we see that the difference between the spectral results for both modes of operation is small. Better to see the difference, we calculate the total flux through each band. The results for the average values of the total spectral flux determined from figure 8 and 9 above are summarised and given in Table 3.

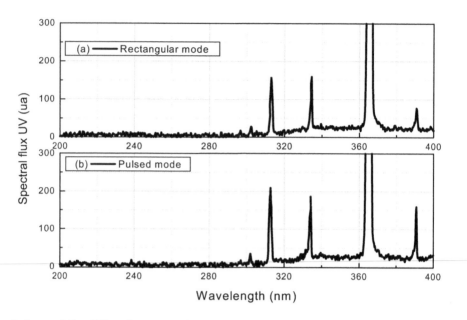

Fig. 8. Spectral flux UV with two supplying modes: (a) rectangular current;
(b) with pulsed current

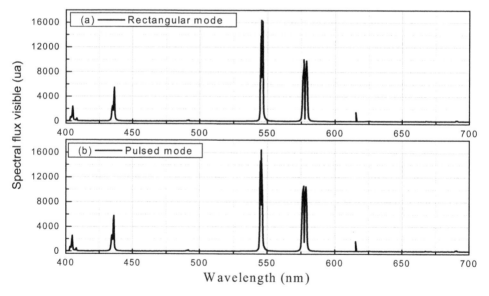

Fig. 9. Spectral flux visible with two supplying modes: (a) rectangular current;
(b) with pulsed current

Spectral Bandwidth (nm)		total UV	Visible
		200-400	400-700
Total flux (u.a)	Rectangular mode	4100	57512
	Pulsed mode (7 pulses per half period)	5510	60296
	Relative progress (%)	34,4	4,84

Table 3. Comparison between the relative total flux of UV and visible radiation bands for two feeding modes of current: rectangular and pulsed

4.2.3 Discussion

We note a clear increase in all the lines measured in the pulsed mode for the same power as in rectangular mode. However, the increase is particularly marked in the ultraviolet band spectrum and limited to the visible (Table 3). We can say that the pulsed mode favors the short wavelengths emission (UV band). This increase is mainly due to rising temperatures in the pulsed mode.

The increase in the UV and visible radiation in pulsed mode compared to the rectangular is confirmed by the results found by (Chammam et al., 2005).

4.3 The radiation produced by medium pressure lamp in pulsed operation

In this part, we present experimental results (electric and spectral) for a medium pressure lamp. This special lamp is provided by the Canadian company Trojan-UV. It is intended for water treatment because it has a broad emission band in the UV and visible spectral range. The geometrical and electrical provided with this lamp are shown in Table 4 below:

Characteristics	values
Inter-electrode length (cm)	25
Diameter (mm)	22
Nominal Arc current (Arms)	6,8
Maximum arc current (Arms)	7,9
Nominal arc voltage (Vrms)	440±5%
Maximum arc voltage (Vrms)	550
Power (W)	3000

Table 4. Electrical and geometrical characteristics of the medium pressure lamp

4.3.1 Electrical measurements

In this part we will present some electrical measurements carried out under pulsed current. To power the lamp at rated power of 3 kW, were overlaid seven pulse of amplitude equal to 4 A on a rectangular current level of 5.5 A in each half cycle of the rectangular current. The pulse duration is about 0.5 ms and the base frequency of the rectangular current is 50 Hz. In figure 10 we represent, the current and the instantaneous power consumed by the lamp in pulsed mode (Bouslimi et al., 2008).

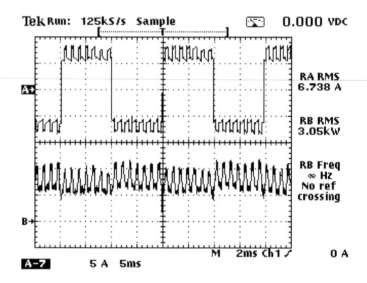

Fig. 10. Instantaneous Current and power in the lamp in pulsed mode, A: Current (5 A/div), B: Power (2 kW/div), Time: 5 ms/div

Note that the instantaneous peak of power in the medium pressure lamp reaches almost twice the level. Thus, it is because the impulses that are causing successive short duration peaks of high power. The radiation produced, called pulsed light, is required by some photochemical applications such as disinfection of wastewater or drinking.

4.3.2 Spectral flux measurements in ultraviolet and visible band

For this lamp, in order to evaluate the influence of pulses on the spectral flux of ultraviolet and visible radiation, spectral measurements are performed with a rectangular and pulsed current. The results obtained for the same power consumed by the lamp are shown in figures (11, 12, 13 and 14).

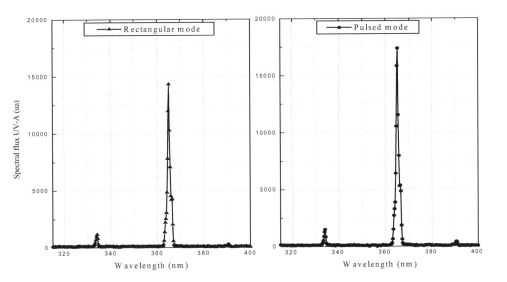

Fig. 11. Spectral flux band UV-A with two feeding modes of current: rectangular and pulsed

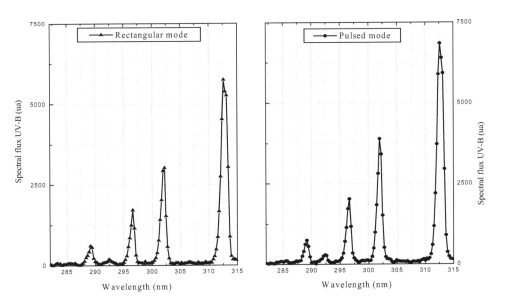

Fig. 12. Spectral flux band UV-B with two feeding modes of current: rectangular and pulsed

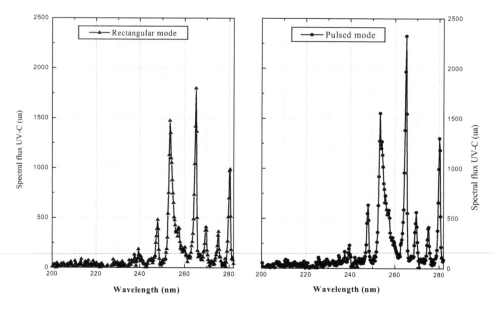

Fig. 13. Spectral flux of UV-C band with two feeding modes of current: rectangular and pulsed

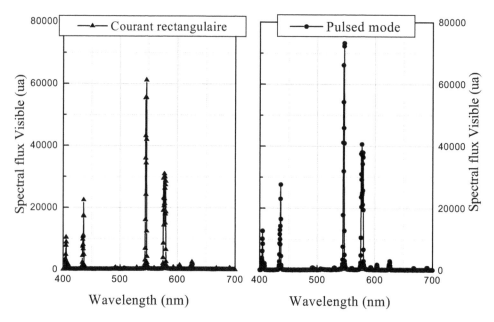

Fig. 14. Spectral flux of visible band with two feeding modes of current: rectangular and pulsed.

Spectral bands (nm)	UVC 200-280	UVB 280-315	UVA 315-400	UV total 200-400	Visible 400-700
Rectangular	11330	16495	29415	57108	308235
Pulsed	14760	20195	35945	70826	380810
Relative increase (%)	30,2	22,4	22,1	24,2	23,5

(Left vertical label: Relative tota flux (u.a))

Table 5. Comparison between the relative total flux of UV and visible radiation bands for two feeding modes of current: rectangular and pulsed (medium pressure lamp)

4.3.3 Discussion of results

In figures (11, 12, 13 and 14) there is a clear increase in the flux of all the spectral lines measured in pulsed mode for the same power in rectangular mode. However, the increase is particularly important for the band UVC spectrum, dominated by the 254 nm line and in particular the molecular line 265 nm, very used to destroy bacteria. Increases in the UVA, UVB and visible, important, too, are substantially identical (about 23%). The increase of the radiation is mainly due to the increase of the electron temperature in the medium pressure discharge lamp. Note that for this lamp the increase is important both in the UV than in the visible.

5. Conclusion

In this work, we have exposed the UV radiation and its applications in photochemistry. The mechanism of UV disinfection and the biological effects are also presented. Some discharge lamps and their power suppliers are showed.

In a great part of this work, we have showing some experimental results carried out on two types of mercury lamp, considered as UV sources: high and medium pressure.

An attempt to raise the efficacy and to improve the performance was made by going to pulse operation instead of operating the arc on a rectangular wave power supply. It is possible with this method to increase the efficacy to sufficiently high values.

The spectral flux results obtained highlight and evaluate the effectiveness of the pulsed current on the radiation production in the ultraviolet and the visible part of the spectrum.

We also note that the improvement of the production of radiation considered, interested many photochemical applications and field lighting.

The applications of the pulsed supply with short duration and sharp dismounted front are considered as relatively recent techniques. It allows us to study in the future, the dynamic behavior of the discharge lamps and their instantaneous effects on the microorganisms in various water treatments (drinking water, waste water, seawater for aquaculture and shellfish culture).

6. References

Bouslimi, L., Chammam, A., Ben Mustapha, M., Stambouli, M. & Cambronne, J.P. (2009a). Simulation and Experimental Study of an Electronic Pulsed Power Supply for HID Lamps Intended for Photochemical Applications. *International Review of Electrical Engineering (IREE)*. Vol. 4, No.5, Part A, (September-October 2009), pp. 799-808, ISSN 1827- 6660

Bouslimi, L., Chammam A., Ben Mustapha M., Stambouli, M., Cambronne, J.P., & G.Zissis. (2009b). Electric and spectral characterization of a high pressure mercury lamp used in the photochemical treatment, *International Journal on Sciences and Techniques of Automatic control & computer engineering (IJ-STA)*, Vol. 3, No.2, (December 2009), pp. 1064-1071, ISSN 1737-7749.

Bouslimi, L., Chammam, A., Ben Mustapha M., Stambouli, M., & Cambronne, J.P.,(2009c). The experimental study of the current pulse duration on the HID lamps luminance, *PES/SSD ' 09 - Power systems*. Djerba,Tunisia, 23-26 March 2009

Bouslimi, L., Chammam, A. et al. (2008). The experimental study of a pulsed power supply used for high power discharge lamp: Application to a 3 KW MP lamp used in waste water treatment, *EVER MONACO ' 2008, International Conference*, Mars 27-30, 2008

Brates, N. & Wyner E F. (1987). Pulsed operation of a high pressure sodium lamp, *J. Illum. Eng. Soc*, Vol. 19, pp. 50-66 (1987).

Chalek, CL. & Kinsinger R E. (1981). A theoretical investigation of the pulsed high pressure sodium arc, *J. Appl. Phys*, Vol. 52, p. 716 (1981).

Chammam, A., Elloumi, H., Mrabet, B., Charrada, K., Stambouli, M., & Damelincourt, J J. (2005). Effect of a pulsed power supply on the spectral and electrical characteristics of HID lamps, *J. Phys. D : Appl. Phys.*, Vol.38, No.8, p. 1170

Craun, G.F. (1990). Causes of water borne out breaks in the United States. In : *AWWA Water Quality Technology Conference*, San Diego, California, Nov. 11-15, 1990

Dunn, J., Bushnell, A., Ott., & Clark, W., (1997a). Pulsed white light food processing, *Cereal Foods World*, Vol.42, No.7, 1997, pp. 510-515.

Dunn, J., Burgess, D. & Leo, F., Investigation of pulsed light for terminal sterilization of WFI filled blow/fill/seal polyéthylène containers, Parenteral Drug Assoc. (1997b). *J. of Pharm. Sci. & Tech.*, Vol. 51, No. 3, 1997, pp. 111-115.

Dunn, J., Clark, RW. ,Asmus ,JF. , Pearlman , JS. , Boyer, K., Painchaud , F. & Hofmann , GA. (1990). Methods for Aseptic Packaging of Medical Devices, *U.S. Patent 4*, pp. 910-942, 1990

Fine, F. & Gervais, P. (2004). Efficiency of pulsed UV light for microbial decontamination of food Powders, *Journal of food protection*: vol. 67, N° 4, 2004, pp. 787-792.

Groocock, N.H. (1984). Disinfection of drinking water by ultraviolet light. *J. of the Institute of Water Engineers and Scientists*, 1984 ; vol. 38, No.20, pp. 163-172.

Hecht, S., (1920). The relation between the wave-length of light and its effect on the photosensory process. *The Journal of General Physiology*, 1920, pp. 375-390

Jagger, J., (1967). Introduction to Research in Ultraviolet Photobiology, *Prentice-Hall, Englewood Cliffs*, NJ, 1967, pp. 53-59.

Meulemans, C.C.E. (1986). The basic principles of UV-sterilization of water, In: *Ozone + Ultraviolet Water Treatment*, Aquatec Amsterdam, 1986.Paris: International Ozone Association, 1986: B.1.1-B.1.13.

Mimouni, A., (2004). Inactivation microbienne par lampes flash ou lumière pulsée, *La Lettre - Traitements de surfaces* - n°10 + 1, pp. 21-25, juillet 2004

Mrabet, B., Elloumi, H., Chammam, A., Stambouli, M., & Zissis, G., (2006). Effect of a pulsed power supply on the ultraviolet radiation and electrical characteristics of low pressure mercury discharge, *Plasma Devices and Operations* .Vol.14, No. 4,pp. 249-259, December (2006).

O'Brian, W.J., Hunter, G.L., Rosson, J.J., Hulsey, R.A. & Carns, K.E. (1995). Ultraviolet system design : past, present and future. In : *Proceedings Water Quality Technology Conference*, AWWA, pp. 271-305, New Orleans, LA., Nov. 12-16, 1995

Phillips, R. (1983). Sources and applications of ultraviolet radiation. New York, New York: *Academic Press Inc.*, 1983

Schenck, G.O. (1981). Ultraviolet Sterilization. In : W. Lorch. *Handbook of Water Purification*. Chichester : *Ellis Horwood Ltd.*, 1981, pp. 530-595.

Severin, B.F., Suidan, M.T. & Engelbrecht, R.S. (1984).Mixing effects in UV disinfection. *Journal WPCF*, 1984, vol. 56, No.7, pp. 881-888.

Sonntag, C. von & Schuchmann, H-P. (1992). UV disinfection of drinking water and by-product formation-some basic considerations. *J Water SRT–Aqua*, vol. 41(2), pp. 67-74, 1992

USEPA. (1996). Ultraviolet light disinfection technology in drinking water application – an overview. *EPA 811-R-96-002* Washington, DC : U.S. Environmental Protection Agency, Office of Ground Water and Drinking Water, 1996.

Weigert, F., (1907). *Ann. Phys.*, Vol. 24, p. 243 (1907).

Wolfe, R.L. (1990). Ultraviolet disinfection of potable water, current technical and research needs, *Envir. Sci. Technol.*, 1990, vol. 24, No.6, pp. 768-773.

Wright, H. B., & Cairns W. L. (1998). Ultraviolet light, *Trojan Technologies Inc*, Available from http://www.bvsde.paho.org/bvsacg/i/fulltext/symposium/ponen10.pdf

Zoeteman, B.C.J., Hrubec, J., de Greef, E. & Kool, H.J. (1982). Mutagenic activity associated with by-products of drinking water disinfection by chlorine, chlorine dioxide, ozone, and UV-irradiation. *Environmental Health Perspectives*, 1982, vol.46, pp. 197-205.

10

The Comparison of the Photoinitiating Ability of the Dyeing Photoinitiating Systems Acting via Photoreducible or Parallel Series Mechanism

Janina Kabatc* and Katarzyna Jurek
University of Technology and Life Sciences,
Faculty of Chemical Technology and Engineering,
Poland

1. Introduction

Light-induced polymerization reaction is largely encountered in many industrial applications. For example, laser direct imaging, graphics arts, holography, and dental materials require irradiation in the visible light region to benefit from laser technologies or simply to avoid UV damaging effects on skin [1]. The basic idea is to readily transform a liquid resin or a soft film into a solid film upon light exposure to form either a coating as developed in the UV curing area or an image as used in the (laser) imaging area. The starting resin is in fact a formulation that consist in an oligomer, a monomer, a photoinitiating system, and various additives depending on the applications (formulation agents, stabilizers, pigments, fillers, etc.).

The imaging technology industries where lasers are very often used currently, appear in high-tech sectors combining photochemistry, organic and polymer chemistry, physics, optics, electronics such as (i) microelectronics – photoresists for the printed circuits, integrated circuits, very large and ultralarge scale integration circuits and laser direct imaging (LDI) technology that allows to write complex relief structures for the manufacture of microcircuits or to pattern selective areas in microelectronic packaging, and so on, (ii) graphic arts – manufacture of conventional printing plates, computer-to-plate technology that directly helps to reproduce a document on a printing plate, and so on (iii) 3D machining (or three-dimensional photopolymerization or stereolithography) which is giving the possibility to make objects for prototyping applications, (iv) optics – holographic recording and information storage, computer generated and embossed holograms, manufacture of optical elements (diffraction grating, mirrors, lenses, waveguide, array illuminators, and display applications), design of structured materials on the nanoscale size.

Great effort is taken at present in the design the new photosensitive systems being able to work in well-defined conditions. As far as the polymerization reactions are concerned in UV curing and imaging areas, they are mostly based on a radical process.

---

* Corresponding Author

In this chapter, we will focus on photosensitive systems that are used in free radical photopolymerization reactions. We will give the most exhaustive presentation of potentially interesting systems developed on a laboratory scale together with the characteristic of their excited-state properties. We will also show how modern time resolved laser spectroscopy techniques allows to probe the photophysical/photochemical properties as well as the chemical reactivity of a given photoinitiating system [2].

2. Properties of photoinitiating system

A photoinitiating system (PIS) consist at least in a photoinitiator (I). Very often, a co-initiator (coI), a radical scavenger (RS) or a photosensitizer S can be added. Basically, a photoinitiating system leads to radicals that can initiate the polymerization (1).

$$I \xrightarrow{h\upsilon} R \bullet \xrightarrow{M} RM \bullet \longrightarrow Polymer \qquad (1)$$

The photoinitiator (I) is usually an organic molecule. Upon excitation by light, (I) is promoted from its ground singlet state S_0 to its first excited singlet state S_1 and then converted into its triplet state T_1 via a fast intersystem crossing. In many cases, this transient T_1 state yields the reactive radicals $R^\bullet$ that can attack a monomer molecule and initiate the polymerization [1, 2].

Radicals of photoinitiators are produced through several following typical processes:

- A photoscission of a C-C, C-S, C-B and C-P bonds (most cleavable compounds are based on the benzoyl chromophore),
- An hydrogen abstraction reaction between (I) and (coI), which plays the role of a hydrogen donor (such as an alcohol, a thiol, etc.); two radicals are formed: one on an donor and another on (I),
- An electron transfer process between (I) and (coI).

The spectral absorption range of photoinitiator is a decisive factor: the wavelength range of the (I) absorption has to match the spectral emission range of the light source. Therefore, when pigmented or colored media are used, a spectral window has to be found to excite. It may happen that the direct excitation of photoinitiator is impossible. In that case, a photosensitizer (S) must be added. The role of sensitizer is to absorb the light and to transfer the excess of energy to the photoinitiator through the well-known energy transfer process. The process is efficient only if the energy level of a donor is higher than that of an acceptor.

The panchromatic sensitization of free radical polymerization under visible light can occur in a presence of the dye alone (one-component) or in a presence of two-, three- or multi-component photoinitiating systems composed of dye molecule (sensitizer) and second compound acting as a co-initiator (either as electron or hydrogen atom donor).

Commonly, visible-light activated initiators are typically two-component initiator systems: a light-absorbing photosensitizer and co-initiator. In this type of photoinitiating system, the photo-excited dye may act as either an electron acceptor (for example, if an amine is used as the second component), or an electron donor (for example, when an iodonium salt is used as the second component) [3]. Athouugh both reaction pathways are known, electron transfer

The Comparison of the Photoinitiating Ability of the Dyeing Photoinitiating Systems Acting via Photoreducible
or Parallel Series Mechanism

217

from an electron donor to the photo-excited dye and the generation of radicals followed by either proton transfer from radical cation of electron donor or bond cleavage in electron donor is more common [3]. The intrinsic characteristics of two-component initiator systems leads to numerous kinetic limitations. For example, since the back electron transfer step is invariably thermodynamically feasible, back electron transfer and radical recombination decrease the potential concentration of free radical active centers. Furthermore, an inefficient radical is often produced simultaneously in this electron transfer/proton transfer reaction step because the dye-based radical is not active for initiation but is able to terminate a growing polymer chain [3]. These cumulative effects significantly limit polymerization kinetics of two-component initiator systems and tend to make visible light polymerization less attractive, than UV photocuring in applications where reaction rate is a primarily consideration [3].

Some dyes absorbing in the visible region have been reported to be photoreduced in the presence of amines [1]. These compounds belong to the families of xanthenes, fluorones, acridines, phenazines, thiazenes, and so on. For example, methylene blue is well known to react from its triplet state with amine to initiate the photopolymerization of acrylates. The photoreduction is accompanied with an important photobleaching of the dye, rendering the photopolymerization of thick samples under visible light. The photobleaching is not so important in the case of xanthenes or fluorones, although the polymerization can be very efficient. Very good efficiencies were reported using thionine, rose bengal, eosin Y, erythrosin, riboflavin, polymethine dyes as photosensitizers, and co-initiators, such as amines, sulfinates, carboxylates, organoborate salts [1]. In the case of amine as co-initiator, the reaction involves a hydrogen abstraction from a amine to semireduced form of a dye. But in the case of organoborate salts acting as a co-initiator, the reaction involves an electron transfer from borate anion to polymethine dye in its excited singlet state. These systems are able to shift the spectral sensitivity of photopolymers up to the red region of the visible spectrum. However, dye/co-initiator systems were not developed significantly in the industry. Very often, dark reactions take place that lead to poor shelf life of the formulation, an effect that was detrimental to their industrial use for a long time. In addition, the conversion of the monomer to polymer was generally limited. Indeed, for most of the industrial applications, conversion of more than 60% have to be reached, a goal that is difficult to achieve with conventional dye/co-initiator photoinitiating systems (PIS) [1].

In the last decade, three-component photoinitiating systems have emerged as an attractive alternative for visible light polymerization based on numerous demonstrations that the kinetic effectiveness of a two-component electron/proton transfer initiator system can be improved by the addition of a third component.

Like the two-component system, the three-component (PIS) include a light absorbing moiety, an electron donor (ED) and an electron acceptor (EA). In such systems, the third component is supposed to scavenge the chain-terminating radicals that are generated by the photoreaction between other two components or produce the additional initiating radicals. This process leads to an increase of the free radical polymerization rate. Therefore, certain additives improve the polymerization efficiency, leading to the development of the so-called three-component photoinitiating systems [3-12]. Three-component initiator systems have consistently been found to be faster, more efficient, and more sensitive than their two-component counterparts [3]. The mechanism involved is rather complex and is based on chemical secondary reactions. It was reported, that different radical intermediates generated during the irradiation and in the subsequent polymerization reaction react with the additive to give new reactive radicals.

The development of new photoinitiating systems remains an interesting challenge. In specific areas, for example in graphic arts or in conventional clear coat and overprint varnish applications, the photoinitiators must exhibit particular properties, among them a high photochemical reactivity leading to high curing speeds.

Kim et al. [4], used the thermodynamic feasibility and kinetic considerations to study photopolymerization initiated with rose bengal or fluorescein as photosensitizer to investigate the key factors involved with visible-light activated free radical polymerization involving three-component photoinitiating systems. Many of the same photosensitizers used for two-component electron-transfer initiating systems may also be used in three-component ones [3]. Examples include coumarin dyes, xanthene dyes, acridine dyes, thiazole dyes, thiazine dyes, oxazine dyes, azine dyes, aminoketone dyes, porphyrins, aromatic polycyclic hydrocarbons, p-substituted aminostyryl ketone compounds, aminotriaryl methanes, merocyanines, squarylium dyes, and pyridinium dyes [3, 13-17].

A number of kinetic mechanisms have been suggested to explain the enhanced kinetics and sensitivity for three-component initiatior systems.

There are few mechanisms of free radicals generation in dyeing three-component photoinitiating systems:

- Photooxidizable series mechanism,
- Photoreducible series mechanism (dye/amine/iodonium salt),
- Parallel series mechanism.

The photoreducible series mechanism is the well-known representative kinetic mechanism for three-component photoinitiating systems. Until now, photoreducible series mechanism for (PIS) containing camphoquinone or methylene blue dye have been reported as a representative kinetic mechanism. However, alternative kinetic mechanisms must be considered since a variety of dyes used in three-component initiator systems impose different thermodynamic and kinetic constraints. For this study, we used three-component photoinitiator systems containing thiacarbocyanine dye (Cy). This dye has excellent attributes that make it attractive for these mechanistic studies. Because this photosensitizer has both reduction potential as well as oxidation potential, the photo-excited dye allows thermodynamically feasible direct interactions with an electron donor as well as with an electron acceptor simultaneously.

In this chapter, the efficiency of the three-component photoinitiating system based on thiacarbocyanine dye to induce visible light polymerization of triacrylate monomer will be described. The ability of both photoinitiating systems formed by Cy/n-butyltriphenylborate/second co-initiator and Cy/1,3,5-triazine derivative/heteroaromatic thiol to initiate polymerization under visible light will be reported.

To understand their efficiency in terms of monomer conversion, the photochemistry of these systems was investigated by means of steady state and time resolved spectroscopy.

The Comparison of the Photoinitiating Ability of the Dyeing Photoinitiating Systems Acting via Photoreducible
or Parallel Series Mechanism

219

2.1 Polymethine dyes as sensitizer in photoinitiating system

Polymethine dyes were first synthesized in 1856 by Greville Williams. Classical polymethine dyes are cationic molecules in which two terminal nitrogen heterocyclic subunits are linked by a polymethine bridge as shown by the general structure 1.

(1)

In the ensuing 150+ years, thousands more cyanines have been synthesized due to demand based on diverse applications of these versatile dyes [18]. As it is known, these dyes present intense absorption and fluorescence bands in the green-red visible region of the electromagnetic spectrum and exhibit high fluorescence quantum yields. The best known application of these dyes is in the laser field, where they showed higher laser efficiency than rhodamine dyes. Besides their use as laser dyes, polymethines have also shown very good performance as sensitizing dyes in free radical photopolymerization, with the idea of using the photopolymers in industrial applications, such as photoimaging, holography, computer-to-plate, and so on. They have been used as sensitizer dye with organoborate or 1,3,5-triazine derivatives as a radical generating reagent. The ion pair composed of cyanine dye cation and an alkyltriarylborate anion was first described by G. B. Schuster et al. [19, 20].

The work of Schuster and co-workers [19, 20] on the photochemistry of cyanine borates led to the preparation of the color-tunable, operating in the visible region commercial photoinitiators [21]. This research group discovered that, photolysis of 1,4-dicyanonaphthalene containing an alkyltriphenylborate leads to one electron oxidation of alkyltriphenylborate salts yielding an alkyltriphenylboranyl radical that undergoes carbon-boron bond cleavage and the formation of free radicals [22].

The laser flash photolysis data allows one to describe the mechanism of the polymerization initiation process. The initiation step of the reaction involves alkyl radical formation as a result of photoinduced electron transfer from borate anion to the excited singlet state of cyanine dye, followed by the rapid cleavage of the carbon-boron bond of the boranyl radical (see Scheme 1).

Scheme 1 summarizes possible primary and secondary processes, which may occur during the free radical photoinitiated polymerization with the use of cyanine borate initiators; where k_{BC} denotes the rate of the carbon-boron bond cleavage, the reverse step is designated as k_{-BC}, and k_{bl} is the rate constant of the free radicals cross-coupling step yielding bleached dye.

As it was mentioned above this chapter reports the use of polymethine dye as a part of a three-component photoinitiating system for radical polymerization in the visible region of the spectrum, together with an alkyltriphenylborate salt and different additives as co-initiators. In the study, we examined the ability of the systems formed by Cy/borate salt,

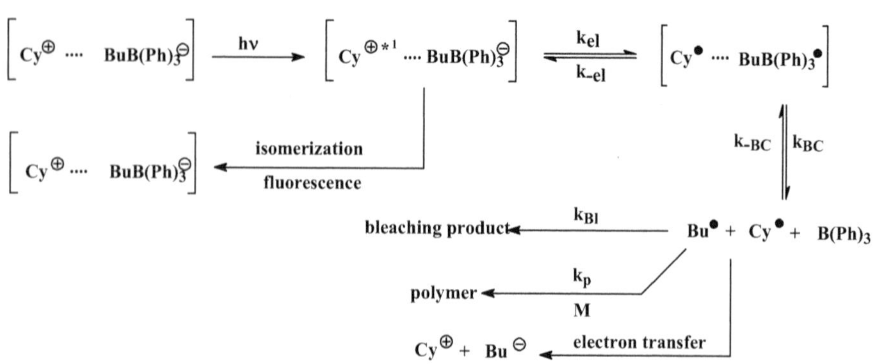

Scheme 1. Primary and secondary processes occuring during the free radical photoinitiated polymerization with the use of cyanine borate photoinitiators.

Cy/borate salt/different derivatives, and Cy/1,3,5-triazine/heteroaromatic tiol to initiate polymerization under visible light (Scheme 2).

Photosensitizer

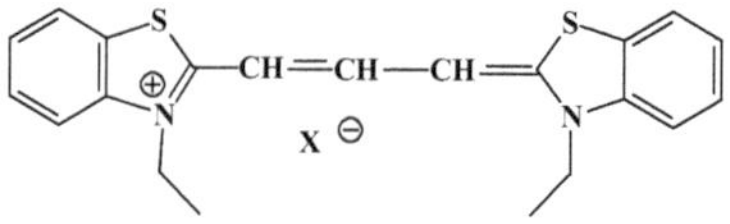

Cy

Co-initiators:

Onium salts

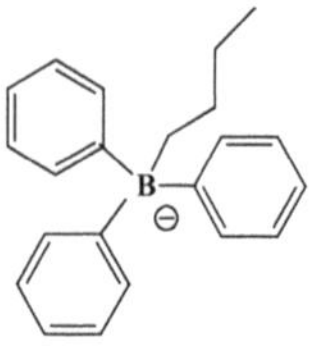

B2 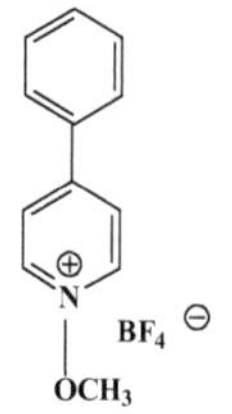   NO 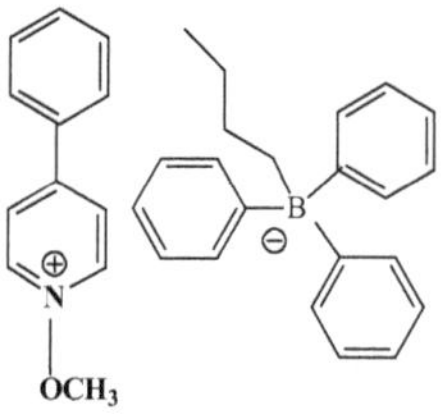   NOB2 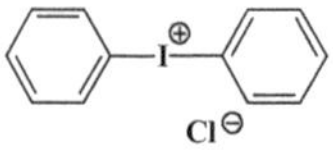   I

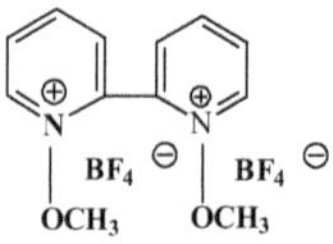

Bp

The Comparison of the Photoinitiating Ability of the Dyeing Photoinitiating Systems Acting via Photoreducible
or Parallel Series Mechanism

221

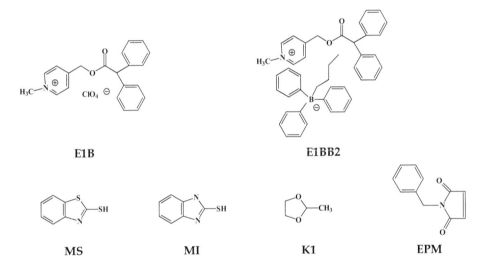

Scheme 2. Compounds used in this study

2.1.1 Kinetic key factors for visible-light activated free radical polymerizations

The efficiency of different combinations of polymethine dye and additives as (PIS) for the polymerization of triacrylate, was evaluated using the differential scanning calorimetry (DSC), under isothermal conditions at room temperature, using a photo-DSC apparatus constructed on the basis of the TA Instruments DSC 2010 Differential Scanning Calorimeter.

The different formulations, in molecular ratio of each component, for dye studied are detailed in Table 1. No significant photopolymerization was detected in the absence of the dye. Figures 1-4 show the corresponding kinetic observed for N,N'-diethylthiacarbocyanine dye, and Table 1 shows the final conversions, polymerization rates and inhibition times for all runs after 5 min of irradiation.

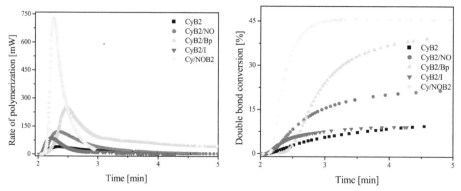

Fig. 1. Comparative study of the photopolymerization of TMPTA/MP mixture (9:1) (2-ethyl-2-(hydroxymethyl)-1,3-propanediol triacrylate/1-methyl-2-pyrrolidinone) using different photoinitiating systems based on the polymethine dye and onium salts.

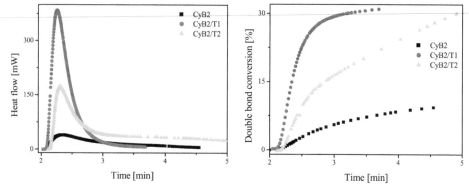

Fig. 2. Comparative study of the photopolymerization of TMPTA/MP mixture (9:1) using different photoinitiating systems based on the polymethine dye and 1,3,5-triazine derivatives.

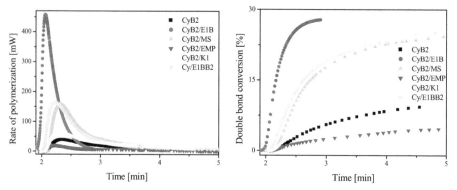

Fig. 3. Comparative study of the photopolymerization of TMPTA/MP mixture (9:1) using different photoinitiating systems based on the polymethine dye and other additives.

The Comparison of the Photoinitiating Ability of the Dyeing Photoinitiating Systems Acting via Photoreducible or Parallel Series Mechanism

223

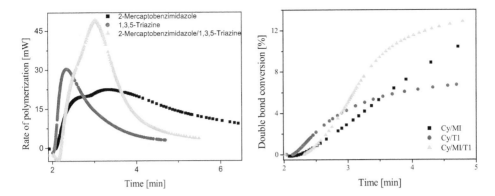

Fig. 4. Comparative study of the photopolymerization of TMPTA/MP mixture (9:1) using photoinitiating systems based on the polymethine dye, heteroaromatic thiol and 1,3,5-triazine derivative.

Co-initiator	Dye [M]	B2 [M]	Other Additives [M]	Molar ratio B2/other additives	Inhibition time [s]	R_P [μmol/s]	Final conversion (%)
B2	5×10^{-3}	5×10^{-3}	0	0	6.5	0.502	10
NO	5×10^{-3}	5×10^{-3}	5×10^{-4}	10	1	1.54	22
NOB2	5×10^{-3}	1×10^{-2}	1×10^{-2}	1	0	9.38	46
Bp	5×10^{-3}	5×10^{-3}	5×10^{-4}	10	0	3.20	39
I	5×10^{-3}	5×10^{-3}	3×10^{-3}	1.67	0.8	1.11	10
T1	5×10^{-3}	5×10^{-3}	5×10^{-2}	0.1	0	4.92	31
T2	5×10^{-3}	5×10^{-3}	1×10^{-2}	0.5	1.4	2.24	30
E1B	5×10^{-3}	5×10^{-3}	1×10^{-2}	0.5	1	5.86	28
E1BB2	5×10^{-3}	1×10^{-2}	1×10^{-2}	1	1.5	2.11	25
MS	5×10^{-3}	5×10^{-3}	5×10^{-2}	0.1	9	2.02	24
K1	5×10^{-3}	5×10^{-3}	1×10^{-1}	5	9	0.84	15
EPM	5×10^{-3}	5×10^{-3}	5×10^{-3}	1	0	0.261	4
T1	5×10^{-3}	0	5×10^{-2}	0	3	0.388	7
MS	5×10^{-3}	0	5×10^{-2}	0	8	0.29	10.7
T1 + MS	5×10^{-3}	0	5×10^{-2}	0	9	0.63	13

Table 1. Molar composition of the samples, corresponding B2/other additive molar ratio, final conversion obtained after 5 min of irradiation, maximum polymerization rate R_P and inhibition time.

It has been reported that to enhance the kinetics of a visible-light activated initiation process, it is important to: (1) retard the back electron transfer and recombination reactions and (2) use the secondary reaction step to consume the nonproductive dye-based radical and thereby regenerate the original photosensitizer (dye) [3]. Figures 1-4 provide a comparison of the visible-light activated free radical polymerizations initiated by two-component initiator systems (CyB2, Cy/MS and Cy/T) with the corresponding three-component PIS. These examples clearly show that the three-component initiator systems produce the highest rates and final conversion as predicted.

Figures 1-3 demonstrate that the photoreducible series mechanism (Cy/B2/second co-initiator) produces the highest conversion and the fastest rates of polymerization. In such photoinitiating system, since Cy* reacts directly with borate salt (ED), this kinetic pathway can prevent photon energy wasting steps such as back electron transfer [23]. But it is well known, that in a case when stable alkyl radical (initiating radical) is formed as a result of carbon-boron bond cleavage in boranyl radical (product of primary photochemical reaction) the back electron transfer process does not occur. Therefore, in the EA-based secondary reaction step, the dye-based radical can be consumed and photosensitizer (Cy) can be regenerated.

The parallel-series mechanism (Figure 4 (Cy/thiol/triazine)) showed intermediate reaction kinetics because this kinetic pathway simultaneously involves both the photoreducible and photooxidizable mechanisms in the primary photochemical reaction. These results are also supported by Table 1 which illustrates that the photoreducible series mechanisms (Cy/B2/second co-initiator) produced the highest reaction kinetics and photo-sensitivity then the alternative kinetic pathway.

The comparison of Cy/MS and Cy/T systems also illustrates the importance of preventing of back electron transfer reaction. Grotzinger and coworkers reported that when 1,3,5-triazine derivative accepts an electron, it produce 1,3,5-triazine radical anion which fragments to produce an active, initiating 1,3,5-triazynyl radical and a chloride anion [12]. Thus, triazine (T) accepts an electron from Cy*, and the product obtained undergoes a rapid unimolecular fragmentation reaction that limits back electron transfer. Because of the reduced back electron transfer between the Cy* and T, Cy/T system leads to the generation of higher concentrations of active centers than Cy/MS system (however, complete bleaching of the dye in the photochemical reactions results in the low conversion in the two-component systems and the conversion reaches < 10 %.

On the other hand, the excited dye wastes photon energy in an electron transfer process between dye and co-initiator because of the back electron transfer competes with separation of gemine radical pair. It has generally been reported than only 10% of the absorbed light energy may be used for photo-induced electron transfer in the bimolecular organic electron transfer reaction [23]. Hence the Cy/MS initiator system only reached ~ 10 % of final conversion. The Cy/MS/T three-component initiator system produced enhanced conversion about 13 %.

As expected, this behavior is strongly dependent on the composition of the photoinitiating system. The photoinitiating ability of the (PIS) under study depends mostly on the nature of the co-initiator. The use of diphenyliodide or N-phenylethylmaleimide in the CyB2 photoinitiating system leads to poor and slow conversion of the monomer.

On the other hand, the Cy/B2/second co-initiator photoinitiating systems produced dramatically enhanced conversion ranging from 15 to 46 % because of effective retardation of the recombination reaction step and consumption of the dye-based radical to regenerate

The Comparison of the Photoinitiating Ability of the Dyeing Photoinitiating Systems Acting via Photoreducible
or Parallel Series Mechanism

225

of the original photosensitizer (dye) in the secondary reaction step. The enhanced conversion relative to the two-component initiator system also arises from production of two radicals: an active initiating butyl radical and an active alkoxy, triazinyl, picolinium ester, thiyl or phenyl radical. These results are supported by Table 1, which illustrates the reaction rate as well as the final conversion of monomer with photoinitiating systems under study. The data clearly indicate that the three-component initiator system (Cy/B2/second co-initiator) is the most effective in overall radical active centre production as well as the rate of initiation from the onset of polymerization.

On the contrary, the system Cy/T1 exhibits a good reactivity with both higher rate of polymerization and final conversion. However, the best results were obtained for the three-component system CyB2/T1 and CyB2/T2. The addition of 1,3,5-triazine derivative to the CyB2 system increased the polymerization rate as well as the final conversion of the triacrylate compared with the two-component systems.

Finally, all these kinetic results provide very useful information in terms of the selection criteria for each component of photoinitiating system. Because once photosensitizer with both reduction and oxidation potentials is selected, the kinetic pathway is controlled by selection of an electron donor or an electron acceptor based on the thermodynamic feasibility, thereby influencing the conversion and rate of polymerization kinetic data.

As before, these kinetic differences of two kinetic pathways are ascribed to differences in the efficiency of retarding back electron transfer as well as regenerating photosensitizer through the secondary reaction step.

2.1.2 Excited state reactivity

Because polymethine dye tested exhibits medium fluorescence quantum yield (Table 2), the fluorescence quenching by co-initiators was first studied.

	Cy
λ_{max} [nm]	556
ε_{max} [mol^{-1}dm^{3}cm^{-1}]	113 000
E_s [kJ/mol]	203
ϕ_f	0.05
τ_0 [ps]	139, 392
E_{ox} [V/SCE]	1.0
E_{red} [V/SCE]	-1.34

Maxiumum absorption wavelength λ_{max}, molar extinction coefficient ε_{max}, singlet state energy E_s, fluorescence quantum yield ϕ_f, singlet state lifetime τ_0, half-wave oxidation and reduction potentials E_{ox} and E_{red}, respectively.

Table 2. Photophysical and Electrochemical Properties of Polymethine Dye

The quenching rate constants k_q of the singlet excited state by co-initiators tested were determined in ethyl acetate:1-methyl-2-pyrrolidinone mixture (4:1) (Table 3), and showing values close to the diffusion rate constant ($k_q = 2 \times 10^{10}$ M^{-1}s^{-1}).

	k_q [M^{-1}s^{-1}]	ΔG_{el} [kJ/mol]
B2	9.5×10^{11}	-1.93
NO	6.15×10^{10}	-64.64
Bp	1.96×10^{10}	-67,.06
T1	3.6×10^{10}	-27.02
E1B	7.42×10^{9}	-26.44
Thiol	2.78×10^{10}	-12.54

Table 3. Fluorescence Quenching Data k_q and Gibbs Free Energy ΔG_{el} Changes for Thiacarbocyanine Dye with Co-initiators Tested.

2.1.3 Thermodynamics of photo-induced electron transfer reaction

Before investigating the kinetic mechanisms for efficient design of photoinitiator systems, thermodynamic feasibility for electron transfer reactions must be verified. The Rehm-Weller equation was used to predict the thermodynamic feasibility for electron transfer reaction as shown below [24]. In this study, N,N'-diethylthiacarbocyanine dye was selected as photosensitizer because allows thermodynamic feasibility for direct simultaneous interaction with an electron donor as well as with an electron acceptor previously described. B2 or MS are used as electron donor (ED) and NO, Bp, I, T1, T2, E1B, K or EMP is used as (EA).

Because of the redox properties of the dye (Table 2) and the co-initiators, the mechanism for the quenching of sensitizer's excited state likely involves an electron transfer process. The values of the Gibbs free energy change for the photoinduced electron transfer ΔG_{et} on excited state is given by the Rehm Weller equation (2) [24].

$$\Delta G_{el} = E_{ox} - E_{red} - E^* + C \tag{2}$$

where:

E_{ox} and E_{red} are the half-wave oxidation and reduction potentials for the acceptor (Cy; E_{red} = -1.34 V/SCE) and the donor (B2; E_{ox} = 1.16 V/SCE), respectively, and E^* is the energy of the excited state. The coulombic term C is usually neglected in polar solvents.

The ΔG_{et} values are very useful for determining the potential kinetic pathway. As can be seen in Table 3, the calculated values for the intermolecular singlet electron transfer reactions are favorable, indicating that the dye can be reduced in the presence of the electron donors, such as: B2 or heteroaromatic thiol or oxidized with 1,3,5-triazine derivative.

From these results, one can conclude that the carbocyanine dye reacts with the co-initiators mainly through the quenching of the first excited singlet state. The reaction proceeds through the formation of a geminate radical pair, which can recombine through a back electron transfer process or separate into free radicals. The latter process explains the formation of the dye-based radical when alkyltriphenylborate salt is used as a quencher, or the radical cation of the dye when 1,3,5-triazine is used instead.

These results can lead to two initiator systems with two corresponding thermodynamically feasible kinetic pathways, which are (i) photoreducible series mechanism: Cy/B2/onium

The Comparison of the Photoinitiating Ability of the Dyeing Photoinitiating Systems Acting via Photoreducible
or Parallel Series Mechanism

227

salt, Cy/B2/N-methylpicolinium ester, Cy/B2/acetal, Cy/B2/thiol, Cy/B2/triazine, (ii) parallel-series mechanism: Cy/thiol/triazine.

2.2 Mechanism of free radicals formation

From the transient absorption spectra obtained during laser flash photolysis, the ground state photobleaching of sensitizer under addition of borate salt, heteroaromatic thiol or 1,3,5-triazine derivative can be observed at 420 nm. Laser flash photolysis experiments were carried out in acetonitrile solution, exciting at 355 nm. Accordingly, it can be seen in Scheme 3 that the depletion increases with increasing concentration of borate salt, as a consequence of the formation of the dye-based radical and boranyl radical.

2.2.1 Photoreducible series mechanisms

As illustrated in Scheme 3, the kinetic pathway involves electron transfer and carbon-boron bond cleavege from borate salt to the photo-excited dye (Dye*) and produces an active initiating radical (such as butyl radical) as the primary photochemical reaction. The second onium salt (N-alkoxypyridinium or diphenyliodonium salt), as an electron acceptor, consumes an inactive radical and produces another active radical (alkoxy or phenyl), thereby regenerating the original dye in the secondary reaction step. The regenerated (PS) may re-enter the primary photochemical reaction. This kinetic pathway is designed as a photoreducible series mechanism. It is the well-known representative kinetic mechanism for three-component initiators. In this mechanism, the second co-initiator increases the photopolymerization kinetics in two ways: (1) it consumes an inactive dye-based radical (Dye*) and produces an active initiating radical, thereby regenerating the original (PS) (dye), and (2) it reduces the recombination reaction of dye-based radical and boranyl radical.

Unfortunately, the latter species can not be observed under our experimental conditions. The initiating radicals in this case could come mainly from the boranyl radical, which undergoes rapid and irreversible fragmentation as a result of carbon-boron bond cleavage. It should be noted that this reaction will compete with the back electron transfer from dye-based radical (Dye*) to boranyl radical within the gemine radical pair; as well with the recombination of the both radicals.

As stated above, when carbocyanine dye is used with borate salt as co-initiator the excited singlet state is quenched with the rate close to the diffusion rate constant, observing an increase in the signal of dye-based radical: as borate salt acts as an electron donor, the electron transfer reaction of sensitizer excited state and B2 leads to (Dye*) and (B2*) radicals (Schemes 1 and 3). Monitoring the dye radical formation at 420 nm leads to the observation of a increasing absorbance of the dye-based radical, in line with the results obtained for heteroaromatic thiol. This demonstrates that the reaction between carbocyanine dye-excited state and borate salt behaves similary to that of thiol. From all these results, the low conversion observed in the photopolymerization for Cy/MS photoinitiating system could be explained by a low quantum yield of radical formation from MS$^{\bullet+}$.

Turning now to the study of the three-component system, transient absorption spectroscopy at 420 nm shows that the signal of (Dye*) formed from the interaction Dye/B2 (with excess of borate salt with respect to other additives) decreases under addition of N-alkoxypyridinium

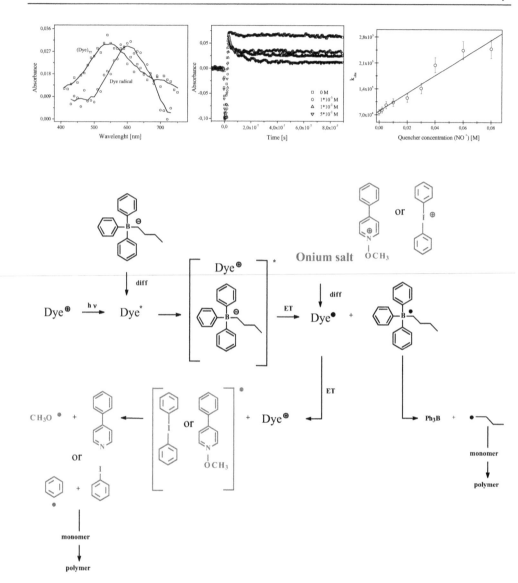

Scheme 3. Mechanism of the reactions occurring in three-component photoinitiating systems based on carbocyanine dye, borate salt and other onium salt. Inset: Left: Transient absorption spectra recorded 100 ns after laser flash (355 nm) for dye in MeCN (squares) and 500 ns after lash for dye in presence n-butyltriphenylborate salt presented dye-based radical formation. In the midst of: Kinetic traces for dye-based radical decay at 610 nm in the presence of various amount of N-methoxy-p-phenylpyridinium salt. The concentration of quencher is marked in Figure. Right: The Stern-Volmer plot of the fluorescence quenching of cyanine dye-based radical by onium salt.

The Comparison of the Photoinitiating Ability of the Dyeing Photoinitiating Systems Acting via Photoreducible
or Parallel Series Mechanism

229

salt, iodonium salt, 1,3,5-triazine derivative, N-methylpicolinium ester or other co-initiators (Schemes 3-6). This indicates that onium salt, triazine or other co-initiators (with the exception of heteroaromatic thiol) react with the Dye-based radical (Dye•) formed from the interaction of dye excited state with borate salt (Schemes 1-6). At the same time, the photobleaching of sensitizer ground state is lowered when second co-initiator is added to the Dye/B2 system. This means that the reaction of second co-initiator with (Dye•) leads to recorvery of the dye ground state. The reaction is expected to proceed through an electron transfer process from dye-based radical to second co-initiator. From the value of the oxidation potential of (Dye•) (E_{ox} = 1.0 V/SCE), the free energy of the electron transfer reaction between (Dye•) and second co-initiator is estimated to be in a range from –0.13 eV to 0.08 eV (e.g. –12.54 kJ·mol^{-1} to 7.72 kJ·mol^{-1}), this value would lead to a fast rate constant of interactions (Schemes 3-6).

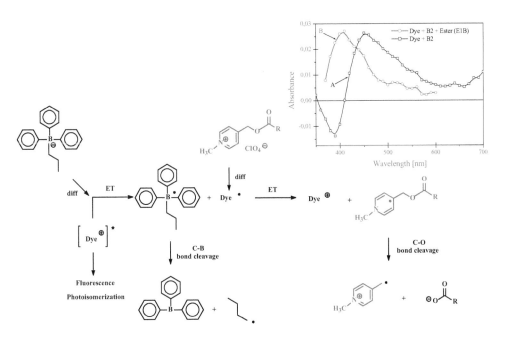

Scheme 4. Mechanism of the reactions occurring in three-component photoinitiating systems based on carbocyanine dye, borate salt and N-methylpicolinium ester. Inset: Transient absorption spectra of: (A) cyanine dye in a presence of borate salt recorded 50 ns after laser puls (squares) presented dye-based radical formation and (B) for cyanine dye in presence of equimolar ratio of tetramethylammonium n-butyltriphenylborate and N-methylpicolinium perchlorate recorded 100 ns after laser puls (circles) presented N-methylpicolinium ester radial formation.

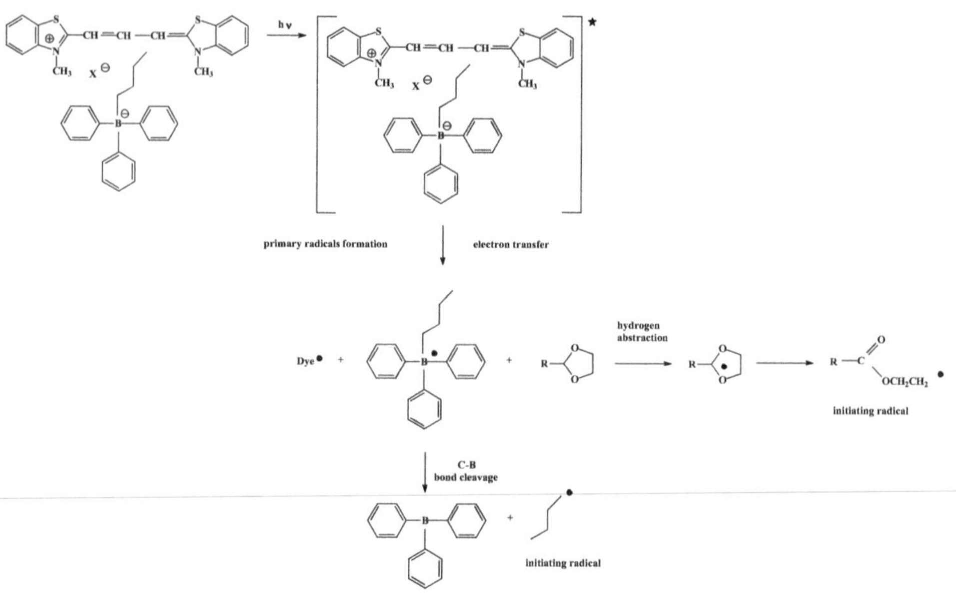

Scheme 5. Mechanism of the reactions occurring in three-component photoinitiating systems based on carbocyanine dye, borate salt and cyclic acetal.

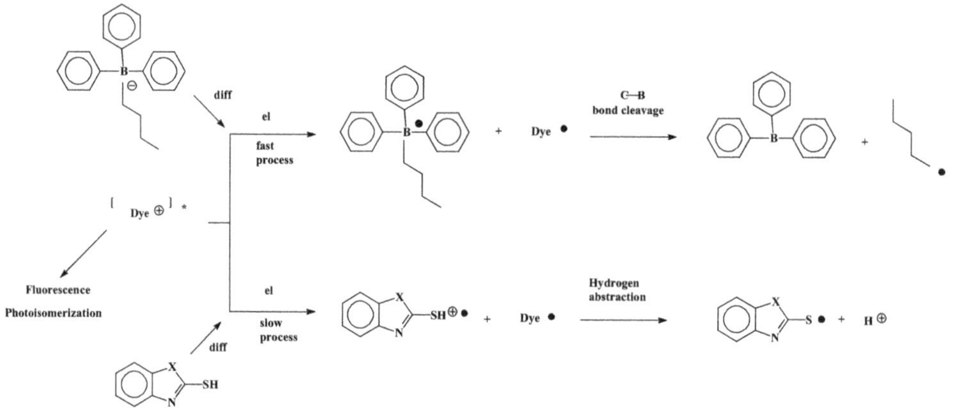

Scheme 6. Mechanism of the reactions occurring in three-component photoinitiating systems based on carbocyanine dye, borate salt and heteroaromatic thiol.

The Comparison of the Photoinitiating Ability of the Dyeing Photoinitiating Systems Acting via Photoreducible
or Parallel Series Mechanism

231

Similarly, if the deactivation of the excited state of the dye proceed through a photoinduced electron transfer with heteroaromatic thiol or triazine (Schemes 6 and 7), the dye-based radical and ($Dye^{\bullet+}$) are formed, respectively. In this case, one can assume that the oxidation and reduction potentials of dye are 1.0 V/SCE and –1.34 V/SCE, respectively. But the oxidation potential of thiol and reduction potential of triazine are in the range from 0.69 V/SCE to 0.90 V/SCE and –0.84 V/SCE, respectively. This leads to the calculation of ΔG_{el} values of –12.54 and –27.02 kJ/mol for thiol and triazine, respectively, showing that this reaction is exergonic and to be feasible.

In summary, when N,N'-diethylthiacarbocyanine dye is used as photosensitizer, the kinetic pathway is seen for the three-component initiator system composed of onium salt, picolinium ester, cyclic acetal, 1,3,5-triazine derivative or maleimide as second co-initiator enhances photopolymerization kinetics as previously described. As an example, the Cy/B2/NO photoinitiator system may be used to explain photoreducible series mechanism (Scheme 3). Because Cy/NO is not a thermodynamically feasible system, the primary photoinduced electron transfer reaction only proceeds between photo-excited dye and borate salt. Then, subsequent electron transfer involves the electron acceptor (NO) in a secondary reaction step.

2.2.2 Parallel-series mechanism

On the other hand, under conditions where photosensitizer (dye) has both reduction and oxidation potentials, the photoexcited dye may act as both an electron donor and an electron acceptor, resulting in a parallel-series mechanism [3]. In this kinetic pathway, the electron transfer between the excited dye molecule and an electron donor competes with an electron transfer between the excited dye and an electron acceptor, as the primary photochemical reaction. The Cy/thiol/1,3,5-triazine photoinitiating system provides example of the combined parallel-series mechanism. Because Cy/MS initiator system is thermodynamically feasible (which did produce free radical active centers as a two-component initiator system; Figure 4) and Cy/T system is also thermodynamically feasible (free radical active centers were also produced), the Cy/MS/T initiator system may engage in the parallel-series mechanism.

For such system composed of polymethine dye/1,3,5-triazine derivative, the photobleaching of the ground state increases with increasing concentration of triazine. This indicates that the photochemical reaction between Dye-excited state and T yields to the formation of transient species. According to the electron transfer reaction, the radical anion ($T^{\bullet-}$) is easily detected at 510 nm (Scheme 7).

It can be seen from Scheme 7 that ($T^{\bullet-}$) is formed within the laser pulse, as a consequence of its formation mainly in the electron transfer process from the excited singlet state of dye to triazine. This leads to the formation of radical cation of dye and the radical anion of triazine. The latter species afterwards loses chloride anion to give the initiating radical ($T^{\bullet-}$), as was demonstrated for other triazine derivatives in presence of rose bengal [12]. Interesingly, the recorded cyclic voltammogram for T1 in acetonitrile (Figure 5 right) exhibits an irreversible reduction wave at –0.84 V/SCE and a noticeable oxidation wave at 1.270 V/SCE indicating a cleavage process within the radical anion ($T^{\bullet-}$). It is expected that the chloride anion is expelled in a fast time scale, preventing the (Dye•+)/(T•−) system to undergo a back electron transfer process.

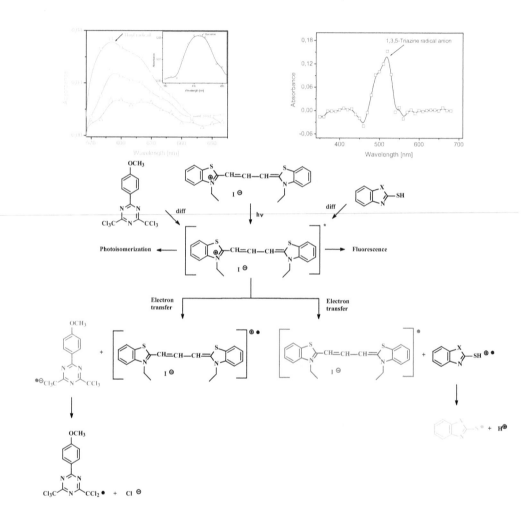

Scheme 7. Mechanism of the reactions occurring in three-component photoinitiating systems based on carbocyanine dye, heteroaromatic thiol and 1,3,5-triazine derivative. Inset: Left: Transient absorption spectra of cyanine dye in a presence of 2-mercaptobenzothiazole (MS) recorded: 1 μs (squares), 4 μs (circles) and 10 μs (triangles) after laser pulse presented the thiyl and dye-based radicals formation. In: Transient absorption spectra of cyanine dye in a presence of 2-mercaptobenzothiazole (MS) recorded 100 ns after laser pulse (circles) in acetonitrile solution. Right: Transient absorption spectra of cyanine dye in a presence of 2,4-bis-(trichloromethyl)-6-(4-methoxy)phenyl-1,3,5-triazine (T) recorded 50 ns after laser pulse presented the 1,3,5-triazinyl radical formation.

The Comparison of the Photoinitiating Ability of the Dyeing Photoinitiating Systems Acting via Photoreducible
or Parallel Series Mechanism

233

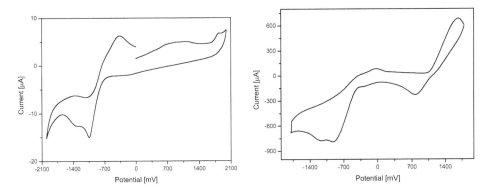

Fig. 5. Cyclic voltammograms of Dye (Cy) and 1,3,5-triazine derivative (T) in 0.1 M
tetrabutylammonium perchlorate solution in dry acetonitrile as the supporting electrolyte.

At 510 nm (Scheme 7) it is possible to observe an increase of $(T^{\bullet-})$ signal, which clearly
evidences the electron transfer process from sensitizer to triazine.

2.3 Photoinitiation efficiency

From all these experiments, it turns out that the photoreactions from the excited state of the
carbocyanine dye are very efficient with both borate salt and second co-initiator. These
photoreactions lead to the formation of initiating species, and therefore, to the conversion of
monomer. A rough estimate of the diffusion rate constant can be given by equation 3:

$$k_d = \frac{8RT}{3\eta} \tag{3}$$

This leads to the value of k_d = 1.84 × 10⁶ M⁻¹s⁻¹ for the monomer used. Consequently, the
quantum efficiency of dye excited state deactivation by a given quencher Q will depend on
k_d × [Q] for most of the photoreactions reported in Table 3. Therefore, the relative efficiency
of the corresponding photochemical processes will be mainly dependent on the
concentration of the co-initiators.

In the case of the three-component photoinitiating systems, the highest concentration of
additive makes the excited state quenched by second co-initiator, leading to initiating
radical (after electron transfer process) and the dye-based radical (Dye$^{\bullet}$). This latter is able to
react with second co-initiator leading to the recorvery of the dye ground state and
additional initiating species (Schemes 3-7). The fact that three-component photoinitiating
systems have higher efficiences than two-component ones is in good agreement with the
expected reaction of second co-initiator with the dye-based radical. The combination of co-
initiators B2 and others have clearly a beneficial effect on the photopolymeryzation process.

By contrast, in a case of three-component photoinitiating system composed of
dye/thiol/triazine, the deactivation of the (Cy) excited state will be mainly governed by the
photoreaction with both thiol and 1,3,5-triazine derivative (Scheme 7). This leads to the
formation of initiating radicals and radical cation (Dye$^{\bullet+}$).

It was show that three-component photoinitiating systems acting via photo-reducible series mechanism producess the highest rates of polymerization and final conversion of monomer (Figure 6).

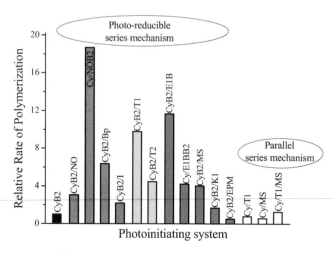

Fig. 6. The comparison of the photoinitiating ability of the dyeing photoinitiating systems acting via photoreducible series mechanism and parallel series mechanism.

3. Conclusions

In this chapter, we have characterized two different kinetic mechanisms using thermodynamic feasibility and key kinetic factors with three-component visible light photoinitiating systems containing thiacarbocyanine dye as a photosensitizer. We used the Rehm-Weller equation to verify the thermodynamic feasibility for the photoinduced electron transfer reaction. Based on this, we have suggested two different kinetic mechanisms, which are (i) photoreducible series mechanism (Cy/B2/second co-initiator) and (ii) parallel series mechanism (Cy/thiol/triazine). In addition, based on experimental kinetic data, we have evaluated two kinetic pathways. The photo-DSC kinetic experiments revealed that the photoreducible series mechanism produced the highest rates of polymerization and final conversion of monomer values. It was found, that three-component PIS showed the best performance. Laser spectroscopy studies allowed the understanding the processes that may explain the behavior observed in terms of photopolymerization. The sensitizer reacts mainly throught a singlet electron transfer mechanism from the borate salt or heteroaromatic thiol to the dye and from the dye to the triazine derivative. Beneficial side-reactions were shown to limit the photobleaching of the dye, resulting in higher final monomer conversion.

Although, these two kinetic pathways presented here can not govern the detailed interactions in all initiator mechanisms, this approach will provide useful information for selection criteria for each component, as well as provide a straightforward manner for classifying the photopolymerization process.

4. Acknowledgment

This work was supported by The Ministry of Science and Higher Education of Poland (MNiSW) (grant No N N204 219734).

5. References

[1] Tarzi, O.I.; Allonas, X.; Ley, C.; Fouassier, J.P. (2010). *Journal of Polymer Sciences: Part A: Polymer Chemistry*, Vol. 48, pp. 2594-2603.

[2] Fouassier, J.P.; Allonas, X.; Lalevée, J.; Dietlin C. (2010). *Photochemistry and Photophysics of Polymer Materials*, Ed. Allen, N.S., John Wiley & Sons, Inc,pp. 351-420.

[3] Kim, D.; Stansbury, J.W. (2009) *Journal of Polymer Science: Part A: Polymer Chemistry*. Vol. 47, pp. 887-898.

[4] Kim, D.; Stansbury, J. W. (2009). *Journal of Polymer Science: Part A: Polymer Chemistry*. Vol. 47, pp. 3131-3141.

[5] Padon, K.S.; Scranton, A. B. (2000). *Journal of Polymer Science: Part A: Polymer Chemistry*. Vol. *38*, pp. 2057-2066.

[6] Kabatc, J.; Pączkowski, J. (2005). *Macromolecules*. Vol. *38*, pp. 9985-9992.

[7] Kabatc, J.; Zasada, M.; Pączkowski, J. (2007). *Journal of Polymer Science: Part A: Polymer Chemistry*. Vol. *45*, pp. 3626-3636.

[8] Kabatc, J.; Pączkowski, J. (2009). *Journal of Polymer Science: Part A: Polymer Chemistry*. Vol. 47, pp. 576-588.

[9] Kabatc, J. (2010). *Journal of Polymer Science: Part A: Polymer Chemistry*. Vol. *48*, pp. 4243–4251.

[10] Kabatc, J. (2010). *Polymer*. Vol. *51*, pp. 5028-5036.

[11] Cavitt, T. B.; Hoyle, C. E.; Kalyanaraman, V.; Jönsson, S. (2004). *Polymer*. Vol. *45*, pp. 1119-1123.

[12] Grotzinger, C; Burget, D.; Jacques, P.; Fouassier, J.P. (2003). *Polymer*. Vol. 44, pp. 3671-3677.

[13] Oxman, J.D.; Ubel, F.A.; Larson, G.B. (1989) *U.S. Patent 4,828,583*.

[14] Palazzotto, M.C.; Ubel, F.A.; Oxman, J.D.; Ali, Z.A. (2000) *U.S. Patent 6,017,660*.

[15] Oxman, J.D.; Jacobs, D.W.; Trom, M.C.; Sipani, V.; Ficek, B.; Scranton, A.B. (2005) *Journal of Polymer Science: Part A: Polymer Chemistry*. Vol. 43, pp. 1747-1756.

[16] Oxman, J.D.; Jacobs, D.W. (1999) *U.S. Patent 5,998,495*.

[17] Oxman, J.D.; Jacobs, D.W. (2000) *U.S. Patent 6,025,406*.

[18] Mishra, A.; Behera, R.K.; Behera, P.K.; Mishra, B.K.; Behera, G.B. (2000). *Chem Rev*. Vol. 100, p. 1973

[19] Chatterjee, S.; Davis, P.D.; Gottschalk, P.; Kurz, M.E.; Sauerwein, B.; Yang, X.; Schuster, G.B. (1990). *J Am Chem Soc*. Vol. 112, p. 6329.

[20] Chatterjee, S.; Gottschalk, P.; Davis, P.D.; Schuster, G.B. (1988). *J Am Chem Soc*. Vol. 110, p.2326.

[21] For examples of cyanine dyes see: (a) Gottschalk, P.; Neckers, D.C.; Schuster, G.B. (1980). *US Patent 4 772 530*; (1987). *Chem Abstr* 107:187434n; (1988). *US Patent 4 842 980*; (1987). *Chem Abstr* 107:187434n; (b) Gottschalk, P. (1989). *US Patent 4 874 450*; (c) Weed, G.; Monroe, B.M. (1992). *US Patent 5 143 818*.

[22] Lan, L.Y.; Schuster, G.B. (1986). *Tetrahedron Lett.* Vol. 27, p. 4261.
[23] Gould, I.R.; Farid, S. (1993). *J Phys Chem.* Vol. 97, pp. 13067-13072.
[24] Rehm, D.; Weller, A. (1970). *Isr J Chem.* Vol. 8, pp. 259.

11

Solar Photochemistry for Environmental Remediation – Advanced Oxidation Processes for Industrial Wastewater Treatment

Anabela Sousa Oliveira[1,2], Enrico Mendes Saggioro[3], Thelma Pavesi[3],
Josino Costa Moreira[3] and Luis Filipe Vieira Ferreira[1]

[1]*Centro de Química-Física Molecular and Institute of Nanosciences and Nanotechnology,
Instituto Superior Técnico, Universidade Técnica de Lisboa, Lisboa,*
[2]*Centro Interdisciplinar de Investigação e Inovação,
Escola Superior de Tecnologia e Gestão, Instituto Politécnico de Portalegre, Portalegre,*
[3]*Centro de Estudos da Saúde do Trabalhador e Ecologia Humana,
Escola Nacional de Saúde Pública, Fundação Oswaldo Cruz, Rio de Janeiro,*
[1,2]*Portugal*
[3]*Brazil*

1. Introduction

Photochemistry is the chemistry induced by light. Being the sun the most abundant and widespread light (and consequently energy) source on earth, it is obvious that solar light can also induce chemical reactions. There are several classes of organic pollutants (organic dyes, pharmaceuticals, polycyclic aromatic hydrocarbons, polychlorinated pesticides, polychlorinated dibenzodioxins, dibenzofurans and biphenyls) that by the seriousness of the risks they pose to environment and human health are considered priorities for environmental monitoring by the most important environmental agencies. In this chapter we will show how solar light can be advantageously used for environmental remediation, leading to the destruction of environmentally relevant molecules, especially when they are present in industrial wastewaters. In fact, solar light can greatly contribute to the remediation (going from the partial decomposition to the complete destruction) of those environmental pollutants. This solar remediation action can be effective either through direct photolysis and photodegradation (light induced chemical bond cleavage leading to the formation of smaller compounds) or as being the photon source that triggers the processes of their photocatalytic degradation (solar photocatalysis through advanced oxidation processes).

Advanced Oxidation Processes (AOPs) are an emergent and promising methodology for the degradation of persistent environmental pollutants, refractory to other environmental decontamination / remediation treatments. AOPs are methods of advanced photocatalysis that use the highly oxidant and non selective hydroxyl radicals, which are able to react with almost all classes of organic compounds leading to their total (complete) mineralization or to the formation of more biodegradable intermediates. The method has the advantage that it can be applied to a large set of different matrixes and that decontamination occurs through pollutants degradation instead of their simple phase transfer. These methodologies become even more attractive when they use the sunlight as energy source, and they are generally identified as Solar Photocatalysis. The most representative solar photocatalysis treatments are semiconductor photocatalysis (Titanium dioxide, TiO_2, is the most used semiconductor) and photo-Fenton. Using TiO_2 or photo-Fenton, highly oxidant hydroxyl radicals are produced to promote the degradation of environmental contaminants.

Photochemistry is not only the aim of solar photocatalysis in what regards source energy and basic mechanism of action but it also furnishes powerful analytical tools for the study, monitoring and understanding of the main photodegradation processes and theirs mechanisms of action. So, many environmental pollutants and effluents which contain them had already their photolytic and photocatalytic photodegradation mechanisms investigated and have been treated through direct photolysis and/or solar activated advanced oxidation technologies in different environmental segments. Although most of the studies reported in literature are performed in laboratorial conditions, most often with artificial irradiation simulating solar light, there are also studies performed in pilot plants at industrial scale. In fact, there are several technological solutions using solar light and photocatalysis being applied in remediation / detoxification pilot plants. Those technologies (that use batch or continuous reactors) as well as the chemical substances or type of effluents that have already been treated with or without solar concentration capabilities will be revised in this chapter.

In our opinion solar photochemistry through advanced oxidation process is an elegant application of fundamental photochemistry that is close to reach a wide industrial use and that in this way well deserves to be included in a reference work on photochemistry and their most relevant applications in modern world. Despite the impressive volume of data published in the last 30 years, the questions related to the actual trends and future involvement of advanced oxidation processes in environmental remediation applications is a hot topic as reflected by the increasing number of publications on the filed in the most recent years.

1.1 Treatment of industrial wastewaters

Generally the term wastewater refers to any residual fluid released into the environment and that contains polluting potential. The equivalent term effluent, which means to spill, derives from the latin *effluente*.

In the last decades the growing environmental awareness led to the implementation of national, international and communitary legislation (Simonsen, 2007) which prohibits or severely restricts the discharge into the environment of untreated industrial effluents containing various classes of substances (a list of controlled or restricted organic pollutants is found in Metcalf & Eddy, 2003, chapter 2, pages 99 – 104). Therefore, particular attention has been devoted to the development of methodologies for industrial wastewater treatment able to destroy or reduce the concentration of restricted chemicals within the allowed legal limits (Davis & Cornwell, 1998; Eckenfelder, 2000; Kiely, 1998; Metcalf & Eddy, 2003; Tchobanoglous at al., 1986; Nevers, 2000).

Effluents discharges into the environment may be liquid and gaseous. Domestic sewage and several different industrial effluents are the main sources of liquid effluents. After treatment, these effluents are generally released into water bodies (rivers or sea), being usually designated as wastewaters (industrial or domestic). Deficient or incomplete wastewater treatment can lead to surface and groundwater contamination. The extensive use of chemicals such as pesticide, fertilizers, pharmaceuticals, detergents, etc. and soil deposition of urban solid residues are the most important causes of water contamination (Tchobanoglous, et al., 1993). Because of the risk posed to public health by consumption of contaminated water, special care must be taken in water source preservation and water treatment.

Water and wastewater treatment sequences consists of several different mechanical, physical, chemical and biological treatments that frequently include harrowing, filtration, flocculation, sedimentation, sterilization and chemistry oxidation of organic pollutants, among others. After physical treatment (filtration and sedimentation) the water still has considerable amounts of organic matter (including organic contaminants), which, in general, can be efficiently degraded under biological treatment (Davis & Cornwell, 1998; Eckenfelder, 2000; Kiely, 1998; Metcalf & Eddy, 2003). The gaseous effluents are often released to the air through tall chimneys and their main treatments include masking, adsorption on active carbon, contact liquid method, combustion and biological treatment (Davis & Cornwell, 1998; Kiely, 1998; Nevers, 2000). The methods for the treatment of water and wastewater (Davis & Cornwell, 1998; Eckenfelder, 2000; Kiely, 1998; Metcalf & Eddy, 2003) and gases (Davis & Cornwell, 1998; Eckenfelder, 2000; Nevers, 2000) are deeply revised in the literature.

However, the treatment of wastewater containing some organic substances cannot be achieved by traditional processes, because they resist to biological degradation (biorecalcitrant or persistent organic pollutants - POPs) or they are not completely removed by traditional treatment.

1.2 Treatment of wastewaters containing persistent organic pollutants

Nowadays, the persistent organic pollutants (POPs) are a matter of great importance, because they cannot be eliminated by the ordinary water or wastewater treatments (Davis & Cornwell, 1998; Eckenfelder, 2000; Kiely, 1998; Metcalf & Eddy, 2003; Nevers, 2000).

POPs are xenobiotic chemicals of natural or anthropogenic origin witch accumulated in the environment and biota, due to theirs highly refractory chemical structures and physical-chemical properties. Structurally they are polycyclic conjugated compounds (polycyclic aromatic hydrocarbons) or they have a high number of halogen atoms, especially chlorine or bromine (pesticides, polychlorinated dibenzodioxins – PCDDs -, polychlorinated dibenzofurans – PCDFs -, polychlorinated biphenyls – PCBs -, brominated flame retardants, etc). Because most POPs are semi-volatile they suffer long range transport and can be found anywhere, even in distant regions where they have never been produced or released. POPs have a lipophylic and hydrophobic characters and so they consequently bioaccumulate in fatty tissues of organisms and are capable of bioaccumulating or biomagnificating into food chains, reaching extremely high concentrations (in comparison with their environmental concentrations) on the top species (Baird, 1999). Many of these compounds are biologically actives possessing mutagenic and/or carcinogenic or even endocrine disruption properties. Although several of them have natural sources, the fast industrial development since the late nineteenth century lead to an enormous increase either on the quantity and on the diversity of the persistent organic pollutants from anthropogenic origin present in the environment. Conjugation of their above mentioned characteristics determines that these compounds represent a high risk to public and environmental health.

Several of those substances have already been classified as prioritary substances for environmental monitoring (see Baird, 1999, chapter 7, pages 293 to 379). Dibenzodioxins, dibenzofurans, polychlorinated biphenyls and organochlorinated pesticides join the list of priority organic pollutants of World Health Organization (WHO), United Nations Environmental Program (UNEP) and other Environmental Protection Agencies (Kiely, 1998; Metcalf & Eddy, 2003). The Stockholm Convention regulates this matter worldwide. This Convention presents a list of POPs (originally 12 substances: aldrin, dieldrin, endrin, chlordane, PCDDs, PCDFs, BHC, DDT, heptachlor, mirex, PCBs, toxaphene). Nowadays there are other under consideration: HCH, chlordecone, hexabromobiphenyl, hexa and heptabromobiphenyl ether, pentachlorobenzene tetra and pentabromodiphenyl ether e perfluorooctanosulfonic acid and its salts) which production, use and trading are banned or severely restricted (United Nations Environment Programme, 2005; Stockholm Convention on Persistent Organic Pollutants, 2005; Oliveira, et al., 2004, 2008, 2011). There are many other synthetic substances that have been identified as priority pollutants for environmental monitoring by the United States Environmental Protection Agency (USEPA) based on theirs probable or confirmed carcinogenic, mutagenic, teratogenic or acute toxicity characteristics. Among them we can mention volatile organic compounds, agricultural fertilizers and chlorinated residues resultant from disinfection processes at water public supply systems. Many of those substances can either be found in the air (as is the case of the volatile organic compounds) or in surface and groundwater and they reach the reception media through domestic or industrial wastewater systems or due to drain-off from agriculture (as appends with pesticides and fertilizers). There are also several substances (i.e. dyes, pharmaceuticals, etc) some of them specially synthesized to be resistant to degradation and conventional wastewater treatment processes are not able to remove them efficiently (Eckenfelder, 2000). Although these substances are not classified as prioritary pollutants, their negative impact in aquatic life and the changes of physical-chemical characteristics of the water bodies even when present in low concentrations make the control of their concentration very important.

Once the use and discharge of bioactive organic substances in the different environmental segments is not easy to eliminate and appears extremely difficult to control its essential to develop new powerful, clean and safe environmental remediation technologies for their treatment especially for the biorecalcitrant organic pollutants. One of the new most promising technologies available uses hydroxyl radical, a highly reactive chemical species that can attack and destroy organic molecules and is denominated advanced oxidation processes (Eckenfelder, 2000; Metcalf & Eddy, 2003).

2. Advanced oxidation processes

Advanced oxidation processes (AOPs) is the common name of several chemical oxidation methods used to remediate substances that are highly resistant the biological degradation. Although oxidation can be total, frequently a partial oxidation is sufficient to decrease the toxicity of the biorecalcitrant compound enabling their final treatment by conventional biological treatment. The complete oxidation leads to mineralization and yields CO_2, H_2O

and inorganic ions. The partial oxidation can be enough to decrease toxicity enabling biological degradation, but is essential to verify if the intermediary products formed are not more toxic than the parent compound under treatment. In the last 30 years several books (Bahnemann, 1999; Halmann, 1995; Pelizzetti & Serpone, 1989; Schiavello, 1988) and reviews (Byrne, et al., 2011; Dusek, 2010; Gogate & Pandit; 2004a, 2004b; Legrini, et al., 1993; Linsebigler, et al, 1995) were published on the subject. Blake, 2001 contains more than 1200 references on the subject. AOPs can remediate all different types of organic pollutants in liquid, gaseous or solid media, reason why they are used on the remediation of contaminated waters, liquid or gaseous effluents and also on the treatment of different hazardous wastes namely on contaminated soils. Some of the above mentioned reviews present comprehensive compilations of the substances and residues already mineralized using different advanced oxidation processes (Blake, 2001, Legrini, et al., 1993).

2.1 Theory of advanced oxidation

Although different advanced oxidation processes use several different reaction systems, all of them have the same chemical characteristic: i.e., the production and use of hydroxyl radicals (OH•) (Eckenfelder, 2000; Metcalf & Eddy, 2003). Hydroxyl radicals are highly reactive species that are able to attack and destroy even the most persistent organic molecules that are not oxidized by the oxidants as oxygen, ozone or chlorine (Eckenfelder, 2000). Table 1 shows oxidation potential of the hydroxyl radical and compares it with others commons oxidants used in chemical oxidation (Fox & Dulay, 1993).

Oxidizing agent	Oxidation potential, Volt
Fluorine	3.06
Hydroxyl radical	2.80
Atomic oxygen	2.42
Ozone	2.08
Hydrogen peroxide	1.78
Hypochlorite	1.49
Chlorine	1.36
Chlorine dioxide	1.27
Molecular oxygen	1.23

Table 1. Oxidation potential of most common oxidizing agents.

Hydroxyl radical is the most powerful oxidant after fluorine; it is able to initiate several oxidation reactions leading to complete mineralization of the original organic substances and their subsequent degradation products. Hydroxyl radical reacts will all classes of organics mainly by hydrogen abstraction:

$$OH• + RH \rightarrow R• + H_2O \qquad (1)$$

Hydrogen abstraction produces organic radicals able to react with molecular oxygen and originating peroxyl radicals.

$$R• + O_2 \rightarrow RO•_2 \rightarrow \qquad (2)$$

Electrophilic additions may also occur (Legrini et al., 1993).

$$OH\bullet + PhX \rightarrow HOPhX\bullet \qquad (3)$$

Electron transfer reactions,

$$OH\bullet + RX \rightarrow RX\bullet^+ + OH^- \qquad (4)$$

and reactions between hydroxyl radicals,

$$2\,OH\bullet \rightarrow H_2O_2 \qquad (5)$$

Hydroxyl radical is characterized by a non-selective attack; this is an extremely useful characteristic for an oxidant to be used on environmental remediation. Other relevant and important characteristics are the existence of several possible pathways for hydroxyl radical production and the fact that all reactions occur at normal temperature and pressure. AOPs advantageously promote complete degradation of pollutants being remediated while classical treatments usually only transfer target pollutants to another phase, leading to the production of secondary residues (slugs) that require further treatment or deposition. Therefore, the advanced oxidation process is a good method for environmental decontamination (Linsebigler et al., 1995).

AOPs versatility is favoured also by the existence of various pathways to produce hydroxyl radicals, which enables a high adaptability to any specific environmental remediation problem. The advanced oxidation process can degrade all types of organic compounds in water therefore they are widely used in industrial wastewater remediation.

2.2 Technologies used in the production of hydroxyl radicals

Advanced oxidation processes enclose several different treatments options: as ozone, hydrogen peroxide, ultraviolet radiation, ultrasound, homogeneous and heterogeneous photocatalysis, photocatalytic disinfection and also their combination (Hoffmann et al., 1995). The use of hydroxyl radicals to promote chemical oxidation it is the common feature of all AOPs. Table 2 shows several of the chemical oxidation technologies available.

Processes with ozone	Processes without ozone
Ozone at high pH (8-10)	$H_2O_2 + UV$
Ozone + UV	Photocatalysis (UV+ photocatalyst)
Ozone + H_2O_2	Ultrasound
Ozone + H_2O_2 + UV	Oxidation supercritical
Ozone + TiO_2	H_2O_2 + UV + iron salts (Foto-Fenton)
Ozone + TiO_2 + H_2O_2	H_2O_2 + iron salts (Fenton reagent)
Ozone + Ultrasound	

Table 2. Technologies used in the production of hydroxyl radicals.

The AOPs classification is frequently based on the use or not of ozone on the production of hydroxyl radicals. The classification can also be based on the use or not of irradiation and on the number of phases (homogeneous or heterogeneous). Ozone + UV, ozone + hydrogen peroxide, ozone + UV + hydrogen peroxide and hydrogen peroxide + UV are the most used commercial processes (highlighted in italic in Table 2) and its use will be analyzed below (Metcalf & Edie, 2003).

2.2.1 Ozone + UV

The hydroxyl radical production is achieved by ozone irradiation (O_3) with ultraviolet radiation (UV) according to:

$$O_3 + h\upsilon_{UV} \rightarrow O_2 + \text{singlet oxygen} \tag{6}$$

$$\text{Singlet oxygen} + H_2O \rightarrow OH\bullet + OH\bullet \text{ (in moistured air)} \tag{7}$$

$$\text{Singlet oxygen} + H_2O \rightarrow OH\bullet + OH\bullet \rightarrow H_2O_2 \text{ (in water)} \tag{8}$$

Ozone photolysis in air with moisture produces two hydroxyl radicals by each ozone molecule while when the same reaction occurs in water the hydroxyl radicals produced suffer rapid recombination and hydrogen peroxide is readily formed. Due to this later reaction, the process in water is not economically viable, since a lot of energy as to be imparted to the system to keep the adequate hydroxyl radical concentration. However, ozone + UV process is efficient on degradation in gaseous phases. The efficiency is even higher if the compounds also undergo direct photolysis by ultraviolet radiation. Figure 1 presents a scheme of an advanced oxidation treatment unity using ozone and UV radiation (Ikehata & El-Din, 2004; Mills & Hunte, 1997).

2.2.2 Ozone + hydrogen peroxide

For compounds that do not efficiently absorb ultraviolet radiation the yield of the degradation processes can be increased adding hydrogen peroxide once the latter when in contact with ozone undergoes an additional reaction, further promoting hydroxyl radical formation.

$$H_2O_2 + 2\,O_3 \rightarrow OH\bullet + OH\bullet + 3\,O_2 \tag{9}$$

Figure 2 presents a scheme of an advanced oxidation unity using ozone and hydrogen peroxide.

2.2.3 Hydrogen peroxide + UV and ozone + UV + hydrogen peroxide

UV irradiation of water with hydrogen peroxide also leads to hydroxyl radical formation (Galindo et al., 2000, Ikehata & El-Din, 2006).

However, frequently the process is not economically viable due to the low absorption extinction coefficient of hydrogen peroxide; this fact determines the use of high concentrations of hydrogen peroxide so that hydroxyl radicals are produced in the adequate amount. The combination of the later process with ozone promotes a better efficiency on the use of UV radiation. Figure 3, presents a scheme of an advanced oxidation unit using ozone, hydrogen peroxide and UV radiation.

Although the technologies presented above have reached commercial application, especially in industrial wastewater treatment and water disinfection all of them use high amounts of expensive reagents (hydrogen peroxide and ozone) and consume a lot of energy on UV radiation generation. So, their application is restricted to processes where more economic alternatives are not viable.

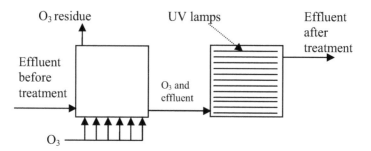

Fig. 1. Scheme of advanced oxidation treatment unit using ozone and UV radiation.

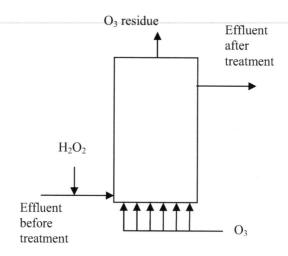

Fig. 2. Scheme of advanced oxidation unit using ozone and hydrogen peroxide.

$$H_2O_2 + h\upsilon_{UV} \rightarrow OH\bullet + OH\bullet \qquad (10)$$

Taking in account their cost, advanced oxidation processes can be used in integrated treatment systems for water, and domestic or industrial wastewaters where prior to biological treatment an advanced oxidative degradation of toxic and refractory substances is performed (Mantzavinos & Psillakis, 2004; Oller et al., 2007; Zapata et al., 2010). Frequently, the primary attack promoted by hydroxyl radical is sufficient to produce less toxic compounds that can already undergo biological treatment. This integrated treatment scheme can lead effectively to global treatment cost reduction. At different times of the advanced oxidative treatment it is advisable to perform toxicity tests with different microorganisms commonly used in biological treatment. These toxicity tests will help to determine the moment that the advanced oxidation process can already be substitute by the biological treatment (Fujishima et al., 2000).

However, because AOPs are essential for treatment of resistant substances in wastewater the most recent research efforts were on the development of more efficient energy processes. AOPs that do not need UV irradiation to activate hydroxyl radical production and that alternatively can use sunlight (wavelengths greater than 300 nm) for the same propose are extremely attractive (Byrne et al., 2011; Malato et al., 2002, 2009).

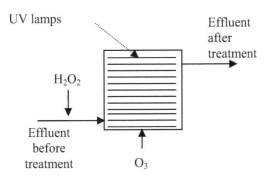

Fig. 3. Scheme of advanced oxidation unit using ozone, hydrogen peroxide and UV light.

3. Advanced oxidation processes with sunlight

The degradation of persistent organic pollutant using advanced oxidation processes with sunlight as energy source have as great advantage their lower costs. There are two advanced oxidation processes that enable the use of sunlight as energy source: heterogeneous photocatalysis using semiconductors and homogeneous photocatalysis using photo-Fenton processes (Fujishima et al., 2000; Pirkanniemi & Sillanpaa, 2002).

We can compare the solar emission spectra (starting at 300 nm) with the absorption spectra of titanium dioxide and of Fenton reactant (Malato et al., 2002). Heterogeneous photocatalysis activated by sunlight uses near ultraviolet solar spectrum (wavelength under 380 nm) and homogeneous photocatalysis by photo-Fenton uses a larger portion of solar spectrum (wavelength up to 580 nm). Both processes are efficient in the photodegradation of persistent organic pollutants, they are a innovative way of using a renewable energy and they are very promising technologies in what regards environmental remediation (Gogate & Pandit, 2004a, 2004b).

Photocatalysis is the combination of photochemistry and catalysis, a process where light and catalysis are simultaneously used to promote or accelerate a chemical reaction. So, photocatalysis can be defined as "catalysis driven acceleration of a light induced reaction". Direct light absorption is one of photocatalysis bigger advantages compared to thermally activate catalytic processes.

Nowadays, photocatalysis appears as an excellent tool for final treatments of samples containing persistent organic pollutants (POPs) when compared to classical treatments (Doll & Frimmel, 2005; Hincapié, 2005). In a near future they can turn in one of the most used technologies for POPs remediation.

3.1 Heterogeneous photocatalysis using semiconductors – TiO₂/UV

Heterogeneous photocatalysis is a (sun)light activated process that produces reducing and oxidizing species able to promote mineralization of organic pollutants using a semiconductor (TiO_2, ZnO, etc) as catalyst. The interaction of a photon with the catalyst produces an electron/hole pair on it. Excited electrons can be transferred to chemicals (reduction) into the semiconductor particle environment and at the same time the catalyst accepts electrons of oxidized specie. The neat flux of electrons in both directions is null and the catalyst stays unaltered. The mechanism and the electron/hole generation processes of heterogeneous photocatalysis is addressed in several reviews (Davis & Cornwell, 1998; Eckenfelder, 2000; Kiely, 1998; Metcalf & Eddy, 2003; Nevers, 2000) that also present exhaustive list of organic residues remediated already by the method.

The ability of heterogeneous photocatalysis to eliminate organic pollutants from gaseous or aqueous (Eckenfelder, 2000; Kiely, 1998; Metcalf & Eddy, 2003) streams was largely demonstrated. Polycyclic aromatic hydrocarbons, pentachlorophenol, 2,4,5-trichloro-phenoxyacetic acid (2,4,5-T), 4,4'-DDT, dichlorobiphenyls and dichlorodibenzodioxins were already mineralized by heterogeneous photocatalysis (Blake, 2001; Pelizzetti & Serpone, 1989). Once TiO_2 only uses about 10% of the available solar radiation, several groups are working on improving TiO_2 visible light absorption (Anandan & Yoon, 2003; Chan et al., 2011; Fox & Dulay, 1993; Gupta & Tripathi, 2011; Janus & Morawski, 2010; Linsebigler et al., 1995; Mills & Hunte, 1997; Mourao et al., 2009; Reynaud et al., 2011; Roy et al., 2011; Zhang et al., 2004) to improve photocatalysis efficiency.

3.1.1 Heterogenous photocatalysis mechanism

Semiconductors (as TiO_2, ZnO, CdS, ZnS, etc.) have a typical electronic structure composed by a fully occupied valence band (VB) and an empty conduction band (CB). This typical semiconductor's electronic structure enable they can act as sensitizers of light induced oxidation processes. Photocatalysis action mechanism can be visualized on Figure 4. In this scheme the valence and conduction band of a semiconductor are represented over a spherical semiconductor particle.

The semiconductor (TiO_2) absorbs photons with enough energy to promote an electron from the valence band to the conduction band and on the process an electron/hole pair is formed. The energy of the absorbed photon has to be equal or higher than that of the semiconductor "band-gap". The process can be described in a simple way by the following set of equations:

$$TiO_2 + h\upsilon \rightarrow h^+_{VB} + e^-_{CB} \tag{11}$$

While the electron is promoted to the conduction band a hole is produced in the valence band; this hole has a high oxidative power (+1 to +3.5 V, depending on semiconductor and pH) able not only to oxidize the water absorb on semiconductor surface producing hydroxyl radicals but also able to oxidize hydroxide ions, OH^-, or the substrate itself, RX (Figure 4).

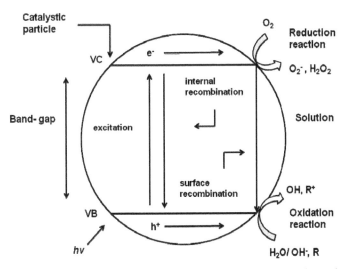

Fig. 4. Heterogeneous photocatalysis on a semiconductor (TiO_2) particle surface.

$$TiO_2 (h^+_{VB}) + H_2O_{\;adsorv} \rightarrow TiO_2 + OH\bullet_{\;adsorv} + H^+ \tag{12}$$

$$TiO_2 (h^+_{\;VB}) + OH^-_{adsorv} \rightarrow TiO_2 + OH\bullet_{\;adsorv} \tag{13}$$

$$TiO_2 (h^+_{\;VB}) + RX_{adsorv} \rightarrow TiO_2 + RX\bullet^+_{\;adsorv} \tag{14}$$

Once formed hydroxyl radicals promote the already mentioned oxidation reactions that degrade the persistent organic pollutants. The electrons promoted to the conduction band are also able to reduce the oxygen available in the surroundings to superoxide radicals. Presence of oxygen is essential in all oxidative degradation processes

$$TiO_2 (e^-_{CB}) + O_2 \rightarrow TiO_2 + O_2\bullet^- \tag{15}$$

When hydrogen peroxide is added the speed of the photodegradation significantly increase because its reaction with TiO_2 provides another hydroxyl radical source.

$$TiO_2 (e^-_{CB}) + H_2O_2 \rightarrow TiO_2 + OH^- + OH\bullet \tag{16}$$

Although the nature of all oxidizing species formed on semiconductor surface after light absorption is controversial, all authors agree that hydroxyl radical is the main oxidizing species formed on semiconductor surface. Effectively all detected intermediary species during photodegradation of polycyclic aromatic hydrocarbons and halogenated organic compounds are typically hydroxylated structures (Bahnemann, 1999; Oliveira et al., 2004b; Xavier et al., 2005). More difficult to clarify is if oxidation proceeds by direct or indirect route, directly by holes or by hydroxyl radical formed from them, bonded to surface or in solution. On the other hand, the strong correlation between speed of degradation and the concentration of pollutants absorbed on catalyst surface also suggests and reinforces the hypothesis that the species responsible for the photodegradation are hydroxyl radicals formed on the surface of the photocatalyst. Laser flash photolysis and electronic paramagnetic resonance proved to be helpful on the elucidation on the nature of intermediary species formed during photocatalytic degradation processes (Bahnemann, 1999; Botelho do Rego and Vieira Ferreira, 2001, Fox & Dulay, 1993, Oliveira et al., 2004b).

3.1.2 Photocatalysts

The ideal semiconductor to be used as photocatalyst must be photoactive, able to absorb ultraviolet and visible radiation, photostable, chip and biologically and chemically inert. TiO_2, ZnO e CdS are the most studied photocatalysts. Titanium dioxide has all above mentioned characteristics of a good photocatalyst and is in fact the photocatalyst with higher fotocatalytic activity on organic matter decomposition. This quality of titanium dioxide made him the reference semiconductor to establish and compare the photocatalytic activity of other semiconductor materials. TiO_2 photocatalysis also obeys to green chemistry key principles (Anastas & Warner, 1998; Hermann et al., 2007).

TiO_2 occurs in three crystals forms: anatase, rutile and brokite. Anatase is the photocatalytic active form. However, different semiconductor batches have present different photocatalytic activities from batch to batch and between different producers. It is also difficult to reproduce the photocatalytic activity between laboratories. Because of that, TiO_2 P25 from Degussa is currently accepted as the standard titanium dioxide. TiO_2 Degussa P25 without any treatment is used on phenol degradation for comparative proposes of the photocatalytic reactors performance. Degradation of 4-chlorophenol is also a standard reaction for certification of titanium dioxide photocatalytic activity.

TiO_2 Degussa P25 is the standard form of TiO_2; it is a powder available commercially with a purity of 99.5% (70:30 anatase : rutile), with a superficial area of 50 ± 15 m^2/g, its not porous and have cubic particles of rounded edges and a average particle diameter of 21 nm. However TiO_2 particles does not exist isolated but as complex irreducible primary aggregates of about 1 μm.

To perform a heterogeneous photocatalytic reaction activated by light it is necessary to use semiconductors with the adequate "band-gap" to be activated by solar energy. TiO_2 have a high bang-gap, of 3.2 eV, being consequently activated only by radiation below 380 nm, i.e., using only 10% of the sunlight spectrum. However, metal oxides with high band-gap, as TiO_2, go on being strongly used on photocatalysis since they are usually resistant to photocorrosion. Although photocatalysts with lower band-gaps present bigger sensitivity to solar spectrum, they are not frequently used because they experience strong photocorrosion, being globally less effective.

Maximizing the efficiency of photocatalysis is an object of great challenge for scientists. Many efforts have been devoted to extend the photocatalytic properties into the visible region (Anandan & Yoon, 2003; Asahi et al., 2001; Augugliaro et al., 2006; Chan et al., 2011; Emeline et al., 2008; Fox & Dulay, 1993; Gupta & Tripathi, 2011; Janus & Morawski, 2010; Linsebigler et al., 1995; Mills & Hunte, 1997; Mourao et al., 2009; Reynaud et al., 2011; Roy et al., 2011; Zhang et al., 2004). Approaches such as doping the TiO_2 with transition metal ions or the deposition of a noble metal on semiconductor particles have been successfully used. Inclusion of iron into TiO_2 particles, for example, has been effectively used in the degradation of chlorinated organic compounds. Additionally nanotechnology is providing new insights in this subject with several classes of nanoscale materials (some of them including already titanium or other catalysts) that are being already evaluated as functional materials for water purification (Biswas & Wu, 2005; Wang et al., 2008; Xu et al., 2011): metal-containing nanoparticles, carbonaceous nanomaterials, zeolites and dendrimers (Savage & Diallo, 2005). The use of light to activate such nanoparticles opens up new ways to design green oxidation technologies for environmental remediation (Kamat & Meisel, 2002, 2003; Savage & Diallo, 2005). Due to their high-specific surface area, nanoparticles exhibit enhanced reactivity when compared with their bulk counterparts by several reasons such as the proportion of surface sites at edges or corners, the presence of distorted high-energy sites, contributions of interfacial free energies to chemical thermodynamics, the effects of altered surface regions, and quantum effects. Nanoparticles can be easily deposited or anchored onto various surfaces or used as a tailored film. These facilities can improve the adsorption of desirable chemicals, such as organics and heavy metals onto film surfaces. (Biswas & Wu, 2005; Kamat & Meisel, 2002, 2003). TiO_2 nanoparticules have been extensively studied for oxidative and reductive transformation of organic and inorganic species present as contaminants water (Wang et al., 2008; Xu et al., 2011). Ashasi et al. (2001) synthesized N-doped TiO_2 nanoparticles that were capable of photodegraded methylene blue under visible light and Bae & Choi (2003) have synthesized visible light-activated TiO_2 nanoparticles based on TiO_2 modified by ruthenium-complex sensitizers and Pt deposits.

3.2 Homogeneous photocatalysis - photo-Fenton process

Fenton's reagent is another extremely useful source of hydroxyl radicals and a potent oxidant of organic compounds. It was first described at the end of the XIX century and consists in a process in homogeneous phase, in which an aqueous hydrogen peroxide solution and Fe^{2+} (ferrous) ions, in acidic conditions (pH = 2-4), generate hydroxyl radicals, in a process that is not activated by light (Nogueira et al., 2007; Pignatello et al., 2006):

$$Fe^{2+} + H_2O_2 \rightarrow Fe^{3+} + OH^- + OH\bullet \tag{17}$$

When Fenton process occurs under solar radiation, degradation rate increases significantly. Although the oxidizing power of Fenton reaction was known for more than one hundred years, only recently it was discovered that Fenton reaction can be accelerated by irradiation with ultraviolet or visible light (λ <580 nm), making it a photocatalytic process (Fe^{2+} is regenerated). The so-called photo-Fenton reaction produces additional hydroxyl radicals and leads to the photocatalyst reduction by light (Nogueira et al., 2007; Pignatello et al., 2006):

$$Fe^{3+} + H_2O + h\upsilon_{UV-Vis} \rightarrow Fe^{2+} + H^+ + OH\bullet \tag{18}$$

The great advantage of this process when compared with heterogeneous photocatalysis with TiO_2 it is its sensibility to light up to 580 nm. When compared with TiO_2 photocatalysis, this process allows a more efficient use of sunlight. The contact between the pollutant and the oxidizing agent is more effective once the process occur in homogeneous phase. The disadvantages of the photo-Fenton process are the treatment aggressivity due to low pH required (usually below 4), the high consumption of hydrogen peroxide and the need of removing iron at the end of the treatment.

The use of sunlight instead of artificial light for photo-Fenton activation, besides increasing the efficiency, also significantly decreases the cost of treatment. Therefore photo-Fenton is a great advance towards industrial implementation of photocatalysis processes (Brillas et al., 2009; Nogueira et al., 2007; Pignatello et al., 2006). Foto-phenton's process ability to treat water containing various pollutants (Kavitha & Palanivelu, 2004; Soon & Hameed; 2011; Umar et al., 2010) was already proved.

4. Industrial units of wastewater treatment by photocatalysis

4.1 Solar collectors for photochemical processes

The use of light activated advanced oxidation processes requires the development of dedicated photochemical solar technology that include the design of efficient solar photons collection technologies and the direction of those photons to the appropriate reactor in order to promote the photodegradation of the persistent organic pollutants to be remediated. For solar photochemical processes it is more interesting the collection of photons with high-energy and low wavelength, since typically the majority of the photocatalysis processes use solar radiation in the ultraviolet (300-400 nm). The exception is photo-Fenton process, which uses all sunlight below 580 nm. Usually radiation with wavelengths higher than 600 nm does not have any utility for photocatalysis processes. Solar flux at ground level is about 20 to 30 W per square meter, so sun approximately provides 0.2 to 0.3 moles of photons per square meter per hour (Bahnemamm, 1999; Malato et al., 2002).

The equipment that makes the efficient collection of photons is the solar collector. This equipment represents the largest source of operating costs of a photocatalysis unit for treatment of effluents. Solar collectors can be classified into three types according to the level of solar concentration achieved, which is usually directly related to the temperature reached by the system. So we have solar collectors which are non concentrators, moderately concentrators or highly concentrators. They can also be called concentrators of low (<150 °C), medium (150 - 400 ° C) and high (> 400 ° C) temperature.

The non concentrating collectors or low temperature collectors are static and usually consist of flat plates directed towards the sun with a certain inclination, depending on the geographic location. Its main advantage is their great simplicity and low cost.

The moderately concentrating collectors or medium temperature collectors concentrate the sun 5 to 50 times; to achieve this concentration factor the equipment must be able to continuously follow the sun. The parabolic and holographic are such type of collectors. Parabolic collectors have a parabolic reflecting surface which concentrates the radiation in a tubular collector located at the focus of the parabola and can have uni or biaxial movement. Holographic collectors consist of reflective surfaces like convex lenses that deflect radiation and concentrate it in a focus.

The highly concentrating collectors or high temperature collectors have a punctual focus rather than a linear one and they are based on a paraboloid with solar tracking. These reactors ensure solar concentrations from 100 to 10,000 times, reason why they require high precision optical elements.

As the temperature plays no role in photochemical reactions, their technological applications typically only use low and medium temperature collectors, which have a much more economical construction. An important difference between these two reactors is that the first type of concentrators uses both direct and diffuse radiation while the concentrators collectors only use direct radiation. In terms of collector and reactor design itself, the systems have much in common with that of conventional thermal collectors; however, as the effluent to be cleaned must be directly exposed to sunlight, the absorber must be transparent to the photons. As temperature is not important, the systems are not thermally isolated. Most photocatalytic remediation systems involve wastewaters (Duran et al., 2010, 2011; Malato et al., 2007a ,2007b; Marugan et al., 2007; Miranda-Garcia et al., 2011; Navarro et al., 2011; Oyama et al., 2011; Vilar et al., 2009; Zayani et al., 2009), but the appropriated technology for gas phase photocatalytic processes is also possible (Lim et al., 2009).

4.2 Reactors with solar collectors

Several types of solar reactors for effluents decontamination have been tested. We describe below the four most commonly used.

4.2.1 Thin film fixed bed reactor

This reactor, whose simplified diagram is presented in Figure 5, is one of the first non-concentrators solar reactors, so it can use the total solar radiation (direct and diffuse) for the photocatalytic process. Quantities of direct and diffuse radiation reaching earth are nearly identical, so concentrator reactors, by not using the diffuse radiation, profit from only half of the maximum radiation available.

The most important part of the reactor is a tilted fixed dish (0.6 m x 1.2 m) coated with a thin film of photocatalyst, typically Degussa P25 TiO_2, which is continuously washed with a film of about 100 μm from the wastewater to be treated at a rate of 1 to 6.5 liters per hour (Bahnemamm, 1999; Malato et al., 2002).

4.2.2 Parabolic trough reactor

This reactor directly concentrates sunlight by a factor from 5 to 50. Tracking of solar radiation is done by a single or dual motors system that allow the continued alignment of the solar concentrator with the sun and various reactors can be connected in series or in parallel. In the parabolic trough reactor, the reflector has a parabolic profile and the tube where the photocatalytic reaction takes place is in its focus, in this way, only the light that enters parallel in the reflector can be focused on the reaction tube (Figure 6). This type of reactor is being used in solar decontamination circuits installed in the United States (Albuquerque, Sandia National Laboratories, and California, Lawrence Livermoore Laboratories) and Spain (Plataforma Solar de Almeria), Figure 7 (Navntoft et al., 2009).

The concentrated radiation is focused into a tube containing an aqueous suspension of TiO_2 and the effluent to be treated. In fact, only about 60% of the radiation collected is effectively used, the rest being lost by various causes.

At Almeria reactors, the total volume of effluent are about 400 liters, but the illuminated tube of about 180 meters contains only about half of that volume of effluent, which moves at speeds between 250 to 3500 liters per hour (Bahnemamm, 1999; Malato et al., 2002).

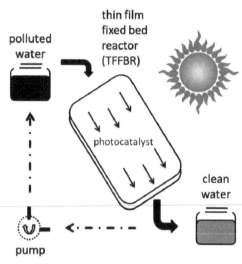

Fig. 5. Simplified diagram of a thin film fixed bed reactor.

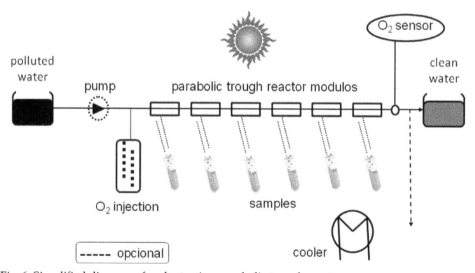

Fig. 6. Simplified diagram of a plant using parabolic trough reactors.

4.2.3 Compound parabolic collecting reactor

This reactor, whose simplified diagram is presented in Figure 7, is an open reactor without solar concentration. Basically this reactor differs from the conventional open parabolic reactor in the form of the reflectors. These reactors are static collectors with a reflective surface that surrounds a circular reactor, as shown in Figure 7. They had shown to provide better efficiency in the treatment of low pollutant concentration effluents. This reactor is an effective combination of the two reactors types described above (Bahnemamm, 1999; Malato et al., 2002).

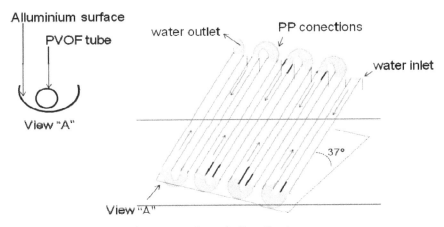

Fig. 7. Simplified diagram of a ompound parabolic collecting reactor.

4.2.4 Double skin sheet reactor

This type of reactor without concentration consists of a transparent box with an internal structure similar to that shown in Figure 8, through which is pumped the suspension containing the pollutant and the photocatalyst. It has the advantage of using the total radiation and be very simple to operate (Bahnemamm, 1999; Malato et al., 2002).

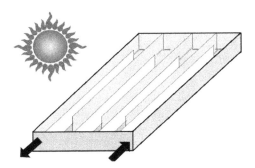

Fig. 8. Simplified diagram of a double sheet reactor.

4.3 Industrial units

Figure 9 shows a diagram of a photocatalytic installation that can be alternatively used for heterogeneous TiO_2 photocatalysis or for homogeneous photo-Fenton photocatalysis (or any other of the treatments previously described). In both cases the catalyst (TiO_2 or iron) must be separated at end of treatment to be recycled and reused.

The surface area of the solar collector depends essentially on the effluent to be treated, mandatorily of the type and concentration of the contaminant, and on solar irradiation conditions and the location where treatment plant will be installed. The lifetime of the catalysts depends on the type of effluent to be treated and of the desired treatment final quality. In the end, the toxicity of the treated effluent must always be evaluated.

The project of an industrial effluent decontamination plant by photocatalysis requires a careful selection of the type of reactor to use, the arrangement of the reactor at the installation (series or parallel), the operation mode of the photocatalyst (fixed or suspended), the system for recycling catalysts and flow velocity, among others. The concentration of photocatalyst is also a key parameter and must be adjusted according to the following basic principles: for suspensions of TiO_2, the speed of reaction is maximum for concentrations among 1 to 2 grams of TiO_2 per liter of effluent to be treated, when the optical path is small (1-2 cm maximum). When the optical path is substantially higher, the appropriate concentration of photocatalyst is several hundred milligrams per liter. Anyway, when TiO_2 concentrations is too high there is an internal filter effect and the rate of photodegradation decreases due to excessive opacity of the solution, which itself inhibits the illumination of the photocatalyst.

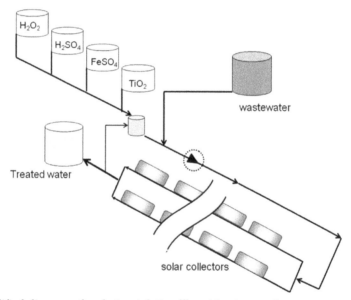

Fig. 9. Simplified diagram of a photocatalytic effluent treatment plant.

If the treatment plant is intended for treatment of a specific effluent, it does not need to be versatile and will be very similar to that shown in Figure 10. On the other hand, if an installation needs to treat various types of wastes, it must have the versatility to adapt to the optimal photodegradation conditions of the various types of effluents, e.g., having different types of solar collectors and reactors (as is the case of Almeria solar platform) and the project will be much more complicated.

5. Applications of photocatalysis on the treatment of industrial effluents

Solar driven AOPs proved to be an excellent environmental remediation method to destroy persistent organic compounds not treatable by biological processes. In many cases, they allow the degradation of several persistent organic toxic pollutants decreasing the toxicity of the effluents released into the environment. These methods are particularly suitable for treating recalcitrant substances including those requiring special attention (hazardous or controlled ones). Several organochlorinated substances (dioxins, PCBs, etc) are persistent and sufficient toxic to disturb the environmental health and must be degraded prior to their environmental release. Without being exhaustive on the list of applications and systems to be treated, we will refer some examples that we consider most significant.

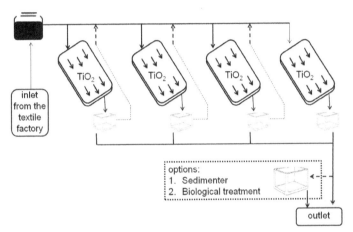

Fig. 10. Simplified diagram of a unit for dyes degradation with thin film fixed reactors.

Water is essential for life and therefore a key resource for humanity. Although it may seem that the water is very handy on our planet that is not true. Of the entire planet's water, 97.5% of the water is salty, among the remaining 2.5%, 70% is frozen and the rest is largely inaccessible in underground aquifers or as soil moisture. In fact, less than 1% of world potable water is available for immediate human consumption and even that is not uniformly distributed around the globe. For this reason, methodologies such as advanced oxidation processes that allow the maintenance of water quality are essential (Andreozzi et al., 1999; Chong et al., 2010; Comninellis et al., 2008; Matilainen & Sillanpaa, 2010). The problematic of water treatment and industrial wastewater treatment are inseparable issues since these industrial effluents constitute a major source of water contamination and are usually discharged into the environment in aqueous media.

The classical treatment processes of drinking water include treatment with ozone and filtration through granular activated carbon beds. Photocatalysis emerged as a promising tool for the treatment of water (and for the degradation of persistent substances even when they are present in low concentrations or complex matrices). So the advanced oxidation processes have been widely reported as an appropriate remediation methodology of all kinds of biorecalcitrants pollutants in water and industrial wastewater and their application to large-scale treatment facility is already being implemented, as discussed in section 4.

Typical examples of water pollutants that were efficiently mineralized by photocatalysis are effluents from industries containing dyes (Guillard et al., 2003), pesticides (Burrows et al., 2002; Marinas et al., 2001) and the effluents from the paper industry (Peiro et al., 2001). Various applications are also known for the decontamination of waste gases (Fu et al., 1996; Hay & Obee, 1999) including those involving self-cleaning surfaces (Hashimoto & Watanabe, 1999).

5.1 Industrial effluents containing dyes

The dyes are common industrial residues present in wastewaters of different industries, ordinarily in textile dyeing process, inks, and photographic industries, among others. The environmental aspects of the use of dyes, including their degradation mechanisms in various environment compartments, have been a target of increasingly interest. It is estimated that nearly 15% of world production of dyes is lost during synthesis and dyeing process. Concomitantly, the major problem related with dyes is the removal of their colour from effluents. The non treated effluents frequently are highly colored and then particularity susceptible to public objection when disposed in water bodies. The dye concentration in residual waters can be smaller than others contaminants, but because of its high molar absorption coefficients they are visible even in very low concentrations. So, methodologies of effluents discoloration became very relevant. The oxidation processes are very much used in treatment of dye containing effluents (Khataee & Kasiri, 2010; Oliveira et al., 2008, 2011; Rauf & Ashraf, 2009; Saggioro et al., 2011; Soon & Hameed, 2011). Figure 10 presents a pilot unit commonly used on the degradation of dyes with thin film fixed bed (Guillard et al., 2003).

5.2 Effluents containing pesticides and pharmaceuticals

The photodegradation and mineralization of pesticides and pharmaceuticals has been widely studied because of the danger they represent to the environment and also due to the highly recalcitrant nature of some of these compounds. For a comprehensive review of pesticide degradation see Blake, 1999 or some of the reviews listed here (Atheba et al., 2009; Bae & Choi, 2003; Felsot et al., 2003). Either titanium photocatalysis or Fenton parent methodologies usually promote rapid destruction of persistent pesticides.

Municipal water recycling for industrial, agricultural, and non-potable municipal may contain several different pharmaceuticals including antibiotics, hormones and other endocrine disruptors, sulphonamides, antipyretics, etc. Those are present in municipal sewage, largely as a result of human use and/or excretion. Much of the concern regarding the presence of these substances in wastewater and their persistence through wastewater treatment processes is because they may contribute to directly or indirectly affect the environmental and human health (Exall, 2004; Vigneswaran & Sundaravadivel, 2004).

In spite of the variable removal of antibiotics during conventional waste water treatment processes, many of these chemicals are often observed in secondary treated effluents. Conventional water and wastewater treatment are inefficient for substantially removing many of these compounds. While there appears to be no standard treatment for removal of all residual pharmaceuticals under conventional treatment processes, there is a strong opinion that advanced oxidation processes can be used for the effective removal of these compounds (Auriol et al., 2006; Dalrymple et al., 2007; Gueltekin & Ince, 2007; Homem & Santos, 2011; Khetan & Collins, 2007; Santos et al., 2009) . Research has show that advanced oxidative processes, which generate very active oxidative species such as the hydroxyl radicals, are promising tools for the destruction of pharmaceuticals compounds (Gültekin, et al., 2007; Le-Minh et al., 2010).

6. Conclusion

Advanced oxidation processes offer a consistent path to the treatment of recalcitrant substances that can not be treated by conventional effluents treatments. Either TiO_2 mediated photocatalysis or Fenton related methodologies offer feasible alternatives for the treatment of dyes, organochlorinated substances (pesticides, dioxines, furanes, PCBs, etc.) and pharmaceutical products, enabling the decomposition of such substances. Those methods, which are very attractive from the point of view of sustainable and green chemistry because they can use solar light as energy source, are being increasingly tested in several treatment plants (some of them pilot plans) with the help of solar collecting technology.

7. Acknowledgements

The authors thank to Fundação para a Ciência e Tecnologia (FCT, Portugal) for financial support through projects PTDC/QUI/65510/2006 and PTDC/QUI/70153/2006. J.C. Moreira and E.M. Saggioro thanks Faperj and ENSP/FioCruz for Master grants. A.S. Oliveira thanks M.E.A.M.R. Vieira Ferreira for critical reading of initial version of this chapter.

8. References

Anandan, S.; Yoon, M. (2003). Photocatalytic activities of the nano-sized TiO_2-supported Y-zeolites. *Journal of Photochemistry and Photobiology C:Photochemistry Reviews*, Vol. 4, No. 1, pp. 5-18.

Anastas, P.T; Warner, J.C. (1998). *Green Chemistry: Theory and Practice*, Oxford University Press, London.

Andreozzi R; Caprio V; Insola A.; Marotta, R. (1999). Advanced oxidation processes (AOP) for water purification and recovery. *Catalysis Today*, Vol. 53, No. 1, pp. 51-59.

Atheba, P.; Robert, D.; Trokourey, A.; Bamba, D.; Weber, J.V. (2009). Design and study of a cost-effective solar photoreactor for pesticide removal from water. *Water Science and Technology*, Vol. 60, No. 8, pp. 2187-2193.

Asahi, R; Morikawa, T; Ohwaki, K; Aoki, K; Taga, Y. (2001). Visible-light photocatalysis in nitrogen-doped titanium dioxides. *Science*, Vol. 293, No. 5528, pp. 269-271.

Augugliaro, V.; Litter, M.; Palmisano, L.; Soria, J. (2006). The combination of heterogeneous photocatalysis with chemical and physical operations: A tool for improving the photoprocess performance. *Journal of Photochemistry and Photobiology C: Photochemistry Reviews*, Vol. 7, No. 4, pp. 127-144.

Auriol, M.; Filali-Meknassi, Y.; Tyagi, R.; Adams, C.; Surumpalli, R. (2006). Endocrine disrupting compounds removal from wastewater, a new challenge. *Process Biochemistry*, Vol. 41, No. 3, pp. 525-539.

Bahnemann, D. (1999). *Photocatalytic detoxification of polluted waters*, The Handbook of Environmental Chemistry (Springer-Verlag), Berlin, Germany.

Baird, C. (1999). *Environmental Chemistry* (2nd ed. WH Freeman and Company), New York, EUA.

Bae, E; Choi, W. (2003). Highly enhanced photoreductive degradation of perchlorinated compounds on dye-sensitized metal/TiO_2 under visible light. *Environmental Science Technology*, Vol. 37, No. 1, pp.147-152.

Biswas P.; Wu, C.Y. (2005). Nanoparticles and the environment, *Journal Air and Waste Management Association*, Vol. 55, pp. 708-46.

Blake, D.M. (2001). *Bibliography of Work on the Heterogeneous Photocatalytic Removal of Hazardous Compounds from Water and Air* (National Renewable Energy Laboratory), Golden, Colorado.

Botelho do Rego A.M., Vieira Ferreira L.F. (2001). Photonic and electronic spectroscopies for the characterization of organic surfaces adsorbed on surfaces. In: *Handbook of Surfaces and Interfaces of Materials*, Nalwa H.S. (Ed.), Academic Press, New York, Vol.2, Ch.7, 275-313.

Brillas, E.; Sires, I.; Oturan M.A. (2009). Electro-Fenton Process and Related Electrochemical Technologies Based on Fenton's Reaction Chemistry. *Chemical Reviews*, Vol. 109, No. 12, pp. 6570-6631.

Burrows, H.D; Canle, L.M; Santaballa, J.A; Steenken, J. (2002). Reaction pathways and mechanisms of photodegradation of pesticides. *Journal of Photochemistry and Photobiology B: Biology*, Vol. 67, No. 2, pp. 71-108.

Byrne, J.A..; Fernandez-Ilbanez, P.A.; Dunlop, P.S.M.; Alrousan, D.M.A.; Hamilton, J.W.J. (2011). Photocatalytic Enhancement for Solar Disinfection of Water: A Review. *International Journal of Photoenergy*, Vol. 2011, pp. 1-12.

Chan, S.H.S.; Wu, T.Y.; Juan, J.C.; The, C.Y. (2011). Recent developments of metal oxide semiconductors as photocatalysts in advanced oxidation processes (AOPs) for treatment of dye waste-water. *Journal of Chemical Technology and Biotechnology*, Vol. 86, No. 9, pp. 1130-1158.

Chong, M.N.; Jin, B.; Chow, C.W.K.; Saint, C. (2010). Recent developments in photocatalytic water treatment technology: A review. *Water Research*, Vol. 44, No. 10, pp. 2997-3027.

Comninellis, C.; Kapalka, A.; Malato, S.; Parsons, S.A.; Poulios, I.; Mantzavinos, D. (2008). Advanced oxidation processes for water treatment: advances and trends for R&D. *Journal of Chemical Technology and Biotechnology*, Vol. 83, No. 6, pp. 769-776.

da Silva, A.M.T. (2009). Environmental Catalysis from Nano-to Macro-Scale. *Materiali in Tehnologije*, Vol. 43, No. 3, pp. 113-121.

Dalrymple, O.K.; Yeh, D.H.; Trotz, M.A. (2007). Removing pharmaceuticals and endocrine-disrupting compounds from wastewater by photocatalysis. *Journal of Chemical Technology and Biotechnology*, Vol. 82, No. 2, pp. 121-134.

Davis, M.L; Cornwell, D.A. (1998). *Introduction to Environmental Engineering* (3rd ed. McGraw Hill), ISBN 0-07-115234-2, Nova York, EUA.

Doll, T.E.; Frimmel, F.H. (2005). Removal of selected persistent organic pollutants by heterogeneous photocatalysis in water. *Catalysis Today*, Vol. 101, pp. 195-202.

Duran, A.; Monteagudo, J.M.; Carnicer, A.; Ruiz-Murillo, M. (2011). Photo-Fenton mineralization of synthetic municipal wastewater effluent containing acetaminophen in a pilot plant. *Desalination*, Vol. 270, No. 1-3, pp. 124-129.

Duran, A.; Monteagudo, J.M.; San Martin, I.; Aguirre, M. (2010). Decontamination of industrial cyanide-containing water in solar CPC pilot plant. *Solar Energy*, Vol. 84, No. 7, pp. 1193-1200.

Dusek, L. (2010). Purification of Wastewater Using Chemical Oxidation Based on Hydroxyl Radicals. *Chemickelisty*, Vol. 104, No. 9, pp. 846-854.

Eckenfelder, W.W. (2000). *Industrial Water Pollution Control* (3rd ed. McGrawHill), ISBN 0-07-116275-5, Boston, EUA.

Exall, K. (2004). A review of water reuse and recycling, with reference to Canadian practice and potential: 2. Applications. *Water Research Journal of Canada*, Vol. 39, No. 1, pp. 13-28.

Felsot, A.S.; Racke, K.D.; Hamilton, D.J. (2003). Disposal and degradation of pesticide waste. *Reviews of Environmental Contamination and Toxicology*, Vol. 177, pp. 123-200.

Fox, M.A.; Dulay, M.T. (1993). Heterogeneous Photocatalysis. *Chemical Review*, Vol. 93, No.1, pp. 341-357.

Fu, X.; Zeltner, W.A; Anderson. (1996). Semicondutor Nanoclusters. *Elsevier*, Vol. 103, pp. 445-461.

Fujishima, A.; Rao, T.; Tryk, D.A. (2000). Titanium dioxide photocatalysis. *Journal Photochemical and Photobiology C: Photochemistry Reviews*, Vol. 1, pp. 1-21.

Galindo, C.; Jacques, P.; Kalt, A. (2000). Photodegradation of the aminoazobenzene acid orange 52 by three advanced oxidation processes: UV/H_2O_2 UV/TiO_2 and VIS/TiO_2 - Comparative mechanistic and kinetic investigations. *Journal of Photochemistry and Photobiology A: Chemistry*, Vol. 130, No. 1, pp. 35-47.

Gogate, P.R.; Pandit, A.B. (2004a). A Review of imperative technologies for wastewater treatment I: Oxidation technologies at ambient conditions. *Advanced Environmental Research*, Vol. 8, pp. 501-551.

Gogate, P.R.; Pandit, A.B. (2004b). A Review of imperative technologies for wastewater treatment II: Hybrid methods. *Advanced Environmental Research*, Vol. 8, pp. 553-597.

Guillard, C.; Disdier, J.; Monnet, C.; Dussaud, J.; Malato, S.; Blanco, J.; Maldonado, M.I.; Herrmann, J.M. (2003). Solar efficiency of a new deposited titania photocatalyst: chlorophenol, pesticide and dye removal applications. *Applied Catalysis B: Environmental*, Vol. 46, No. 2, pp. 319-332.

Güitekin, I; Ince, N.H. (2007). Synthetic endocrine disruptors in the environment and water remediation by advanced oxidation processes. *Journal of Environmental Management*, Vol. 85, pp. 816-832.

Gupta, S.M.; Tripathi, M. (2011). A review of TiO_2 nanoparticles. *Chinese Science Bulletin*, Vol. 56, No. 16, pp. 1639-1657.

Halmann, M.M. (1996). *Photodegradation of Water Pollutants*, ISBN 0849324599, Florida, EUA.

Hashimoto, K.; Wataneble, W. (1999). *TiO₂ photocatalysis*, Fundamentals and Applications, BKC, Tokyo.

Hay, S.O.; Obee, T.N. (1999). The Augmentation of UV Photocatalytic Oxidation with Trace Quantities of Ozone. *Advanced Oxidation Technologies*, Vol. 4, No.2, pp. 209-218.

Herrmann, J.M.; Duchamp, C.; Karkmaz, M.; Hoai, B.T.; Lachheb, H.; Puzenat; Guillard. (2007). Environmental green chemistry as defined by photocatalysis. *Journal of Hazardous Materials*, Vol. 146, No. 3, pp. 624-629.

Hincapié, M.; Maldonado, M.I.; Oller, I.; Gernjak, W.; Sánchez-Perez, J.A.; Ballestros, M.M.; Malato, S. (2005). Solar photocatalytic degradation and detoxification of EU prioritary substances. *Catalysis Today*, Vol. 101, pp. 203-210.

Hoffmann, M.R.; Martin, S.T.; Choi, W.; Bahnemann, D.W. (1995). Environmental Applications of Semiconductor Photocatalysis. *Chemical Review*, Vol. 95, pp. 69-95.

Homem, V.; Santos, L. (2011). Degradation and removal methods of antibiotics from aqueous matrices - A review. *Journal of Environmental Management*, Vol. 92, No. 10, pp. 2304-2347.

Ikehata, K.; El-Din, M.G. (2004). Degradation of recalcitrant surfactants in wasterwater by ozonation and advances oxidation processes: A review. *Ozone-Science & Engineering*, Vol. 26, No. 4, pp. 327-343.

Ikehata, K.; El-Din, M.G. (2006). Aqueous pesticide degradation by hydrogen peroxide/ultraviolet irradiation and Fenton-type advanced oxidation processes: a review. *Journal of Environmental Engineering and Science*, Vol. 5, No. 2, pp. 81-135.

Janus, M.; Morawski, A.W. (2010). Carbon-modified TiO₂ as Photocatalysts. *Journal of Advanced Oxidation Technologies*, Vol. 13, No. 3, pp. 313-320.

Kamat, P; Meisel, D. (2002). Nanoparticles in Advanced Oxidation Processes. *Current Opinion in Colloid & Interface Science*, Vol. 7, pp. 282-7.

Kamat, P; Meisel, D. (2003). Nanoscience opportunities in environmental remediation. *Comptes Rendus Chimie*, Vol. 6, pp. 999-1007.

Kavitha, V.; Palanivelu, K. (2004). The role of ferrous ion in Fenton and photo-Fenton processes for the degradation of phenol. *Chemosphere*, Vol. 55, No. 9, pp. 1235-1243.

Khataee, A.R.; Kasiri, M.B. (2010). Photocatalytic degradation of organic dyes in the presence of nanostructure titanium dioxide: Influence of the chemical structure of dyes. *Journal of Molecular Catalysis A: Chemical*, Vol. 328, No. 1-2, pp. 8-26.

Khetan, S.K.; Collins, T.J. (2007). Human pharmaceuticals in the aquatic environment: A challenge to green chemistry. *Chemical Reviews*, Vol. 107, No. 6, pp. 2319-2364.

Kiely, G. (1998). *Environmental Engineering* (international edition, McGraw Hill), ISBN 0-07-116424-3, Boston, EUA.

Legrini, O.; Oliveros, E.; Braun, A.M. (1993). Photochemical Porcesses for Water Treatment. *Chemical Review*, Vol. 93, No.2, pp. 671-698.

Le-Minh, N; Khan, S.J; Drewes, J.E; Stuetz, R.M. (2010). Fate of antibiotics during municipal water recycling treatment processes. *Water Research*, Vol. 44, pp. 4295-4323.

Lim, M.; Zhou, Y.; Wang, L.; Rudolph, V.; Lu, G.Q.M. (2009). Development and potential of new generation photocatalytic systems for air pollution abatement: an overview. *Asia-Pacific Journal of Chemical Engineering*, Vol. 4, No. 4, pp. 387-402.

Linsebigler, A.L.; Lu, G; Yates Jr., J.T. (1995). Photocatalysis on TiO$_2$ surfaces: Principles,
mechanisms and selected rules. *Chemical Review*, Vol. 95, pp. 735-768.

Malato, S.; Blanco, J.; Alarcon, D.C.; Maldonado, M.I.; Fernández-Ibáñez, P.; Gernjak, W.
(2007a). Photocatalytic decontamination and disinfection of water with solar
collectors. *Catalysis Today*, Vol. 122, No. 1-2, pp. 137-149.

Malato, S.; Blanco, J.; Maldonado, M,I.; Oller, I.; Gernjak, W.; Perez-Estrada, L. (2007b).
Coupling solar photo-Fenton and biotreatment at industrial scale: Main results of a
demonstration plant. *Journal of Hazardous Materials*, Vol. 146, No. 3, pp. 440-446.

Malato, S.; Fernandez-Ibanez, P.; Maldonado, M.I.; Blanco, J.; Gernjak, W. (2009).
Decontamination and disinfection of water by solar photocatalysis: Recent
overview and trends. *Catalysis Today*, Vol. 147, No. 1, pp. 1-59.

Malato, S.; Blanco, J.; Vidal, A.; Richter, C. (2002). Photocatalysis with solar energy at a pilot-
plan scale: an overview. *Applied Catalysis B: Environmental*, Vol. 37, pp. 1-15.

Mantzavinos, D.; Psillakis, E. (2004). Enhancement of biodegradability of industrial
wastewaters by chemical oxidation pre-treatment. *Journal of Chemical Technology and
Biotechnology*, Vol. 79, No. 5, pp. 431-454.

Marinas, A.; Guillard, C.; Marinas, J.M.; Fernandez-Alba, A.; Herrmann, J.M. (2001).
Photocatalytic degradation of pesticide-acaricide formetanate in aqueous
suspension of TiO$_2$. *Applied Catalysis B: Environmental*, Vol. 34, No. 3, pp. 241-252.

Marugan, J.; Aguado, J.; Gernjak, W.; Malato, S. (2007). Solar photocatalytic degradation of
dichloroacetic acid with silica-supported titania at pilot-plant scale. *Catalysis Today*,
Vol. 129, No. 1-2, pp. 59-68.

Matilainen, A.; Sillanpaa, M. (2010). Removal of natural organic matter from drinking water
by advanced oxidation processes. *Chemosphere*, Vol. 80, No. 4, pp. 351-365.

Metcalf & Eddy. (2003). *Wastewater Engineering – Treatment and Reuse* (4rd ed. McGrawHill),
ISBN 0-07-124140X, Boston, EUA.

Mills, A.; Hunte, S. (1997). An overview of semiconductor photocatalysis. *Journal
Photochemical and Photobiology A: Chemistry*, Vol. 108, pp. 1-35.

Miranda-Garcia, N.; Suarez, S.; Sanchez, B.; Coronado, J.M.; Malato, S.; Maldonado, M.I.
(2011). Photocatalytic degradation of emerging contaminants in municipal
wastewater treatment plant effluents using immobilized TiO$_2$ in solar pilot plant.
Applied Catalysis B: Environmental, Vol. 103, No. 3-4, pp. 294-301.

Mourao, H.A.J.L.; de Mendonça, V.R..; Malagutti, A.R..; Ribeiro, C. (2009). Nanostructures in
photocatalysis: a Review about synthesis strategies of photocatalysts in nanometric
size. *Química Nova*, Vol. 32, No. 8, pp. 2181-2190.

Navarro, S.; Fenoll, J.; Vela, N.; Ruiz, E.; Navarro, G. (2011). Removal of ten pesticides from
leaching water at pilot plant scale by photo-Fenton treatment. *Chemical Engineering
Journal*, Vol. 167, No. 1, pp. 42-49.

Navntoft, C.; Dawidowski, L.; Blesa, M.A.; Fernandez-Ibanez, P.; Wolfram, E.A.; Paladini, A.
(2009). UV-A (315-400 nm) irradiance from measurements at 380 nm for solar water
treatment and disinfection: Comparison between model and measurements in
Buenos Aires, Argentina and Almeria, Spain. *Solar Energy*, Vol. 83, No. 2, pp. 280-
286.

Nevers, N. (2000). *Air Polution Control Engineering* (2rd ed. McGraw Hill), New York, EUA.

Nogueira, R.F.P.; Trovo, A.G.; da Silva, M.R.A. (2007). Fundaments and environmental applications of fenton and photo-Fenton processes. *Química Nova*, Vol. 30, No. 2, pp. 400-408.

Oliveira, A.S.; Fernandes, M.B.; Higarashi, M.M.; Moreira, J.C.; Ferreira, M.E.A.; Ferreira, L.F.V. (2004a). CARCINOGÉNEOS QUÍMICOS AMBIENTAIS: utilização de técnicas de fotoquímica de superfícies na análise destes compostos em amostras ambientais. *Revista de Química Industrial*, Vol.9, No. 4, pp. 15-23.

Oliveira, A.S.; Ferreira, L.F.V.; Silva, J.P.; Moreira, J.C. (2004b). Surface Photochemistry: Photogradation Study of Pyrene Adsorbed onto Microcrystalline Cellulose and Silica. *International Journal of Photoenergy*, Vol. 6, pp. 205-213.

Oliveira, A.S.; Saggioro, E.M.; Barbosa, N.R., Mazzei A.; Ferreira, L.F.V.; Moreira, J.C. (2011). Surface Photocatalysis: A Study of the thickness of TiO_2 layers on the photocatalytic decomposition of soluble indigo blue dye. *Revue de Chimie (Bucarest)*, Vol. 62, pp. 462-468.

Oliveira, A.S.; Ferreira, L.F.V.; Moreira, J.C. (2008). Surface photochemistry techniques applied to the study of environmental carcinogens. *Revue Roumaine de Chimie*, Vol. 53, pp. 893-902.

Oller, I.; Fernandez-Ibanez, P.; Maldonado, M.I.; Estrada, L.P.; Gernjak, W.; Pulgarin, C.; Passarinho, P.C.; Malato, S. (2007). Solar heterogeneous and homogeneous photocatalysis as a pre-treatment option for biotreatment. *Research on Chemical Intermediates*, Vol. 33, No. 3-5, pp. 407-420.

Oyama, T.; Otsu, T.; Hidano, Y.; Koike, T.; Serpone, N. (2011). Enhanced remediation of simulated wastewaters contaminated with 2-chlorophenol and other aquatic pollutants by TiO_2-photoassisted ozonation in a sunlight-driven pilot-plant scale photoreactor. *Solar Energy*, Vol. 85, No. 5, pp. 938-944.

Peiro, A.M.; Ayllon, J.A.; Peral, J.; Domenech, X. (2001). TiO_2 photocatalized degradation of phenol and ortho-substituted phenolic compounds. *Applied Catalysis B: Environmental*, Vol. 30, pp. 359-367.

Pelizzetti, E.; Serpone, N. (1989). *Photocatalysis: Fundamental and Applications*, New York, EUA.

Pignatello, J.J.; Oliveros, E.; MacKay, A. (2006). Advanced oxidation processes for organic contaminant destruction based on the Fenton reaction and related chemistry. *Critical Reviews in Environmental Science and Technology*, Vol. 36, No. 1, pp. 1-84.

Pirkanniemi, K.; Sillanpää, M. (2002). Heterogeneous water phase catalysis as an environmental application: a review. *Chemosphere*, Vol. 48, pp. 1047-1060.

Rauf, M.A..; Ashraf, S.S. (2009). Fundamental principles and application of heterogeneous photocatalytic degradation of dyes in solution. *Chemical Engineering Journal*, Vol. 151, No. 1-3, pp. 10-18.

Reynaud, M.A.G.; Concha, G.M.O.; Cuevas, A.C. (2011). A brief review on fabrication and applications of auto-organized TiO_2 nanotube arrays. *Corrosion Reviews*, Vol. 29, No. 1-2, pp. 105-121.

Roy, P.; Berger, S.; Schmuki, P. (2011). TiO_2 Nanotubes: Synthesis and Applications. *Angewandte Chemie-International Edition*, Vol. 50, No. 13, pp. 2904-2939.

Saggioro, E.M.; Oliveira, A.S., Pavesi, T.; Maia, C.G., Vieira Ferreira, L.F.; Moreira, J.C. (2011). Use of Titanium Dioxide Photocatalysis on the Remediation of model textile wastewaters containing azo dyes. *Molecules*, Vol. 16, pp. 10370-10386.

Savage, N; Diallo, M.S. (2005). Nanomaterials and water purification: Opportunities and challenges. *Journal of Nanoparticle Research*, Vol. 7, pp. 331-342.

Santos, M.S.A.; Trovo, A.G.; Bautitz, I.R.; Nogueira, R.F.P. (2009). Degradation of Residual Pharmaceuticals by Advanced Oxidation Processes. *Quimica Nova*, Vol. 32, No. 1, pp. 188-197.

Simonsen, C.B. (2007). *Essentials of Environmental Law* (Pearson Prentice Hall), ISBN 0-13-228045-0, Upper Saddle River; New Jersey, EUA. 3rd edition.

Schiavello (1988). *Photocatalysis and Environment: Trends and Applications*, Kluwer Academic Publishers, Dordrecht.

Soon, A.N.; Hameed, B.H. (2011). Heterogeneous catalytic treatment of synthetic dyes in aqueous media using Fenton and photo-assisted Fenton process. *Desalination*, Vol. 269, No. 1-3, pp. 1-16.

Stockholm Convention on Persistent Organic Pollutants (2005). 01/12/2011, Available form: http://www.pops.int

Tchobanoglous, G; Theisen, H; Vigil, S.A (1993). *Integrated Solid Waste Management: engineering principles and management issues* (McGraw Hill), ISBN 0-07-112865-4, New York, EUA.

Umar, M.; Aziz, H.A.; Yusoff, M. S. (2010). Trends in the use of Fenton, electro-Fenton and photo-Fenton for the treatment of landfill leachate. *Waste Management*, Vol. 30, No. 11, pp. 2113-2121.

United Nations Environmental Programme (2005). 01/12/2011, Available form: http://www.chem.unep.ch/pops/default.html

Vigneswaran, S.; Sundaravadiel, M. (2004). Recycle and Reuse of domestic wastewater, In: *Wastewater recycle, reuse and reclamation*, Saravanamuthu, V., Eolss Publiseh, Oxford, UK.

Vilar, V.J.P.; Gomes, A.I.E.; Ramos, V.M.; Maldonado, M.I.; Boaventura, R.A.R. (2009). Solar photocatalysis of a recalcitrant coloured effluent from a wastewater treatment plant. *Photochemical & Photobiological Sciences*, Vol. 8, No. 5, pp. 691-698.

Vilhunen, S.; Sillpanpaa, M. (2010). Recent developments in photochemical and chemical AOPs in water treatment: a mini-review. *Reviews in environmental science and bio-technology*, Vol. 9, No. 4, pp. 323-330.

Wang, C.C; Lee, C.K; Lyu, M.D; Juang, L.C. (2008). Photocatalytic degradation of C.I Basic Violet 10 using TiO_2 catalysts supported by Y zeolite: An investigation of the effects of operational parameters. *Dyes and Pigments*, Vol. 76, pp. 817-824.

Xavier, L.F.W.; Moreira, I.M.N.S.; Higarashi, M.M.; Moreira, J.C.; Ferreira, L.F.V.; Oliveira, A.S. (2005). Fotodegradação de hidrocarbonetos policíclicos aromáticos em placas de sílicas impegnadas com dióxido de titânio. *Química Nova*, Vol. 28, pp. 409-413.

Xu, S; Ng, J; Zhang, X; Bai, H; Sun, D.D. (2011). Adsorption and photocatalytic degradation of Acid Orange 7 over hydrothermally synthesized mesoporous TiO_2 nanotube. *Colloids and Surfaces A: Physicochemical and Engineering Aspects*, Vol. 379, pp. 169-175.

Zapata, A.; Malato, S.; Sanchez-Perez, J.A.; Oller, I.; Maldonado, M.I. (2010). Scale-up strategy for a combined solar photo-Fenton/biological system for remediation of pesticide-contaminated water. *Catalysis Today*, Vol. 151, No. 1-2, pp. 100-106.

Zayani, G.; Bousselmi, L.; Mhenni, F.; Ahmed, G. (2009). Solar photocatalytic degradation of commercial textile azo dyes: Performance of pilot plant scale thin film fixed-bed reactor. *Desalination*, Vol. 246, No. 1-3, pp. 344-352.

Zhang, X.H.; Li, W.Z.; Xu, H.Y. (2004). Application of zeolites in photocatalysis. *Progress in Chemistry*, Vol. 16, No. 5, pp. 728-737.

12

Light-Induced Iminyl Radicals: Generation and Synthetic Applications

Miguel A. Rodríguez

Departamento de Química, Unidad Asociada al C.S.I.C., Universidad de La Rioja
Spain

1. Introduction

The photochemistry of the carbon-nitrogen double bond was first explored in depth in the seventies (Padwa, 1977; Pratt, 1977). The low photochemical reactivity of this bond is due to the deactivation of the excited state of the imine by E–Z isomerization processes, which do not have any synthetic utility because of their very low energy barrier for thermal conversion (Padwa & Albrecht, 1974a, 1974b). However, the presence of an electronegative atom on the nitrogen opens up the field of reactivity. Studies on the energies of the N–O bonds in acyloximes show that this bond can easily undergo homolytic cleavage induced by ultraviolet light, a process that leads to iminyl radicals (Okada et al., 1969). The use of radicals has proven to be a useful tool in organic synthesis (Renaud & Sibi, 2001; Togo, 2004; Zard, 2003). In particular, the cyclization of nitrogen-centred radicals, such as aminyl or iminyl radicals (Chart 1), is a valuable procedure for the preparation of nitrogen heterocycles (Fallis & Brinza, 1997; Zard, 2008). It is therefore necessary to have effective methods for the production of these radicals.

$$R^1\text{-}\underset{R^2}{\overset{|}{N}}\bullet \qquad\qquad R^2\overset{R^1}{=}N\bullet$$

Aminyl   **Iminyl**

Chart 1.

The direct way to create nitrogen radicals involves the homolytic cleavage of N–X bonds, while the addition of a radical to an unsaturated nitrogen functional group, such as a nitrile or an imine derivative, constitutes the most widely used indirect method. The cleavage of the bond may be triggered thermally or photochemically. This chapter focuses on the photochemical generation of iminyl radicals and the study of their reactivity. Firstly, the methods for the production of iminyl radicals will be reviewed and the reactivity of this species will be examined, highlighting its ability to be added to unsaturated systems, with particular emphasis on intramolecular cyclization reactions. At this point, the regioselectivity in the formation of five- and six-membered rings will be analysed. In terms of the course of these reactions, both experimental results and theoretical calculations on the reaction mechanism will be discussed. Finally, the last part of the chapter will be devoted to further synthetic applications of these reactions, particularly in the synthesis of different polycyclic heteroaromatic compounds and the preparation of natural products.

2. Methods for the generation of iminyl radicals

Different methods have been developed for the light-induced formation of iminyl radicals. The homolytic cleavage of N–X bonds is a well known and widely used method for the production of this kind of radical, while the use of radical addition to nitriles is much more limited. These two alternatives will be described in detail below.

2.1 Homolytic cleavage of N–X bonds

A range of different unsaturated nitrogen derivatives (C=N–X) have been used as starting materials, where the heteroatom X can be oxygen, nitrogen, sulfur or halogen. Of these, the most widely used approach is the cleavage of the N–O bond and these systems will therefore be the starting point in this section.

2.1.1 Cleavage of N–O bonds

The average bond energies of C=N, N–O and C–O bonds are 147, 53 and 86 kcal/mol, respectively (Petrucci et al., 2011). The application of energy in the form of heat or light to C=N–O–C structures should consistently fragment the N–O bond. To the best of my knowledge, the first reactions to involve the formation of iminyl radicals were the photolysis of oxadiazoles (Newman, 1968a, 1968b; Cantrell & Haller, 1968; Mukai et al., 1969), benzo[c]isoxazole (Ogata et al., 1968), oxadiazolinone (Sauer & Mayer, 1968) and aromatic oxime benzoates (Okada et al., 1969). In the latter case, dimerization of the radical occurred to give the major product (Scheme 1). The homolytic cleavage of similar O-acyl aromatic ketoximes was also observed and this took place in the triplet excited state when the photolysis was conducted in the presence of triplet sensitizers (Yoshida et al., 1975).

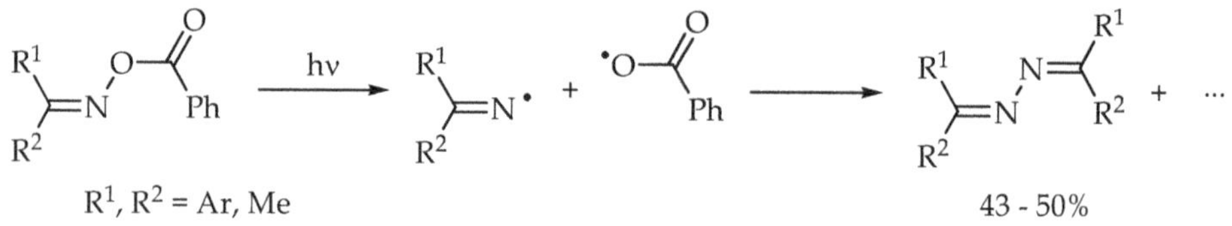

$$R^1, R^2 = Ar, Me \qquad\qquad\qquad\qquad\qquad 43 - 50\%$$

Scheme 1.

Since the nineties, interest in the photochemical generation of iminyl radicals has increased greatly. In a pioneering study by the group of Boivin & Zard (Boivin et al., 1994), modified Barton esters, prepared from O-carboxymethyl derivatives of oximes, were irradiated to give an initial N–O bond cleavage, which was followed by decarboxylation and loss of formaldehyde (Scheme 2), a process that provides a mild and very useful source of iminyl radicals. These radicals can evolve by subsequent intramolecular cyclization to give a five-membered ring and subsequent transfer of a pyridylthiyl group from the starting Barton ester in the absence of an external trap.

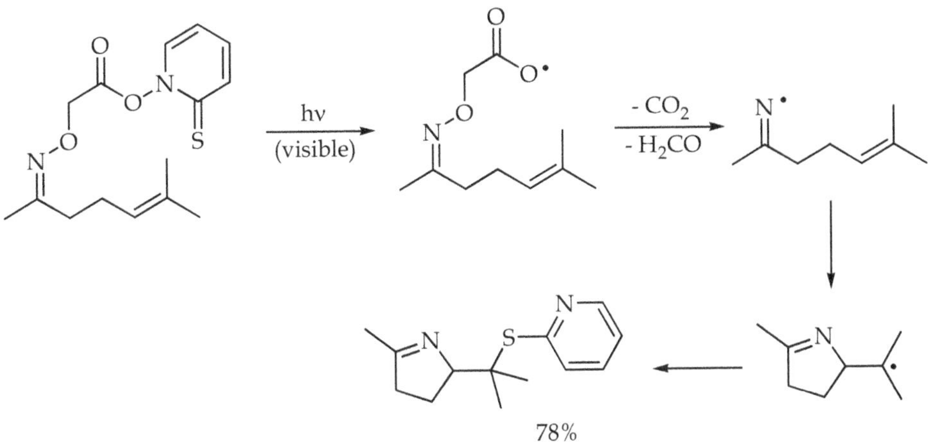

Scheme 2.

A slight modification of this method allowed the homolytic substitution by iminyl radicals at selenium (Fong & Schiesser, 1993). The photolysis of thiohydroxamic esters derived from the *O*-carboxymethyl oxime derivatives of 2-(benzylseleno)benzaldehyde gave 1,2-benzoselenazoles in 70% yield (Scheme 3). Similarly, Barton oxalate esters of oximes were used as starting materials. In this case, the formation of the iminyl radical takes place after a double decarboxylation (Boivin et al., 1994). An analogous thermal decomposition procedure involving decarboxylation and loss of formaldehyde to obtain an iminyl radical has been also described from peresters (Leardini et al., 2001).

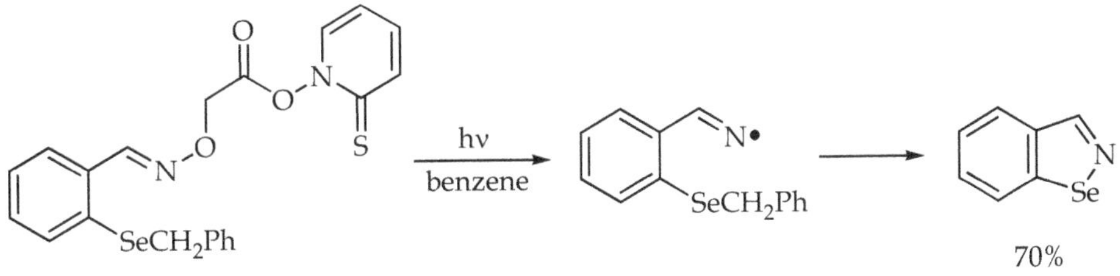

Scheme 3.

Although not used from a synthetic point of view, ketoxime diurethanes also lead to the formation of radicals (Hwang et al., 1999). Of greater applicability is the use of ketoxime xanthates (Gagosz & Zard, 1999), which allow the preparation of a variety of substituted 1-pyrrolines when the appropriate structure is irradiated (Scheme 4). The photoreaction has great versatility and gives yields between 72 and 88%, which exceeds those of the thermal process (Gagosz & Zard, 1999). The authors postulate that iminyl radicals generated by homolysis of the N–O bond eventually escape from the solvent cage and then undergo cyclization. The resulting cyclised radical finally adds to the sulfur atom of the starting oxime xanthate, thus producing the cyclised dithiocarbonate and regenerating the iminyl species to propagate the chain.

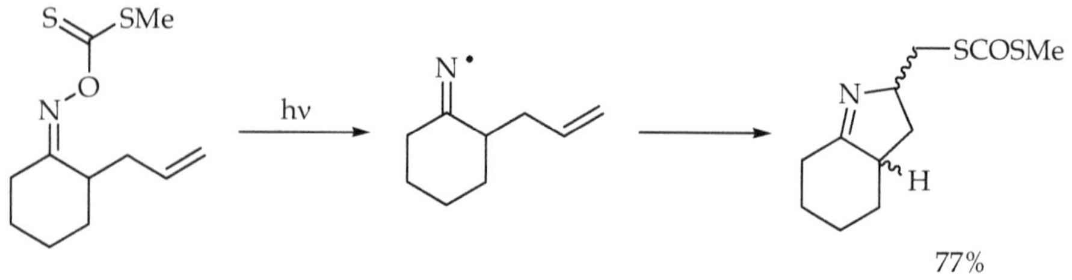

Scheme 4.

One alternative is to use oxime ethers as starting compounds. In a first paper, the group of Narasaka published the thermal treatment of γ,δ-unsaturated O-(2,4-dinitrophenyl)oximes with sodium hydride and 3,4-methylenedioxyphenol, which gave 3,4-dihydro-2H-pyrroles after intramolecular cyclization (Uchiyama et al., 1998). The authors considered that this reaction proceeded with formation of an iminyl radical through an initial one-electron transfer from the sodium phenolate and expected that a similar electron transfer would occur on irradiation of the oxime ether in the presence of a sensitizer. Indeed, irradiation of γ,δ-unsaturated O-aryloximes in the presence of 1,5-dimethoxynaphthalene (DMN) as a sensitizer led to 1-pyrrolines through cyclization of an iminyl radical (Mikami & Narasaka, 2000; 2001). The reaction was carried out in the presence of 1,4-cyclohexadiene (CHD) in order to trap the radical resulting from the intramolecular addition to the alkene (Scheme 5).

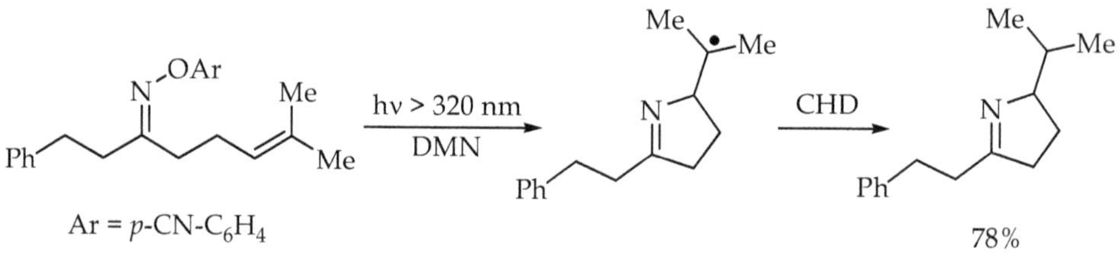

Scheme 5.

The mechanistic aspects of the photosensitized reactions of a series of oxime ethers have been studied by steady-state and laser flash photolysis methods (de Lijser & Tsai, 2004; de Lijser et al., 2007). On the basis of these experiments, the formation of iminyl radicals is rationalized. On the other hand, Narasaka's group also studied the effect of substituting an ether oxime by an acetate oxime (Kitamura et al., 2005; Kitamura & Narasaka, 2008). The sensitized photoreaction in acetonitrile again led to the formation of five-membered rings in good yields (Scheme 6).

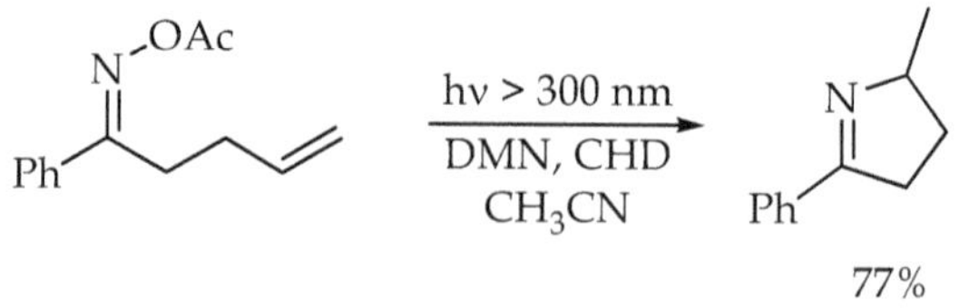

Scheme 6.

The formation of six-membered heterocyclic rings by intramolecular cyclization between an iminyl radical and an olefin was first reported by Rodríguez and co-workers (Alonso et al., 2006). Direct irradiation of 2-vinylbenzaldehyde O-acetyloximes induced N–O bond cleavage and led to the formation of an iminyl radical, which was able to add to a vinyl group to give isoquinolines after aromatisation through the formal loss of a hydrogen atom (Scheme 7). Analysis of the Stern-Volmer plots (Turro, 1991) for the quenching of the photoreactivity of acyloximes in the presence of common triplet-state quenchers shows that both excited states, singlet and triplet, undergo the same N–O fracture (Alonso et al., 2008). Direct and sensitized laser experiments also led to the same conclusion (Lalevée et al., 2002). According to ab initio molecular orbital calculations on the singlet excited states of acyloximes, the oscillator strength for the n–π* $S_0 \rightarrow S_1$ transition should be 0.014, with a λ_{max} of 233 nm, while the π–π* $S_0 \rightarrow S_2$ transition should have an oscillator strength of 0.2, with a λ_{max} of 212 nm, which indicate that S_2 is the spectroscopic state while S_1 is an excited dark state. Relaxation from S_2 leads directly to N–O bond cleavage due to the coupling between the imine π* and the σ* N–O orbitals (Alonso et al., 2008).

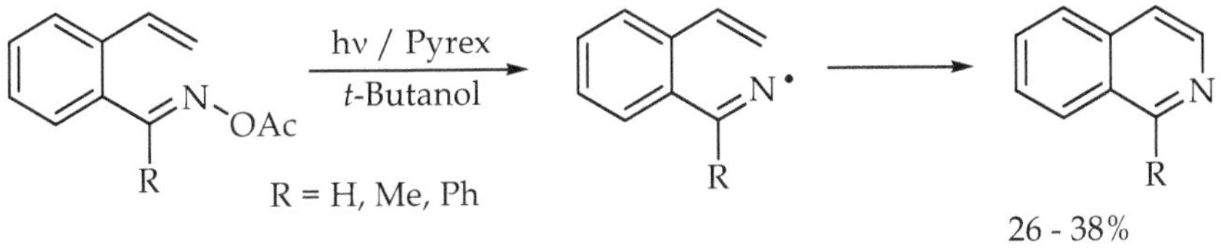

Scheme 7.

The reaction is also effective for carbon-carbon triple bonds (Alonso et al., 2006). In the intramolecular version, the addition of a nitrogen-centred radical should generate an isoquinolyl radical, which may evolve by atom abstraction since the use of 2-propanol-d_7 as solvent led to 4-deuteroquinoline, while the use of methanol-d_1 led to nondeuterated isoquinoline (Scheme 8).

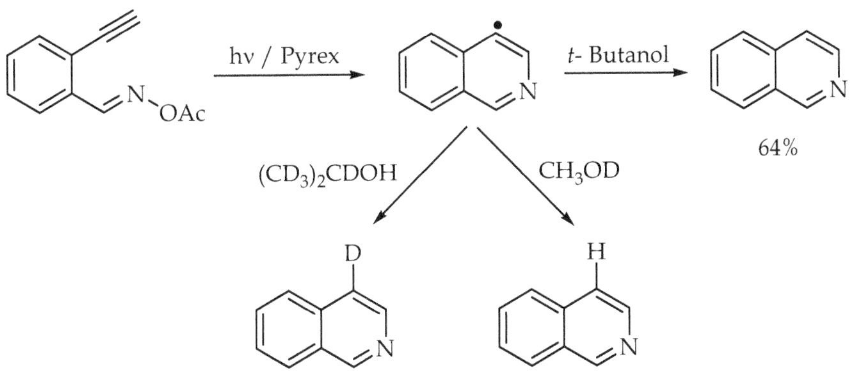

Scheme 8.

The light-induced intermolecular attack of the iminyl radical on a carbon-carbon triple bond has also been reported (Alonso et al., 2006). As shown in Scheme 9, irradiation of benzophenone O-acetyloxime in the presence of tolane or dimethyl acetylenedicarboxylate induces a tandem (cascade) process of intermolecular addition/intramolecular cyclization with formation of isoquinoline in good yields.

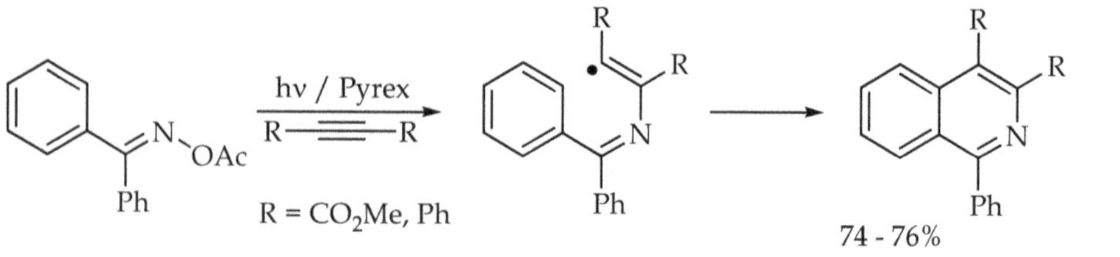

Scheme 9.

Similarly, the iminyl radicals generated by the action of UV light are capable of reacting with alkynyl Fischer carbenes (Blanco-Lomas et al., 2011). The reaction led to the formation of the new carbene complex together with another product, identified as a seven-membered compound (Scheme 10). Based on mechanistic studies, it is proposed that the iminyl radical would attack the alkynyl carbene at the alkynyl carbon (1,4-addition) to form the carbene complex, while 1,2-addition at the carbene carbon should lead to the cyclic structure.

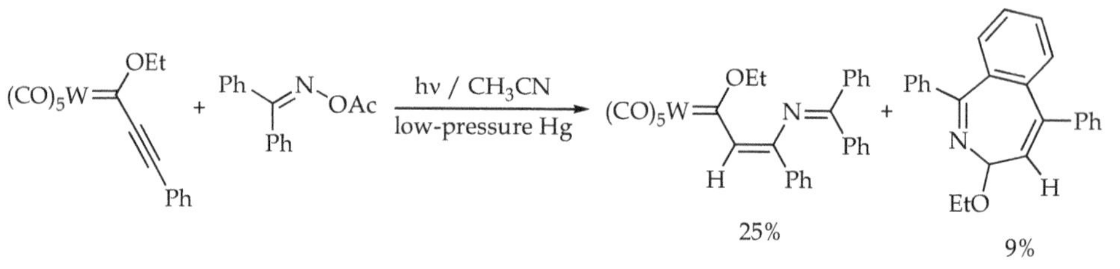

Scheme 10.

An aromatic ring can also be used as an unsaturated system. This kind of attack by iminyl radicals was previously observed when the photolysis of aromatic ketone O-acyloximes took place in aromatic solvents and this process involves a homolytic aromatic substitution (Sakuragi et al., 1976). Starting from appropriately substituted oximes (Scheme 11), different phenanthridines were obtained in good to excellent yields after intramolecular cyclization onto a phenyl ring followed by rearomatisation (Alonso et al., 2006). In the case of aldehyde O-acyloxime (R = H) the iminyl radical partially decomposed into a nitrile (Bird et al., 1976).

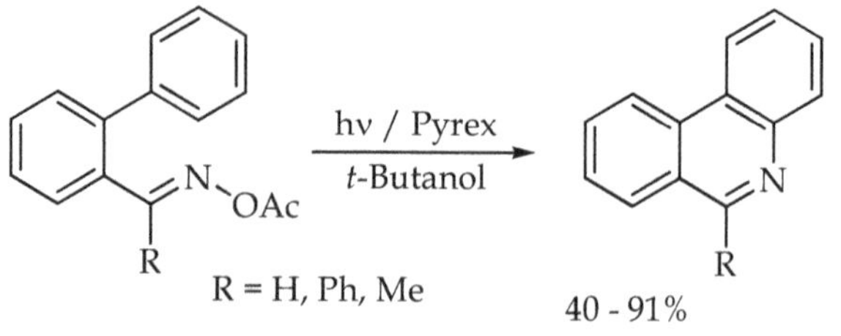

Scheme 11.

Finally, dioxime oxalates are also precursors of iminyl radicals. Irradiation of these compounds in the presence of 4-methoxyacetophenone as a sensitizer allowed the preparation of dihydropyrroles and phenanthridines (Portela-Cubillo et al., 2008).

2.1.2 Cleavage of N–N, N–S and N–Br bonds

Although N–N bonds are generally stronger than N–O bonds, homolysis of N–N bonds also offers an attractive alternative for the generation of iminyl radicals. Irradiation of hydrazones led to the formation of azines (Takeuchi et al., 1972) or, in the presence of a good hydrogen donor, to a mixture of amines and imines (Binkley, 1970) after N–N bond scission. Iminyl radicals are also formed by the reaction of photochemically generated *tert*-butoxyl radicals with primary or secondary alkyl azides (Roberts & Winter, 1979). However, these procedures are of little synthetic utility. In contrast, as shown in Scheme 12, irradiation with a sunlamp of readily accessible thiocarbazone derivatives in the presence of catalytic amounts of hexabutylditin provides adducts that arise from subsequent intramolecular cyclization of an iminyl radical and transfer of an iminodithiocarbonate group (Callier-Dublanchet et al., 1997). The cleavage of the N–N bond seems to be relatively slow and the optimum substituent on the thiocarbazone moiety has yet to be determined.

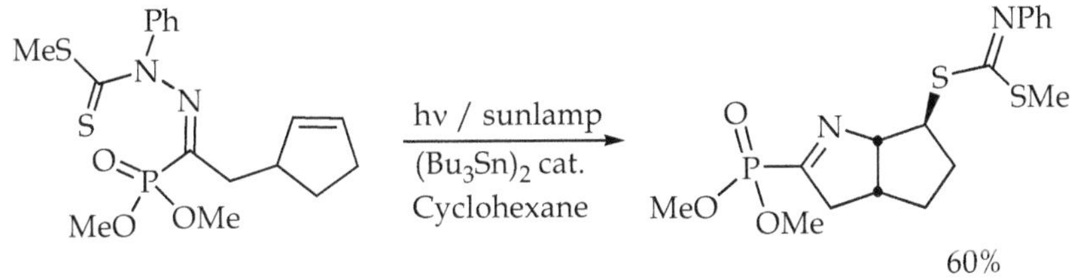

Scheme 12.

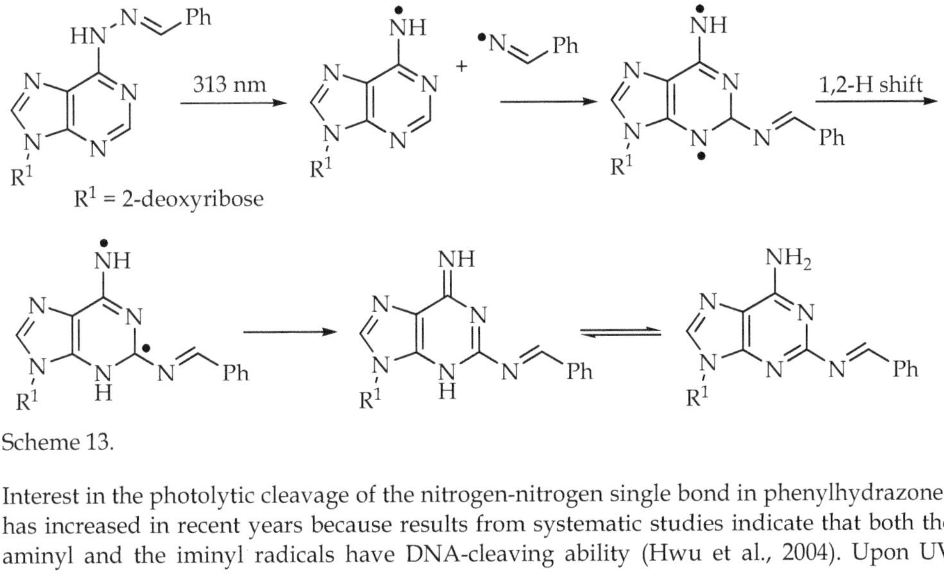

Scheme 13.

Interest in the photolytic cleavage of the nitrogen-nitrogen single bond in phenylhydrazones has increased in recent years because results from systematic studies indicate that both the aminyl and the iminyl radicals have DNA-cleaving ability (Hwu et al., 2004). Upon UV photolysis, modified 2'-deoxyadenosine containing a photoactive phenylhydrazone moiety undergoes efficient homolytic cleavage to give aminyl and iminyl radicals (Kuttappan-Nair et al., 2010). Both radicals evolve by recombination as the main pathway (Scheme 13). Alternatively, it is reasonable to propose that these radicals react with other DNA base moieties, leading to intra- and interstrand cross-linking.

Although thermal reactions are more commonly used (Zard, 2008; Esker & Newcomb, 1993), sulfanyl imines can also be employed as precursors of iminyl radicals in photochemical reactions (Guindon et al., 2001). Irradiation of benzothiazolylsulfanylimines induces the cleavage of the N–S bond and facilities a tandem process of intramolecular iminyl radical cyclofunctionalization/hydrogen transfer to afford *syn-anti* 1-pyrrolines with high levels of 1,2-induction in both steps (Scheme 14).

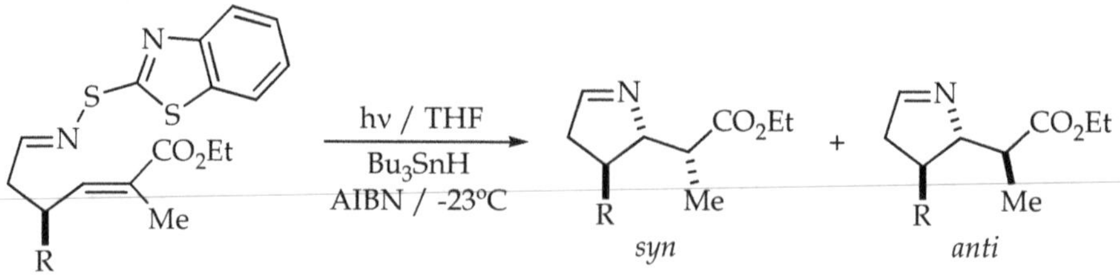

AIBN = Azobisisobutylonitrile

R	syn:anti	Yield (%)
H	1:9	83
Me	1:16	84
OMe	1:7	78

Scheme 14.

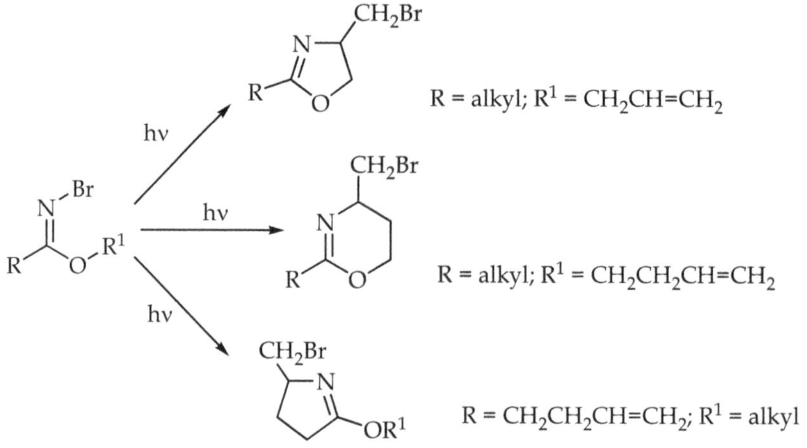

Scheme 15.

The generation of alkoxyiminyl radicals can be addressed by photolysis of *N*-bromo imidates (Glober et al., 1993). These radicals can undergo *exo*-1,5 and *exo*-1,6 cyclization onto olefins on the *O*-alkyl side chains, to give 4,5-dihydrooxazoles or 5,6-dihydro-4*H*-1,3-oxazines, respectively, or *exo*-1,5 cyclization onto an olefin on the iminyl side chain, to yield 2-alkoxy-1-pyrrolines (Scheme 15). The experimental results indicate that dihydrooxazole formation is more favourable than cyclization to 1-pyrrolines, a finding that has been rationalized by semiempirical MNDO molecular orbital calculations.

2.2 Addition of a radical to an unsaturated nitrogen derivative

In the previous section, direct methods of generating an iminyl radical by homolytic cleavage of an N–X bond have been examined. An alternative to these methods is the photochemical formation of another radical and subsequent addition to an unsaturated nitrogen derivative, such as an azide or a nitrile. This indirect approach was first reported by Roberts's group (Cooper et al., 1977). A photochemically generated *tert*-butoxyl radical reacted with primary or secondary alkyl azides to produce iminyl radicals (Scheme 16), which ultimately yielded a ketone after hydrolysis.

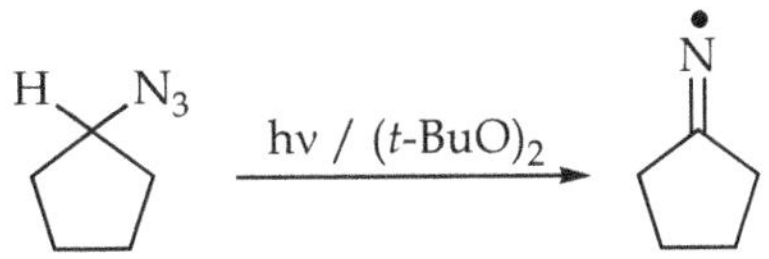

Scheme 16.

5-Bromovaleronitrile was used to establish the rate constant of cyclization of the 4-cyanobutyl radical to the cyclopentiminyl radical (Griller et al., 1979), as determined by kinetic measurements using electron paramagnetic resonance (EPR) spectroscopy, and this was found to be $k_c^{C\equiv N} = 4.0 \times 10^4$ s^{-1} at 80 °C, almost an order of magnitude slower than the analogous 5-hexynyl radical ($k_c^{C\equiv C} = 1.2 \times 10^5$ s^{-1} at 80 °C). The reaction is initiated by photolysis of hexabutylditin (Scheme 17).

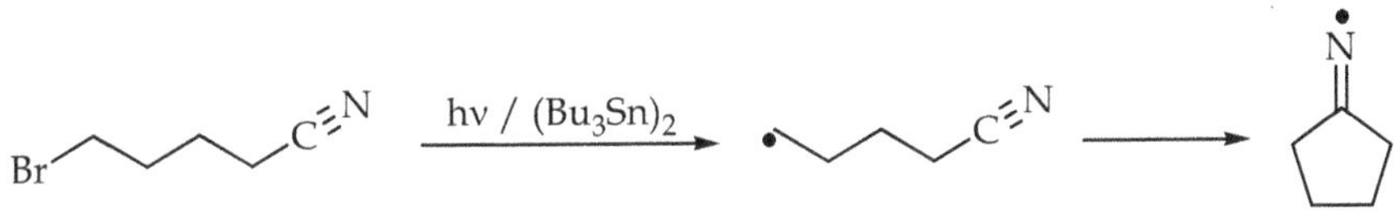

Scheme 17.

With respect to intermolecular reactions, a couple of examples have been reported where photochemically generated radicals add to acetonitrile (Engel et al., 1987; de Lijser & Arnold, 1997). Carbon-centred radicals prefer to add to the carbon of the nitrile to give the iminyl radical (de Lijser & Arnold, 1998), a preference that has been rationalized by a time-resolved infrared study on the photochemistry of *O*-fluoroformyl-9-fluorenone oxime in acetonitrile solution and by ab initio molecular orbital calculations (Bucher et al., 2006). In addition to the N–O bond cleavage, upon laser flash photolysis (266 nm) the short-lived transient fluoroformyl radical and the transient iminyl radical were detected (Scheme 18).

Boryl radicals also add to nitriles. Photochemically formed *tert*-butoxyl radicals were capable of abstracting hydrogen from borohydride anions to form a borane radical anion, which added to cyanides to give iminyl radical adducts (Giles & Roberts, 1983). Similarly, these *tert*-butoxyl radicals reacted with primary amine-boranes to give amine-boryl radicals, which can be intercepted by addition to the CN group (Kirwan & Roberts, 1989).

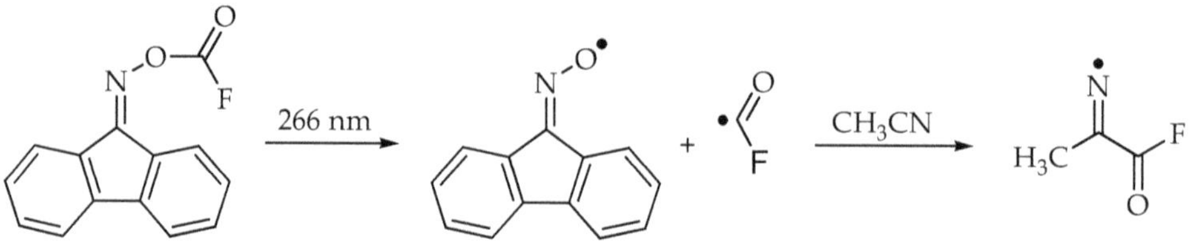

Scheme 18.

3. Characterization, kinetic data and calculations

In order to perform an adequate characterization of iminyl radicals, it is necessary to have valid precursors. In this sense, oxime esters seem to be suitable substrates for these experiments. The photochemistry of 9-fluorenone oxime phenylglyoxilate was investigated by laser flash photolysis at 355 nm, using time-resolved EPR spectroscopy (Kolano et al., 2004). The first step in the reaction would be cleavage of the N–O bond to give, initially, the 9-fluorenoneiminyl radical and the benzoylcarbonyloxy radical (Scheme 19). This process led to the detection, among others, of a set of transient signals, a 1:1:1 triplet centred at 3455.7 G (J_N = 9.7 G) attributed to the iminyl radical. Previous hyperfine splitting values obtained for iminyl radicals were in the range 9.1 to 10.3 G (Griller et al., 1974).

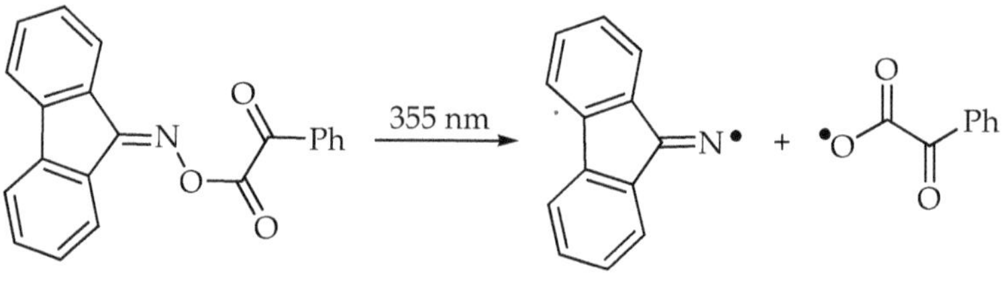

Scheme 19.

The group of Walton has devoted a great deal of attention to iminyl radicals. They used EPR spectroscopy to detect and characterize several N-centred radicals (McCarroll & Walton, 2000). Of particular interest is the study on the pentenyliminyl radicals (Portela-Cubillo et al., 2009). These radicals selectively closed in the 5-*exo* mode, irrespective of the substitution pattern around the C=C double bond of the pentenyl chain (Scheme 20). DFT computations have been used to model the experimental results obtained. The rate constant for the parent compound (R^1 = Ph; R^2 = R^3 = R^4 = H) was $k_c^{5\text{-}exo}$ = 8.8 × 10^3 s^{-1} at 27 °C, while the presence of two phenyl groups at the end of the C=C bond (R^1 = Me; R^2 = H; R^3 = R^4 = Ph) gave rise to a cyclization rate constant of $k_c^{5\text{-}exo}$ = 2.2 × 10^6 s^{-1} at 25 °C (Le Tadic-Biadatti et al., 1997). The rate of cyclization is slower for an iminyl with two H atoms at the terminus of the C=C double bond (R^3 = R^4 = H) than that with two phenyl groups (R^3 = R^4 = Ph), probably due to the higher stability of the phenyl conjugated C-centred radical formed for the later during the cyclization reaction.

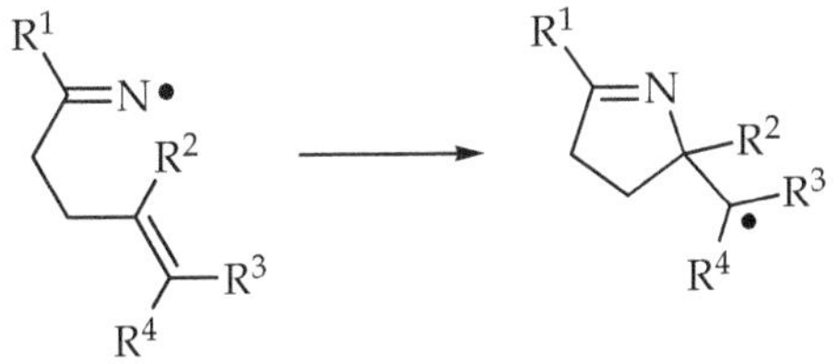

Scheme 20.

The rate constants of other processes involving radicals have also been measured. For example, in the bimolecular combination reactions to form the corresponding azides, in some cases values close to the diffusion controlled limit have been measured, $k_2 \sim 3 \times 10^8$ M^{-1} s^{-1} at -28 °C (Portela-Cubillo et al., 2009), while others range between 10^2 and 10^9 M^{-1} s^{-1} at -35 °C depending on steric factors (Griller et al., 1974). Iminyl radicals also abstract hydrogen from different species. Specifically, the rate constants for the reaction of the iminyl radical shown in Scheme 20 (R^1 = Me; R^2 = H; R^3 = R^4 = Ph) with thiophenols has been estimated at $k_H \sim 10^7$ M^{-1} s^{-1} at 25 °C (Le Tadic-Biadatti et al., 1997).

Iminyl radicals have also been studied by theoretical DFT calculations. Since these species could evolve through cyclization to the five- or six-membered rings, the course of the ring formation has been calculated (Alonso et al., 2008). The choice of B3PW91 was made on the basis of previous results where this functional proved to give satisfactory results in radical chemistry (Pace et al., 2006; Pinter et al., 2007). As can be seen from Table 1 and Figure 1, the calculated relative free energies for the transition states of the cyclization to the five- and six-membered ring when no other factor is present are 12.6 and 16.0 kcal/mol, respectively, at the B3PW91/6-31+G* level, which indicates a preference for the formation of the 1-pyrroline ring. The presence of an aromatic ring as a spacer makes these values 6.4 and 5.6 kcal/mol, respectively, and indicates a small preference for the construction of the six-membered ring (Table 1). These predictions were corroborated by experimental results.

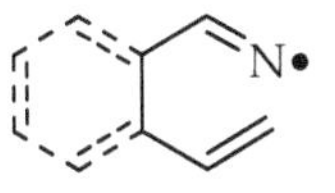	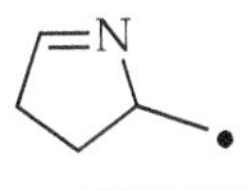	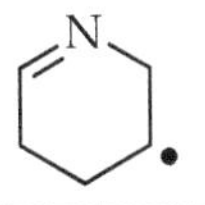	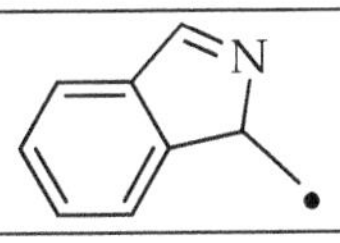	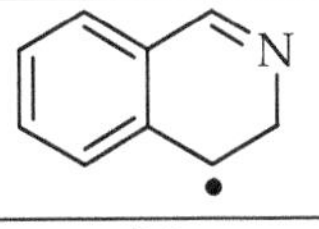
Iminyl Radical	0.0	0.0	0.0	0.0
Transition State	12.6	16.0	6.4	5.6
Product	-5.2	-7.6	-10.6	-27.7

Table 1. Free energies (ΔG, kcal/mol) for the cyclization step to five- or six-membered rings, relative to the corresponding iminyl radical, at the B3PW91/6-31+G* level

DFT calculated structures and energies have also been used to understand the influence of some substituents on the reactivity of iminyl radicals (Alonso et al., 2010). The most significant is the reduction of the energy barrier, from 14.6 to 13.2 kcal/mol at the B3PW91/6-31G* level, when a methyl group is located on the iminic carbon (Figure 2). This change is probably due to electronic effects.

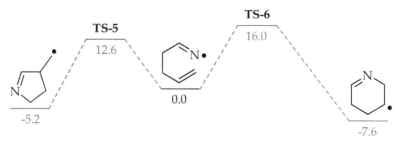

Fig. 1. Free energy diagram (ΔG, kcal/mol) for the cyclization of the simplest iminyl radical, at the B3PW91/6-31+G* level

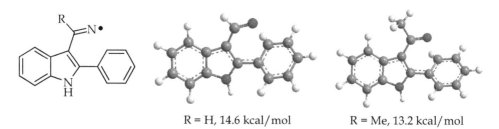

R = H, 14.6 kcal/mol   R = Me, 13.2 kcal/mol

Fig. 2. DFT optimised structures and energy barriers, relative to the corresponding iminyl radical, at the B3PW91/6-31G* level

4. Preparation of polycyclic heteroaromatic compounds and natural products

Direct irradiation of O-acyloximes has also been employed for the preparation of several polycyclic heteroaromatic compounds. As shown in Scheme 21, the use of a combination of reagents with five- and six-membered rings and cyclization of the intermediate iminyl radicals onto phenyl, thiophenyl or pyridinyl rings led to a variety of fused rings with different heteroatoms on the structure in good to excellent yields (48 to 90%) (Alonso et al., 2010). Thus, 2H-pyrazolo[4,3-c]quinoline, thieno[3,2-c]isoquinoline and benzo[c][1,7] naphthyridine derivatives have been prepared.

The versatility of this methodology has been exploited in the preparation of some interesting natural products. On using the appropriate structure (Scheme 22), direct irradiation of O-acyloximes allowed the preparation of several phenanthridine derivatives (Alonso et al., 2010), such as the alkaloid trisphaeridine (R^1,R^2 = OCH$_2$O; R^3 = H) or the vasconine precursor (R^1 = R^2 = OCH$_3$; R^3 = CH$_2$CH$_2$OH), which can also be used to obtain assoanine, oxoassoanine and pratosine (Rosa et al., 1997). Trisphaeridine was also prepared in 59%

yield by irradiation of the corresponding dioxime oxalate in the presence of 4-methoxyacetophenone as a photosensitizer (Portela-Cubillo et al., 2008).

An indirect method to form iminyl radicals by addition of a radical to a nitrile has been used to synthesize heteropolycycles. The process involves a cascade reaction from a photochemically generated radical. The formation of a tetracyclic heteroarene initiated by irradiation of an aromatic disulfide, with cleavage of an S–S bond, is shown in Scheme 23. The intermolecular addition of the sulfanyl radical to isonitriles gave an imidoyl radical, which was able to add to a nitrile to provide an iminyl radical that cyclizes to complete the cascade reaction (Camaggi et al., 1998; Nanni et al., 2000).

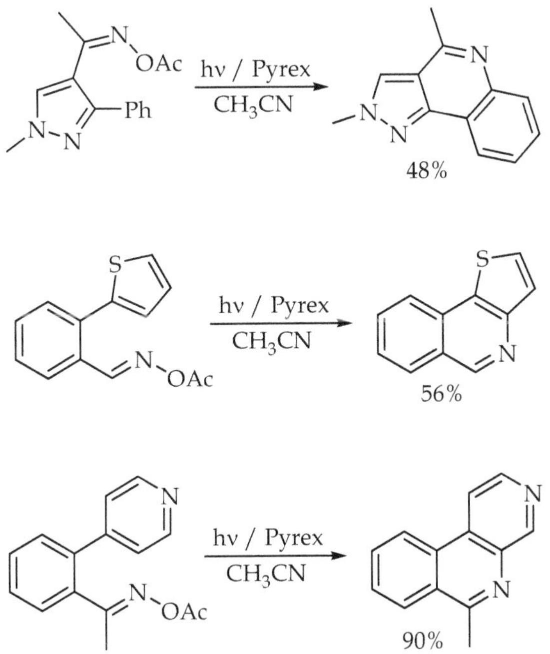

Scheme 21.

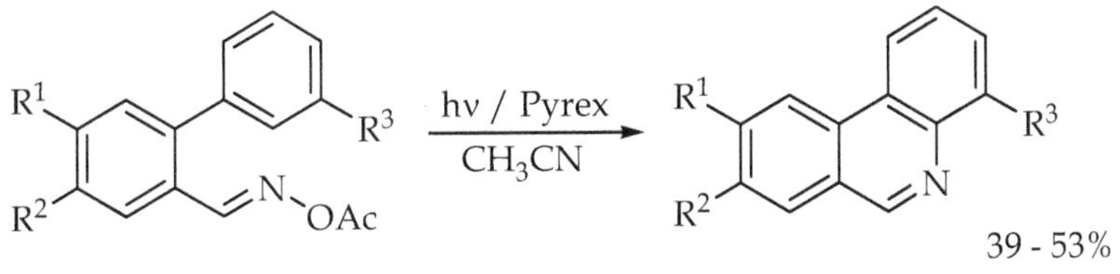

Scheme 22.

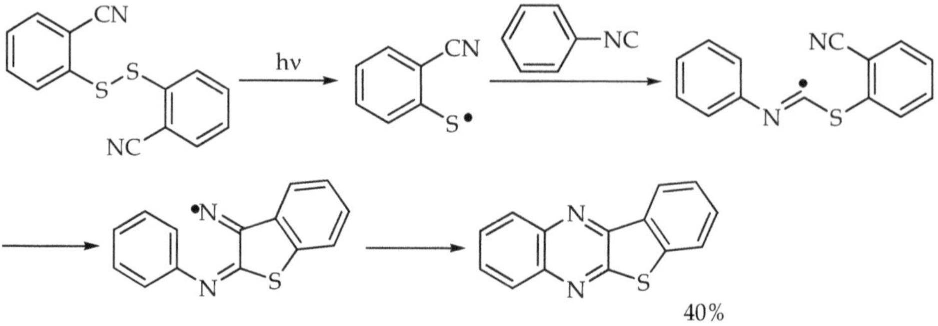

Scheme 23.

The preparation of rings A-D of the alkaloids camptothecin and mappicine has been achieved in good yield using this protocol (Bowman et al., 2001; 2002). After light-induced cleavage of hexamethylditin and abstraction of the iodine atom to give a vinyl radical, subsequent 5-*exo* cyclization onto the nitrile and 6-*exo* cyclization of iminyl radical followed by aromatisation led to the four ring skeleton (Scheme 24).

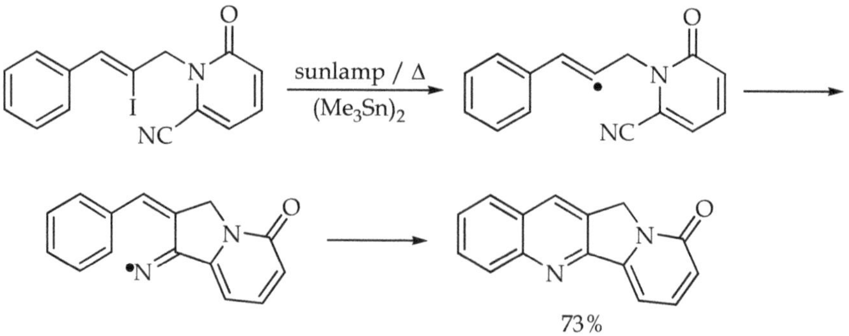

Scheme 24.

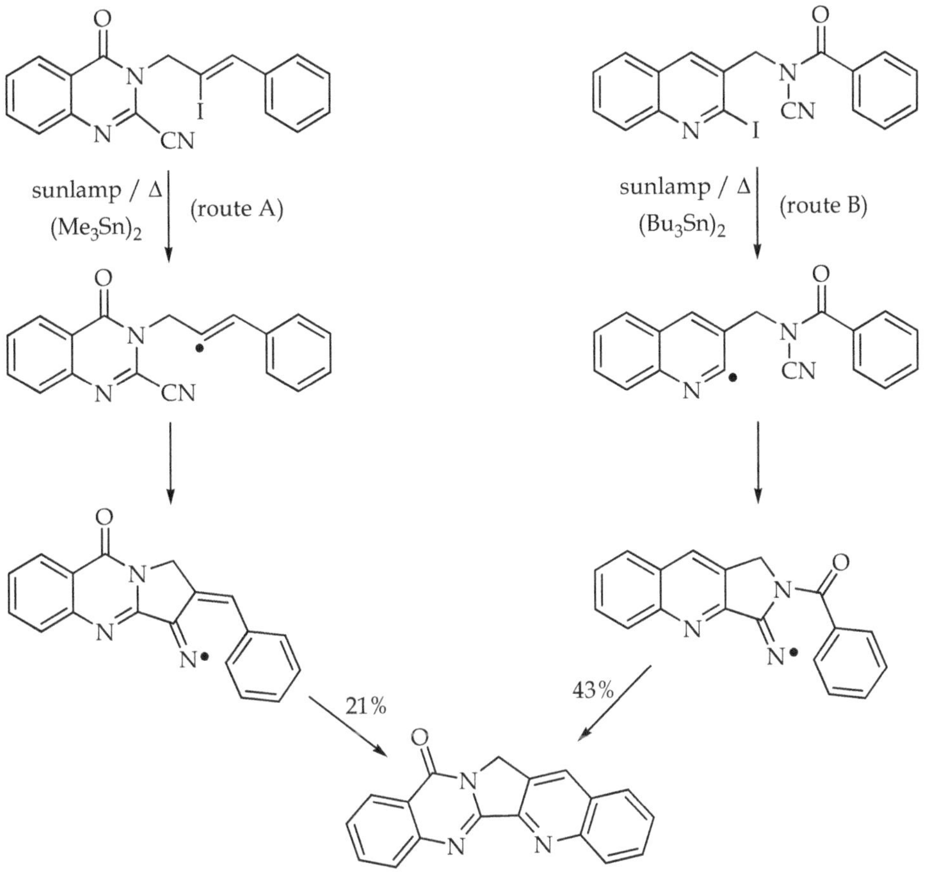

Scheme 25.

The biologically active alkaloid luotonin A has been obtained using cascade radical cyclization *via* an iminyl radical from 4-oxo-3,4-dihydroquinazoline-2-carbonitrile (Bowman et al., 2005). In a similar way, the vinyl radical attacked the cyano group and the resulting iminyl radical cyclized to the alkaloid (Scheme 25, route A). More recently, an alternative cyclization cascade process has been described for the preparation of luotonin A (Servais et al., 2007). As shown in Scheme 25 (route B), under radical conditions the pyrroloquinazoline skeleton is created from N-acyl-N-(2-iodobenzyl)cyanamide.

5. Conclusion

The photochemical generation of iminyl radicals can be performed in a direct way, which involves the homolytic cleavage of N–X bonds, or by an indirect method, which involves the addition of a radical to an unsaturated nitrogen functional group, such as an azide or a nitrile. Regarding the reactivity of this kind of radical, its ability to be added to unsaturated systems (double and triple bonds, aromatic and heteroaromatic compounds) has been revealed. The addition reaction can occur with the formation of five- or six-membered rings. The ease with which these processes take place allows their use in further synthetic applications, particularly in the synthesis of different polycyclic heteroaromatic compounds and in the preparation of natural products.

6. Acknowledgment

Financial support from the Ministerio de Ciencia e Innovación of Spain (CTQ2011-24800) and Universidad de La Rioja (API11/20) is gratefully acknowledged.

7. References

Alonso, R.; Campos, P. J.; García, B. & Rodríguez, M. A. (2006). *Org. Lett.*, 8, 3521

Alonso, R.; Campos, P. J.; Rodríguez, M. A. & Sampedro, D. (2008). *J. Org. Chem.*, 73, 2234

Alonso, R.; Campos, P. J.; Caballero, A. & Rodríguez, M. A. (2010). *Tetrahedron*, 66, 8828

Binkley, R. W. (1970). *J. Org. Chem.*, 35, 2796

Bird, K. J.; Chan, A. W. K. & Crow, W. D. (1976). *Aus. J. Chem.*, 29, 2281

Blanco-Lomas, M.; Caballero, A.; Campos, P. J.; González, H. F.; López-Sola, S.; Rivado-Casas, L.; Rodríguez, M. A. & Sampedro, D. (2011). *Organometallics*, 30, 3677

Boivin, J.; Fouquet, E.; Schiano, A.-M. & Zard, S. Z. (1994). *Tetrahedron*, 50, 1769

Bowman, W. R.; Bridge, C. F.; Cloonan, M. O. & Leach, D. C. (2001). *Synlett*, 765

Bowman, W. R.; Bridge, C. F.; Brookes, P.; Cloonan, M. O. & Leach, D. C. (2002). *J. Chem. Soc., Perkin Trans. 1*, 58

Bowman, W. R.; Cloonan, M. O.; Fletcher, A. J. & Stein, T. (2005). *Org. Biomol. Chem.*, 3, 1460

Bucher, G.; Kolano, C.; Schade, O. & Sander, W. (2006). *J. Org. Chem.*, 71, 2135

Callier-Dublanchet, A.-C.; Quiclet-Sire, B. & Zard, S. Z. (1997). *Tetrahedron Lett.*, 38, 2463

Camaggi, C. M.; Leardini, R.; Nanni, D. & Zanardi, G. (1998). *Tetrahedron*, 54, 5587

Cantrell, T. S. & Haller, W. S. (1968). *Chem. Commun.*, 977

Cooper, J. W.; Roberts, B. P. & Winter, J. N. (1977). *J. Chem. Soc., Chem. Commun.*, 320

de Lijser, H. J. P. & Arnold, D. R. (1997). *J. Chem. Soc., Perkin Trans. 2*, 1369

de Lijser, H. J. P. & Arnold, D. R. (1998). *J. Org. Chem.*, 69, 3057

de Lijser, H. J. P. & Tsai, C.-K. (2004). *J. Phys. Chem.*, 102, 5592

de Lijser, H. J. P.; Rangel, N. A.; Tetalman, M. A. & Tsai, C.-K. (2007). *J. Org. Chem.*, 72, 3057

Engel, P. S.; Lee, W.-K.; Marschke, G. E. & Shine, H. J. (1987). *J. Org. Chem.*, 52, 2813

Esker, J. L. & Newcomb, M. (1993). *Adv. Heterocycl. Chem.*, 58, 1

Fallis, A. G. & Brinza, I. M. (1997). *Tetrahedron*, 53, 17543

Fong, M. C. & Schiesser, C. H. (1993). *Tetrahedron Lett.*, 34, 4347

Gagosz, F. & Zard, S. Z. (1999). *Synlett*, 1978

Giles, J. R. M. & Roberts, B. P. (1983). *J. Chem. Soc., Perkin Trans. 2*, 743

Glober, S. A.; Hammond, G. P.; Harman, D. G.; Mills, J. G. & Rowbottom, C. A. (1993). *Aus. J. Chem.*, 46, 1213

Griller, D.; Mendenhall, G. D.; Van Hoof, W. & Ingold, K. U. (1974). *J. Am. Chem. Soc.*, 96, 6068

Griller, D.; Schmid, P. & Ingold, K. U. (1979). *Can. J. Chem.*, 57, 831

Guindon, Y.; Guérin, B. & Landry, S. R. (2001). *Org. Lett.*, 3, 2293

Hwang, H.; Jang, D.-J. & Chae, K. H. (1999). *J. Photochem. Photobiol. A: Chemistry*, 126, 37

Hwu, J. R.; Lin, C. C.; Chuang, S. H.; King, K. Y.; Su, T.-R. & Tsay, S.-C. (2004). *Bioorg. Med. Chem.*, 12, 2509

Kirwan, J. N. & Roberts, B. P. (1989). *J. Chem. Soc., Perkin Trans. 2*, 539

Kitamura, M.; Mori, Y. & Narasaka, K. (2005). *Tetrahedron Lett.*, 46, 2373

Kitamura, M. & Narasaka, K. (2008). *Bull. Chem. Soc. Japan*, 81, 539

Kolano, C.; Bucher, G.; Wenk, H. H.; Jäger, M.; Schade, O. & Sander, W. (2004). *J. Phys. Org. Chem.*, 17, 207

Kuttappan-Nair, V.; Samson-Thibault & Wagner, J. R. (2010). *Chem. Res. Toxicol.*, 23, 48

Lalevée, J.; Allonas, X.; Fouassier, J. P.; Tachi, H.; Izumitani, A.; Shirai, M. & Tsunooka, M. (2002). *J. Photochem. Photobiol. A: Chemistry*, 151, 27

Leardini, R.; McNab, H.; Minozzi, M. & Nanni, D. (2001). *J. Chem. Soc., Perkin Trans. 1*, 1072

Le Tadic-Biadatti, M.-H.; Callier-Dublanchet, A.-C-; Horner, J. H.; Quiclet-Sire, B.; Zard, S. Z. & Newcomb, M. (1997). *J. Org. Chem.*, 62, 559

McCarroll, A. J. & Walton , J. C. (2000). *J. Chem. Soc., Perkin Trans. 2*, 2399

Mikami, T. & Narasaka, K. (2000). *Chem. Lett.*, 338

Mikami, T. & Narasaka, K. (2001). *C. R. Acad. Sci. Paris, Chimie / Chemistry*, 4, 477

Mukai, T.; Oine, T. & Matsubara, A. (1969). *Bull. Chem. Soc. Japan*, 42, 581

Nanni, D.; Calestani, G.; Leardini, R. & Zanardi, G. (2000). *Eur. J. Org. Chem.*, 707

Newman, H. (1968a). *Tetrahedron Lett.*, 2417

Newman, H. (1968b). *Tetrahedron Lett.*, 2421

Ogata, M.; Kano, H. & Matsumoto, H. (1968). *Chem. Commun.*, 397

Okada, T.; Kawanisi, H. & Nozaki, H. (1969). *Bull. Chem. Soc. Japan*, 42, 2981

Padwa, A. & Albrecht, F. (1974a). *J. Org. Chem.*, 36, 2361

Padwa, A. & Albrecht, F. (1974b). *J. Am. Chem. Soc.*, 96, 4849

Padwa, A. (1977). *Chem. Rev.*, 77, 37

Pace, A.; Buscemi, S.; Vivona, N.; Silvestri, A. & Barone, G. (2006). *J. Org. Chem.*, 71, 2740

Petrucci, R. H.; Herring, F. G.; Madura, J. D. & Bissonnette, C. (2011). *General Chemistry: Principles and Modern Applications* (10th edition), Chapter 10, p 435, Pearson Canada Inc., Toronto, Canada

Pinter, B.; DeProft, F.; VanSpeybroeck, V.; Hemelsoet, K.; Waroquier, M.; Chamorro, E.; Veszpremi, T. & Geerlings, P. (2007). *J. Org. Chem.*, 72, 348

Portela-Cubillo, F.; Scanlan, E. M.; Scott, J. S. & Walton, J. C. (2008). *Chem. Commun.*, 4189

Portela-Cubillo, F.; Alonso-Ruiz, R.; Sampedro, D. & Walton, J. C. (2009). *J. Phys. Chem.*, 113, 10005

Pratt, A. C. (1977). *Chem. Soc. Rev.*, 6, 63

Renaud, P. & Sibi, M. P. (Eds.). (2001). *Radicals in Organic Synthesis*, Wiley-VCH, Weinheim, Germany

Roberts, B. P. & Winter, J. N. (1979). *J. Chem. Soc., Perkin Trans. 2*, 1353

Rosa, A. M.; Lobo, A. M.; Branco, P. S.; Prabhabar, S. & Sá-da-Costa, M. (1997). *Tetrahedron*, 53, 299

Sakuragi, H.; Ishikawa, S.-I-; Nishimura, T.; Yoshida, M.; Inamoto, N. & Tokumaru, K. (1976). *Bull. Chem. Soc. Japan*, 49, 1949

Sauer, J. & Mayer, K. K. (1968). *Tetrahedron Lett.*, 325

Servais, A.; Azzouz, M.; Lopes, D.; Courillon, C. & Malacria, M. (2007). *Angew. Chem. Int. Ed.*, 46, 576

Takeuchi, H.; Nagai, T. & Tokura, N. (1972). *J. Chem. Soc., Perkin Trans. 2*, 420

Togo, H. (2004). *Advanced Free Radical Reactions for Organic Synthesis*, Elsevier, Oxford, U.K.

Turro, N. J. (1991). *Modern Molecular Photochemistry*, p 253, University Science Books, Sausalito, U.S.A.

Uchiyama, K.; Hayashi, Y. & Narasaka, K. (1998). *Chem. Lett.*, 1261

Yoshida, M.; Sakuragi, H.; Nishimura, T.; Ishikawa, S. & Tokumaru, K. (1975). *Chem. Lett.*, 1125

Zard, S. Z. (2003). *Radical Reactions in Organic Synthesis*, Oxford University Press, Oxford, U.K.

Zard, S. Z. (2008). *Chem. Soc. Rev.*, 37, 1603

Permissions

The contributors of this book come from diverse backgrounds, making this book a truly international effort. This book will bring forth new frontiers with its revolutionizing research information and detailed analysis of the nascent developments around the world.

We would like to thank Satyen Saha, for lending his expertise to make the book truly unique. He has played a crucial role in the development of this book. Without his invaluable contribution this book wouldn't have been possible. He has made vital efforts to compile up to date information on the varied aspects of this subject to make this book a valuable addition to the collection of many professionals and students.

This book was conceptualized with the vision of imparting up-to-date information and advanced data in this field. To ensure the same, a matchless editorial board was set up. Every individual on the board went through rigorous rounds of assessment to prove their worth. After which they invested a large part of their time researching and compiling the most relevant data for our readers. Conferences and sessions were held from time to time between the editorial board and the contributing authors to present the data in the most comprehensible form. The editorial team has worked tirelessly to provide valuable and valid information to help people across the globe.

Every chapter published in this book has been scrutinized by our experts. Their significance has been extensively debated. The topics covered herein carry significant findings which will fuel the growth of the discipline. They may even be implemented as practical applications or may be referred to as a beginning point for another development. Chapters in this book were first published by InTech; hereby published with permission under the Creative Commons Attribution License or equivalent.

The editorial board has been involved in producing this book since its inception. They have spent rigorous hours researching and exploring the diverse topics which have resulted in the successful publishing of this book. They have passed on their knowledge of decades through this book. To expedite this challenging task, the publisher supported the team at every step. A small team of assistant editors was also appointed to further simplify the editing procedure and attain best results for the readers.

Our editorial team has been hand-picked from every corner of the world. Their multi-ethnicity adds dynamic inputs to the discussions which result in innovative outcomes. These outcomes are then further discussed with the researchers and contributors who give their valuable feedback and opinion regarding the same. The feedback is then collaborated with the researches and they are edited in a comprehensive manner to aid the understanding of the subject.

Apart from the editorial board, the designing team has also invested a significant amount of their time in understanding the subject and creating the most relevant covers. They scrutinized every image to scout for the most suitable representation of the subject and create an appropriate cover for the book.

The publishing team has been involved in this book since its early stages. They were actively engaged in every process, be it collecting the data, connecting with the contributors or procuring relevant information. The team has been an ardent support to the editorial, designing and production team. Their endless efforts to recruit the best for this project, has resulted in the accomplishment of this book. They are a veteran in the field of academics and their pool of knowledge is as vast as their experience in printing. Their expertise and guidance has proved useful at every step. Their uncompromising quality standards have made this book an exceptional effort. Their encouragement from time to time has been an inspiration for everyone.

The publisher and the editorial board hope that this book will prove to be a valuable piece of knowledge for researchers, students, practitioners and scholars across the globe.

List of Contributors

Xiang-Ming Meng, Shu-Xin Wang and Man-Zhou Zhu
Anhui University, China

Miguel A. Valenzuela, Sergio O. Flores, Omar Ríos-Bern□, Elim Albiter and Salvador Alfaro
Lab. Catálisis y Materiales, ESIQIE-Instituto Politécnico Nacional Zacatenco, México D.F., México

Fabrizio Sordello, Valter Maurino and Claudio Minero
Università degli Studi di Torino, Dipartimento di Chimica Analitica, Torino, Italy

Ji-Jun Zou, Lun Pan, Xiangwen Zhang and Li Wang
Key Laboratory for Green Chemical Technology of Ministry of Education, School of Chemical Engineering and Technology, Tianjin University P.R. China

Maria Teresa Neves-Petersen and Steffen B. Petersen
International Iberian Nanotechnology Laboratory (INL), Braga Aalborg University, Aalborg, Portugal, Denmark

Gnana Prakash Gajula
Materials and Metallurgy Group, Indira Gandhi Centre for Atomic Research, Kalpakkam, India

Takashi Kikukawa and Makoto Demura
Faculty of Advanced Life Science, Hokkaido University, Japan

Jun Tamogami, Kazumi Shimono, Toshifumi Nara and Naoki Kamo
College of Pharmaceutical Sciences, Matsuyama University, Japan

Yukihiro Kimura
Graduate School of Agricultural Science, Kobe University, Japan
Organization of Advanced Science and Technology, Kobe University, Japan

Takashi Ohno, Yasuo Yamauchi and Yong Li
Organization of Advanced Science and Technology, Kobe University, Japan

L. Therese Bergendahl and Martin J. Paterson
Department of Chemistry, Heriot Watt University, Edinburgh, Scotland

Lotfi Bouslimi
Ecole Supérieure des Sciences et Techniques de Tunis, Tunis
Institut Supérieur de Pêche et d'Aquaculture de Bizerte, Tunis

Georges Zissis and Jean Pascal Cambronne
Université de Toulouse, UPS, INPT, LAPLACE (Laboratoire Plasma et Conversion d'Energie), France

Mongi Stambouli and Ezzedine Ben Braiek
Ecole Supérieure des Sciences et Techniques de Tunis, Tunis

Janina Kabatc and Katarzyna Jurek
University of Technology and Life Sciences, Faculty of Chemical Technology and Engineering, Poland

Anabela Sousa Oliveira and Luis Filipe Vieira Ferreira
Centro de Química-Física Molecular and Institute of Nanosciences and Nanotechnology, Instituto Superior Técnico, Universidade Técnica de Lisboa, Lisboa, Portugal
Centro Interdisciplinar de Investigação e Inovação, Escola Superior de Tecnologia e Gestão, Instituto Politécnico de Portalegre, Portalegre, Portugal

Enrico Mendes Saggioro, Thelma Pavesi and Josino Costa Moreira
Centro de Estudos da Saúde do Trabalhador e Ecologia Humana, Escola Nacional de Saúde Pública, Fundação Oswaldo Cruz, Rio de Janeiro, Brazil

Miguel A. Rodríguez
Departamento de Química, Unidad Asociada al C.S.I.C., Universidad de La Rioja, Spain